Girish K. Malhotra
Active Pharmaceutical Ingredient Manufacturing

Also of Interest

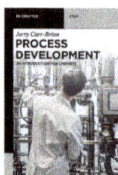

Process Development.
An Introduction for Chemists
Carr-Brion, 2022
ISBN 978-3-11-071786-0, e-ISBN 978-3-11-071787-7

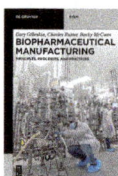

Biopharmaceutical Manufacturing.
Principles, Processes, and Practices
Gilleskie, Rutter, McCuen, 2021
ISBN 978-3-11-061687-3, e-ISBN 978-3-11-061688-0

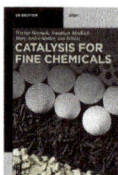

Catalysis for Fine Chemicals
Bonrath, Medlock, Müller, Schütz, 2021
ISBN 978-3-11-057115-8, e-ISBN 978-3-11-057118-9

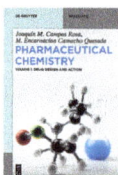

Pharmaceutical Chemistry.
Vol. 1: Drug Design and Action
Campos Rosa, Camacho Quesada, 2017
ISBN 978-3-11-052836-7, e-ISBN 978-3-11-052848-0

Vol 2: Drugs and Their Biological Targets.
Campos Rosa, Camacho Quesada, 2017
ISBN 978-3-11-052851-0, e-ISBN 978-3-11-052852-7

Girish K. Malhotra

Active Pharmaceutical Ingredient Manufacturing

Nondestructive Creation

DE GRUYTER

Author
Girish K. Malhotra
EPCOT International
29150 Bryce Road
Pepper Pike
OH 44124-5701
USA
E-Mail: girish@epcotint.com

ISBN 978-3-11-070282-8
e-ISBN (PDF) 978-3-11-070284-2
e-ISBN (EPUB) 978-3-11-070289-7

Library of Congress Control Number: 2022930254

Bibliographic information published by the Deutsche Nationalbibliothek
The.Deutsche Nationalbibliothek lists this publication in the Deutsche Nationalbibliografie;
detailed bibliographic data are available on the Internet at http://dnb.dnb.de.

To Indu my known and unknown friend for her wisdom and encouragement

Preface

For over 25 years, there has been considerable and at times passionate discussion about "green chemistry" in chemical-related industries, including pharmaceuticals. "Climate change" and "net zero emissions" have become "in vogue." Onus is on the industry to innovate and lower emissions and to become green. However, to make a measurable impact, a significant effort is needed. This has been lacking on industrial scale, especially in pharmaceutical manufacturing. Universities/academics and the regulators can talk about these and may show some pathways. Unless the industry takes the necessary steps to curtail emissions through innovative manufacturing technologies, the stated goal is unlikely to be achieved.

Regulations have to facilitate and promote incorporation of novel processes into the existing practices in order to achieve these goals. However, with the current regulations, and related costs and delays, the industry is hesitant to engage in such challenges. For public health and safety concerns, both the regulators and the industry are also constrained by the current regulations. The regulatory agencies have to rely on the industry for their claims/assertions. Regulators on their part have not made any effort to understand the environmental impact of their guidelines, for example, if a process produces the desired product using excessive solvent or multiple solvents, the industry cannot change the process unless safety and efficacy of the product is proven again. For the new products, an effort to minimize solvent use, that is, emissions, has to be the "mantra" from the onset of the product development. At present, this is not the case.

Pharma companies have to make sure that the technologies that will reduce emissions by 75–80% from the current levels are incorporated in practice as soon as the product efficacy and value are recognized. All of the necessary methods (chemistries, equipment, and processes) have been in existence and have been used for the last 75+ years in the fine/specialty chemicals, the older cousin of the API (active pharmaceutical ingredients), and in their formulations. Pharmaceuticals have not explored and/or adopted many of the available options. Deterrents include speed to market, profitability, and complexity of regulations. These have prevented widespread incorporation of better and innovative technologies. Each improvement in the existing process has to be proven to the regulatory agencies. Companies have to make sure that the efficacy and performance are not compromised. This process is expensive and results in delays resulting in lower profits. In addition, lack of understanding by the regulators also stands in the way of any improvement of manufacturing processes. The necessary methods/equipment that can significantly lower the emissions exist and they are practiced by fine/specialty chemicals. Pharma companies have simply overlooked them.

According to the 2006 Nobel laureate Edmund Phelps, "nondestructive creation" leads to innovation. This methodology involves alternate ways to apply existing methods and teachings to simplify the current practices. Since all of the necessary

https://doi.org/10.1515/9783110702842-202

elements that are needed for API manufacturing technology innovation still exist and are practiced, their incorporation should not be difficult. However, distinctive thinking and divergent execution would be required. Chemists and chemical engineers are familiar with such challenges and have all the tools to apply them. Peter Drucker has said it rightly that "Innovation can be systematically managed – if one knows where to look." However, lack of their application in API manufacturing is startling.

Around Thanksgiving 2019, Mr. Robert Esposito suggested to me to write a book about pharmaceutical manufacturing. Sometimes good things fall in your lap. The goal, starting from API manufacturing and their dosage, was narrowed just to API manufacturing.

Considering the issue of "climate change and global warming," timing for the book could not have been any better. The environmental impact of API manufacturing practices is not acknowledged widely. They contribute the highest per kilo of the product in the chemical manufacturing world. The year 2020 rolled around and the writing started. Global rumbling of an infectious disease was getting louder. In mid-March 2020, the world spun but its inhabitants and their activities came to a screeching halt. With the lockdown, writing became a challenge as references were not easily accessible. Libraries were shut down. However, most of the scientific journals could not and would not let the distribution of knowledge to come to standstill. When Internet became accessible few years ago, we were impressed by what all it offered. With pandemic, Internet came to bat for all of us. It was the best blessing. Online access was offered and the world benefited immensely. Thanks to the "nondestructive creationists" who created the Internet for your vision, creativity, and imagination. We all benefitted and far exceeded the challenge posed. Cuyahoga County Library and the affiliated library systems and others who are my "nondestructive creationists" were a godsend. Again, thanks to all who helped in their own ways.

Chapters review the history and evolution of the API manufacturing and how, in spite of the technological progress, it is still stuck in the stone age. History indicates that the methodologies used for fine/specialty chemical manufacture in the first half of twentieth century are still being used today. Methodologies to move it to the contemporary time are reviewed. They are not earthshakingly complex. These methods do not call for any alternate science or engineering but their application can nondestructively create excellent processes that significantly reduce current environmental emissions. They could even convert batch manufacturing of many products to continuous processes that are inherently more profitable and produce products of higher quality. Lowering emissions from API emissions and their formulations will come from "reimagining" chemical engineering fundamentals that involve physical and chemical properties of chemicals, and their mutual behavior along with the equipment use. It requires exploiting social and mutual behavior of chemicals. Unit processes and unit operations used would be the same but their scale could be different from fine/specialty chemical practices. Scale would be different. Every which way method/s and some may look like repeats are reviewed and options discussed. Chemists and chemical engineers have the creativity and imagination to incorporate these for the chemistries and processes they develop.

It is expected that the reviewed possibilities will invigorate creativity and imagination of chemists and chemical engineers leading to processes that will have significantly lower emissions and carbon footprint. It is very possible that many products could be produced using continuous processes further lowering emissions. Most of the regulators do not have hands-on experience of process development, design, and commercialization, and will have to facilitate innovation. Regulations and filing processes may have to be modified.

Guidance and help came from unexpected sources. Most of them provided intellectual conversations and that helped me a lot. These conversations planted ideas that were filed and polished. Dr. Richard Thomas, Dr. Charles E. Kausch, Mr. James Burton, Dr. Satish Kalhan, MD, and my colleagues at regulatory bodies and the consulting companies offered their counsel and conversation. Due to confidentiality of the relationships, they are not named here but are acknowledged – they know it. Dr. Yusuf K. Hamied of Cipla offered conversation time and suggestions. Thanks to you all.

Drs. Thomas and Kausch, Mr. Lawrence Coven along with my wife Dr. Indu Malhotra proofread, the hardest task, as they have to imagine the writing. They had their respective input and facilitated the task. I am really grateful. Our family and friends endured and I am indebted. I gratefully acknowledge Dr. Karin Sora, Vice President of Science, Technology, Engineering & Mathematics (STEM). Dr. Ria Sengbusch, project editor for chemistry and materials sciences, needs special thanks for her suggestions, guidance, and edits to produce and publish this book. Ms. Tharani Ramachandran, project manager at Integra made this possible. Thanks to you all and cheers!

Girish K. Malhotra, PE
Friday January 28, 2022

Contents

Chapter 6
Active pharmaceutical ingredients (API): innovation, design considerations, and waste reduction —— 177

Chapter 7
Process documentation and operating strategies —— 213

Chapter 1
History of pharmaceuticals

We try to remember that medicine is for the patient. We try never to forget that medicine is for the people. It is not for the profits. The profits follow, and if we have remembered that, they have never failed to appear. The better we have remembered it, the larger they have been.
– George W. Merck Time Magazine August 19, 1952 [1]

1.1 History of pharmaceuticals: their evolution

Going through history, growth and development of the pharmaceutical industry is fascinating from seventeenth century onward. We get to see how various companies will have meaning to history buffs. However, to the current generation, it would be how some companies have appeared and how their names got changed through mergers, demergers, truncations, and change of vision. A short history of some of the companies is presented. Some of the recent mergers (last 5 years) are not included.

1.1.1 Beginning

Different botanical plants were used for remedies of various ailments. Companies started offering soaps, skin care, and distinctive medications for different common diseases. A close look at the companies involved presents an interesting journey. Pharmaceutical industry's current form really started to take place around late sixteenth and early seventeenth centuries [2–4]. Wikipedia has been an excellent source of most of the citations. Table 1.1 is a brief compilation of evolution of the pharmaceutical companies. It is difficult to go into the detailed history of each company. The following narrative is an extremely brief summary. Included references are just "tip of the iceberg." Books can be written about each company's history.

It is interesting to note that some of these companies got their start as a dye and fine chemical business in the seventeenth century. Researchers found that many of the chemicals could selectively kill bacteria, parasites, and disease-causing microorganisms. With evolution of synthetic organic chemistry, chemists were able to convert many of the coal tar based raw materials to disease curing medicines. Their businesses grew to serve the ever-growing needs of the global population. Development of petroleum-based chemicals, continuous learning, and advances of organic chemistry synthesis facilitated the growth of the pharmaceutical industry. A walk through the major company's evolution is a fascinating journey and worth a review as it tells us about their roots. One fact that dominates and it is not widely recognized is that a majority of the small-molecule disease-curing

https://doi.org/10.1515/9783110702842-001

compounds (active pharmaceutical ingredients, API) are organic fine and specialty chemicals and their salts that kill disease-causing bacteria. If they did not have a disease-curing value, they could be an additive of some other values.

An important aspect that cannot be forgotten is that the manufacturing practices used in fine/specialty chemicals found use and value for the manufacture of disease-curing molecules. A need to to develop specific manufacturing practices for API was never an issue as these molecules could be easily synthesized in the equipment that was available and used to manufacture other similar organic molecules.

1.1.2 GlaxoSmithKline

Roots of GlaxoSmithKline (GSK) [5] go back to 1873. Joseph Nathan founded a trading company, Joseph Nathan and Co., in New Zealand. Due to excess milk being produced on the dairy farms, it started producing dried milk in 1904 and sold it under the name "Defiance." They tried to name the product "Lacto" but their trademark was denied. They changed Lacto to Glaxo. It was sold under the slogan "Glaxo builds bonny babies." Due to vitamin A deficiency, they sold cod liver oil as a supplement. Its first pharmaceutical product was vitamin D. It was extracted from cod liver oil and sold.

Tab. 1.1: Pharmaceutical companies from sixteenth century to present.

Company	Current name
Glaxo	GlaxoSmithKline
Smith Kline	
Beecham	
Wellcome Research Labs	
Merck	Merck
Schering Plough	
Pfizer	Pfizer
American Cyanamid (Lederle Labs)	
Warner Lambert	
Parke Davis	
Upjohn	
Wyeth	
Pharmacia	
King Pharmaceuticals	
Hospira	
G. D. Searle	

Tab. 1.1 (continued)

Company	Current name
Bayer AG	Bayer
Merck KGaA Schering	Merck KGaA
Eli Lilly	Eli Lilly
Abbott Labs	Abbott
	AbbVie
Imperial Chemical Industries (Zeneca) Astra	Astra-Zeneca
Bristol Myers Squibb	Bristol Myers Squibb
Ciba Geigy Sandoz	Novartis
Hoffmann-La Roche	Roche
Johnson & Johnson	Johnson & Johnson
Boots	Walgreens
Boehringer Ingelheim	Boehringer Ingelheim
Sanofi	Sanofi
Aventis	
Novo-Nordisk	

Glaxo Laboratories was incorporated as a subsidiary of Joseph Nathan and Co. in the United Kingdom in 1935. However, in 1947 shareholders bought out Joseph Nathan and Co. and reorganized the trading company to make Glaxo, the parent company.

Glaxo acquired Allen & Hanburys [6] in 1958. Glaxo bought Meyer Laboratories in 1978. It expanded its activities in the US market. In 1983, the American arm, Glaxo Inc., moved to Research Triangle Park and Zebulon (US manufacturing) in North Carolina.

Burroughs Wellcome & Company [7] was founded in 1880 in London by the American pharmacists Henry Wellcome and Silas Burroughs. In 1902, the Wellcome Tropical Research Laboratories was established. In the 1920s, Burroughs Wellcome established research and manufacturing facilities in Tuckahoe, New York. It served as the US headquarters until the company moved to Research Triangle Park in North Carolina in 1971. The Nobel Prize-winning scientists Gertrude B. Elion and George H. Hitchings invented drugs such as mercaptopurine [Figure 1.1], which is used for cancer and autoimmune diseases, and are still used many years later.

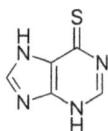

Fig. 1.1: Mercaptopurine (3,7-dihydropurine-6-thione) [8].

In 1959, the Wellcome Foundation bought Cooper, McDougall & Robertson Inc., and expanded to animal health. In 1995, Glaxo and Wellcome merged to form Glaxo Wellcome PLC.

In 1848, Thomas Beecham started selling his Beecham's Pills laxative in England and that created Beecham Group. By the 1960s, Beecham was extensively involved in pharmaceuticals and consumer products such as Macleans toothpaste, Lucozade (an energy drink), and synthetic penicillin research. Horlicks [9] malt milk was a Beecham dietary supplement and it is still commercially available. Glaxo's consumer healthcare business will be divested in 2022 as a separate company.

John K. Smith opened his first pharmacy in Philadelphia in 1830. In 1865, Mahlon Kline joined the business, and 10 years later it was named Smith, Kline & Co. In 1891, it merged with French, Richard and Company, and in 1929 it was named Smith Kline & French Laboratories as it focused more on research. Later, it bought Norden Laboratories, a business doing research into animal health, and in 1963 Recherche et Industrie Thérapeutiques in Belgium to focus on vaccines. Beginning 1969, the company began to expand globally. It bought Allergan, a manufacturer of eye and skincare products, in 1982. It was divested in 1989 and since it has gone through different hands to be finally merged with AbbVie that was created from Abbott Laboratories, discussed later.

In 1982, SmithKline & French merged with Beckman Inc. and changed its name to SmithKline Beckman. In 1988, it bought International Clinical Laboratories, and in 1989 it merged with Beecham to form SmithKline Beecham PLC. The headquarters moved from the United States to England. To expand R&D in the United States, the company bought a new research center in 1995; another opened in 1997 in England at New Frontiers Science Park, Harlow.

In 1999, Glaxo Wellcome had become the world's third largest pharmaceutical company by revenues and the world's largest manufacturer of drugs for the treatment of asthma and HIV/AIDS. Glaxo Wellcome and SmithKline Beecham merged in 2000 forming GSK.

1.1.3 Merck and Merck KGaA

There are two Merck: Merck KGaA, which is Darmstadt, Germany, based, and Merck & Co., which is US based [10, 11].

Roots of both Merck reach back as far as the seventeenth century in Germany. In 1668, Friedrich Jacob Merck, a pharmacist, assumed ownership of the Engel-Apotheke ("Angel Pharmacy") in Darmstadt, Germany. His descendant, Emanuel Merck, took over the pharmacy. Due to his scientific background, he was able to isolate and characterize different alkaloids and that led to discovery of number of drugs. He began the manufacture of these substances "in large quantities" in 1827, touting them as a "Cabinet of Pharmaceutical and Chemical Innovations."

He and his successors progressively built up a chemical-pharmaceutical factory that produced – in addition to raw materials for pharmaceutical preparations – a multitude of other chemicals and medicines. After Wilhelm Adam Sertürner's isolation of morphine from opium (1805), Merck pioneered commercial manufacture of morphine [Figure 1.2] for an expanding global market. Its use increased once the hypodermic syringes were invented.

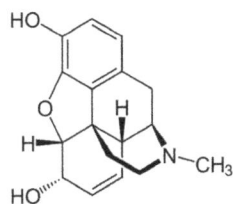

Fig. 1.2: Morphine ((4R,4aR,7S,7aR,12bS)-3-methyl-2,3,4,4a,7,7a-hexahydro-1H-4,12-methanobenzofuro[3,2-e]isoquinoline-7,9-diol) [12].

From 1884 onward, Merck also played a role in the production and marketing of cocaine. Sigmund Freud was an enthusiastic collaborator in Merck's cocaine research. Later, focus shifted to vitamins, a new product category. Vitamin D_3 (Vigantol) was introduced in 1927, followed by vitamin C (Cebion, ascorbic acid) in 1934.

Following the Second World War, Merck was granted permission by the military government to produce drugs, pesticides, food preservatives, reagents, and fine chemicals for laboratory use. Soon afterward, the economic boom set in. This meant double-digit sales for Merck for many years. Products Figure 1.3 included corticoid preparations Fortecortin [13] which is still used today, a cold remedy Nasivin [14], and Gestafortin [15] for hormone therapy.

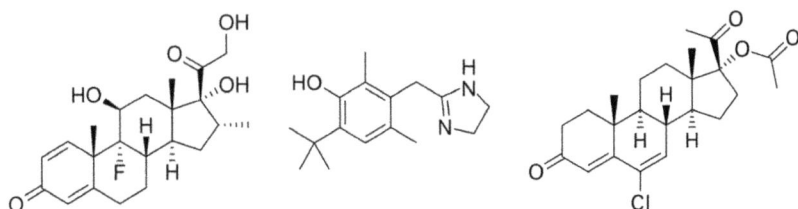

Fig. 1.3: Chemical structures of remedies: Fortecortin,1-dehydro-9α-fluoro-16α methylhydrocortisone [13], Nasivin, 3-(4,5-Dihydro-1H-imidazol-2-ylmethyl)-2,4-dimethyl-6-tert-butyl-phenol [14], Gestofortin Chlormadinone acetate [15].

Merck KGaA today also manufactures pigments and specialty chemicals.

1.1.3.1 Merck

In 1891, Georg(e) Merck of the original Merck family in Germany moved to the United States and set up Merck & Co. in New York. Following the First World War, Merck & Co. was confiscated and set up as an independent company in the United States. Merck started in Germany is known as Merck Sharpe and Dohme (MSD) outside the United States.

Merck started producing the same chemical and medicines its parent company in Germany was producing. In 1933, Merck established a laboratory to develop products vitamins and other products. It synthesized Prontosil [Figure 1.4] which came from sulfanilamide [Figure 1.4], a coal tar [16] by-product. Coal tars are mixture of different phenols, aromatic hydrocarbons, and heterocyclic oxygen, sulfur, and nitrogen compounds. Sulfa drugs and antibiotics, for example, penicillin were launched.

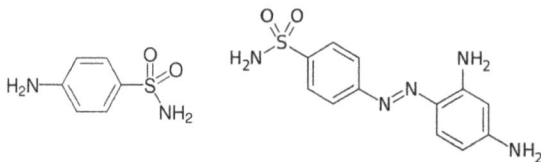

Fig. 1.4: Sulfonamide [4-aminobenzene sulfonamide] [17] and Prontosil [4-(2,4-Diaminophenyl)azo] benzenesulfonamide] [18].

Cortisone [19] was first identified by the American chemists Edward Calvin Kendall and Harold L. Mason while researching at the Mayo Clinic. In 1929, Philip S. Hench and colleagues discovered cortisone's effectiveness in the treatment of rheumatoid arthritis. Cortisone [Figure 1.5] is one of several end products of a process called steroidogenesis. This process starts with the chemical synthesis of cholesterol. Streptomycin and cortisone were commercialized.

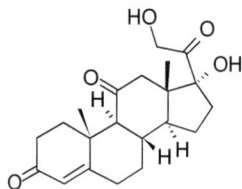

Fig. 1.5: Cortisone [17α,21-Dihydroxypregn-4-ene-3,11,20-trione] [19].

Merck [11] pioneered not only in new drugs but also lowered their selling prices through manufacturing innovation. They lowered the selling price of cortisone from $200 per gram to $16 per gram wholesale.

Schering-Plough Corporation [20] was a US-based pharmaceutical company. It was originally the US subsidiary of the German company Schering AG, founded in 1851 by Ernst Christian Friedrich Schering. In 1971, the Schering Corporation merged with Plough Inc. (founded by Abe Plough in 1908) to form Schering-Plough. Merck & Co. and Schering-Plough merged in 2009 with the new company being called Merck & Co.

Schering-Plough manufactured several pharmaceutical drugs, the most well-known of which were the allergy drugs Claritin and Clarinex, an anticholesterol drug Vytorin, and a brain tumor drug Temodar. These are now available from Merck & Co. All of the active ingredients are fine/specialty chemicals that have a disease-curing value and are formulated to a dispensable dose.

1.1.4 Pfizer

Table 1.1 tells us the storied history of Pfizer [21]. Today it is conglomerate of many companies.

With time, Pfizer grew with the acquisition of Warner Lambert, Parke Davis, Upjohn, Wyeth Labs, Pharmacia, King Pharmaceuticals, G. D. Searle, and Hospira. Lederle Labs was part of American Cyanamid, a specialty chemical company.

Pfizer was founded in New York City in 1849 by German-American Charles Pfizer and his cousin Charles F. Erhart. They launched the chemicals business Charles Pfizer and Company in Williamsburg, Brooklyn. They produced an antiparasitic product called santonin [Figure 1.6]. This was an immediate success. However, it was production of citric acid that really led to Pfizer's growth in the 1880s.

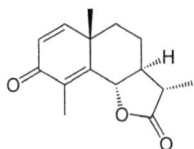

Fig. 1.6: Santonin ((3S,3aS,5aS,9bS)-3,5a,9-trimethyl-3a,5,5a,9b-tetrahydronaphtho[1,2-b]furan-2,8(3H,4H)-dione) [22].

The First World War led to shortage of calcium citrate which Pfizer imported from Italy for the manufacture of citric acid, and the company began a search for an alternative supply. Pfizer chemists learned of a fungus that ferments sugar to citric acid. In 1919, they were able to commercialize the production of citric acid from this source. This led to company's expertise in fermentation technology. These skills were applied by Pfizer to the mass production of the antibiotic penicillin during the Second World War for the need to treat injured allied soldiers.

Penicillin became very inexpensive in the 1940s, and Pfizer searched for new antibiotics with greater profit potential. This company discovered Terramycin (oxytetracycline) in 1950. This changed the company from a manufacturer of fine chemicals to a research-based pharmaceutical company. Pfizer developed a drug discovery program focusing on in vitro synthesis in order to augment its research in fermentation technology.

In 1860, pharmacists John and Frank Wyeth opened a drugstore with a small research lab in Philadelphia. In 1862, they began to manufacture large quantities of commonly ordered medicines.

In 1872, Henry Bowers, a Wyeth employee, developed one of the first rotary compressed tablet machines in the United States. This enabled the mass production of medicines with unprecedented precision and speed. Wyeth sold Anacin (combination of aspirin and caffeine), Phenergan Figure 1.7, and Advil [Figure 1.7]. Anacin was sold globally as an effective pain reliever and was a big seller in the second half of the twentieth century.

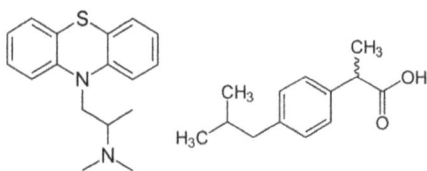

Fig. 1.7: Phenergan ((2-dimethylamino-2-methyl) ethyl-*N*-dibenzoparathiazine) [23] and ibuprofen (Advil) (2-(4-isobutylphenyl) propanoic acid) [24].

1.1.5 Warner Lambert

Warner Lambert [25] was formerly two separate companies. The first company was started in 1856, when William R. Warner founded a drug store in Philadelphia. Warner went on to invent a tablet coating process gaining him a place in the Smithsonian Institution. Tablet coatings facilitate swallowing of the tablets. Jordan Lambert founded Lambert Pharmacal Company of St. Louis. It was famous for Listerine. The two companies merged to form Warner–Lambert. Scientists at Warner–Lambert discovered Lipitor [Figure 1.8] and formed an alliance with Pfizer to bring it to market.

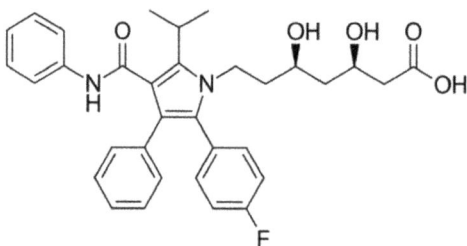

Fig. 1.8: Lipitor generic name atorvastatin ((3*R*,5*R*)-7-[2-(4-fluorophenyl)-3-phenyl-4-(phenylcarbamoyl)-5-propan-2-ylpyrrol-1-yl]-3,5-dihydroxyheptanoic acid) [26].

1.1.6 Parke, Davis

Parke, Davis and Company [27] was founded in Detroit, Michigan. It was once the world's largest pharmaceutical company and is credited with building the first modern pharmaceutical laboratory and developing the first systematic methods of performing clinical trials of new medications. Parke-Davis also produced the broad-spectrum antibiotic chloramphenicol [28], which was a blockbuster product before the discovery of its association with aplastic anemia. It is included on the World Health Organization's (WHO)'s essential medicine list [29]. The starting raw material of chloramphenicol is benzaldehyde (PhCHO), a widely used chemical. Its synthesis is illustrated in Figure. 1.9.

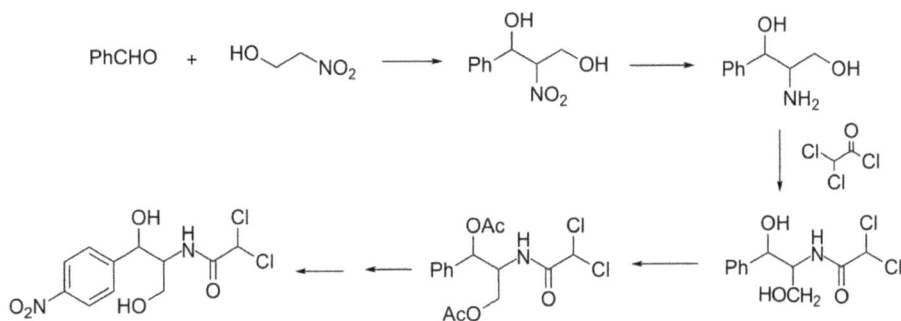

Fig. 1.9: Chemical synthesis of chloramphenicol [2,2-dichlor-N-[(aR,bR)-b-hydroxy-a-hydroxymethyl-4-nitrophenethyl] acetamide] [28].

Another product of the company was pure adrenaline. The compound was patented in 1900 and trademarked as "Adrenalin." Because of the similarity of this name to "Adrenaline," the use of the alternative name "epinephrine" for generics was mandated in the United States and is used to this day. Parke-Davis filed a lawsuit against H. K. Mulford Company alleging Adrenalin patent infringement. The ruling in favor of Parke-Davis is considered crucial to the modern patent law.

1.1.7 Upjohn

Company [30] was a pharmaceutical manufacturing firm founded in 1886 in Kalamazoo, Michigan, by Dr. William E. Upjohn. The company was originally formed to make friable pills, which were specifically designed to be easily digested. These could be "reduced to a powder under the thumb," a strong marketing argument at the time.

Upjohn developed a process for the large-scale production of cortisone. The oxygen atom at the 11th position in this steroid is an absolute requirement for biological activity. There are, however, no known natural sources for starting materials that contain that feature. The only method for preparing this drug prior to 1952 was a lengthy synthesis starting from cholic acid isolated from bile. In 1952, Upjohn biochemists announced that they were able to introduce this crucial oxygen atom by fermentation of the steroid progesterone with a common mold of the genus *Rhizopus*. Over the next several years, a group of chemists developed a process for preparing cortisone from the soybean sterol (stigmasterol). The microbiological oxygenation is a key step in this process.

Subsequently, Upjohn together with Schering biochemically converted cortisone into the more potent steroid prednisone by a bacterial fermentation [30]. Upjohn's most well-known drugs before the acquisition by Pfizer were Xanax, Halcion, Motrin, and Rogaine.

Before its acquisition by Pfizer, Hospira was the world's largest producer of generic injectable pharmaceuticals [31]. The difference between the solid dose pharmaceutical and the injectable dose pharmaceutical is that the solid dose is taken orally and the latter is injected in the blood stream or body tissue. It manufactured generic acute-care and oncology injectables, as well as integrated infusion therapy and medication management systems. Before 2004, Hospira was part of Abbott. Sodium thiopental is an anesthetic discovered by Abbott Laboratories in the 1930s. Hospira manufactured the drug after splitting off from Abbott under the brand name Pentothal.

G.D. Searle became part of Monsanto, a chemical company, and later through Pharmacia, it became a part of Pfizer. One of its famous products is aspartame, an artificial sweetener.

1.1.8 Bayer AG

In 1863, Friedrich Bayer and his partner, Johann Friedrich Weskott, founded a dyestuffs factory in Barman, Germany, which later came to be known as Bayer [32]. Fuchsine [Figure 1.10], a dye, and aniline [Figure 1.11] became the company's most important products.

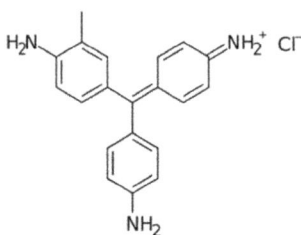

Fig. 1.10: Fuchsine (4-[(4-aminophenyl)-(4-imino-1-cyclohexa-2,5-dienylidene)methyl]aniline hydrochloride [33].

Fig. 1.11: Aniline [34].

More than 2,000 years ago, Hippocrates discovered that chewing willow bark reduced fever and relieved pain [35] but had gastric irritation. French chemist Charles Frederic Gerhardt in 1853 produced a less irritating compound by combining sodium salicylate (sodium helps neutralize the salicylic acid) with acetyl chloride. He lost interest in this compound. Felix Hoffmann, German chemist at Friedrich Bayer & Co, modified Gerhardt's structure to produce acetyl salicylic acid [Figure 1.12].

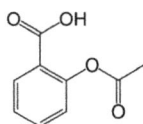

Fig. 1.12: Acetyl salicylic acid (aspirin) [36].

In 1899, Bayer's trademark "Aspirin" was registered worldwide for its brand of acetylsalicylic acid, but it lost its trademark status in the United States, France, and the United Kingdom after the confiscation of Bayer's US assets and trademarks during the First World War and because of the subsequent widespread usage of the word.

As of 2011, approximately 40,000 tons of aspirin were produced each year and 10–20 billion tablets were consumed in the United States alone for prevention of cardiovascular events. Aspirin is on the WHO's essential medicines list [29] and considered one of the most important medications needed in a basic health system. Its crossed logo "BAYER" name is a world-famous one and is stamped on each aspirin tablet.

Heroin (diacetylmorphine), Figure 1.13, now illegal as an addictive drug, was introduced as a nonaddictive substitute for morphine, and trademarked and marketed by Bayer as a cough suppressant and over-the-counter treatment for other common ailments, including pneumonia and tuberculosis. Bayer's scientists were not the first to make heroin, but the company led the way in commercializing it.

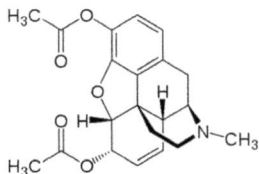

Fig. 1.13: Heroin (diacetylmorphine) [37].

In 1903, Bayer licensed the patent for the hypnotic drug diethylbarbituric acid from its inventors. Systematic investigations of the effect of structural changes on potency and duration of action at Bayer led to the discovery of phenobarbital in 1911 and the discovery of its potent antiepileptic activity in 1912. Phenobarbital was among the most widely used drugs for the treatment of epilepsy through the 1970s and is on the WHO's list of essential medications.

In 1916, Bayer's scientists discovered suramin [Figure 1.14], an antiparasite drug, for sleep sickness and is still sold under the brand name Germanin. It is also on the WHO's list of essential medicines.

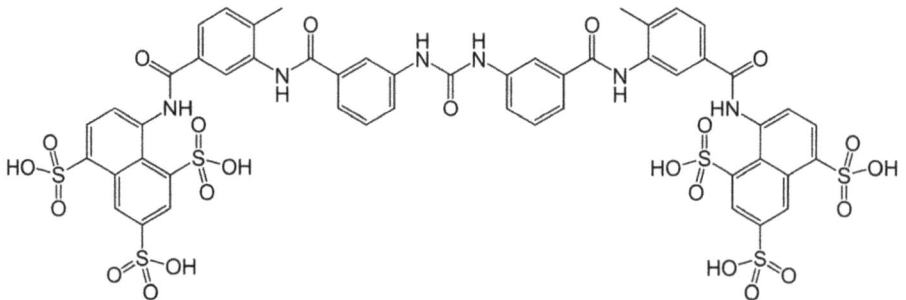

Fig. 1.14: Suramin (8,8'-[ureylenebis[m-phenylenecarbonylimino(4-methyl-m-phenylene)carbonylimino]]di-1,3,5-naphthalenetrisulfonic acid) [38].

In 1925, Bayer became a part of IG Farben [39]. Six chemical companies were BASF, Bayer, Hoechst, Agfa, Chemische Fabrik Griesheim-Elektron, and Chemische Fabrik vorm. Weiler Ter Meer made this conglomerate. Its history is a fascinating read.

Bayer's scientists discovered prontosil, the first commercially available antibacterial drug. Antibiotic ciprofloxacin [40] [Figure 1.15] and Yaz [Figure 1.16] [41] (drospirenone) birth control pills were also introduced. Ciprofloxacin was widely used for Anthrax treatment in 2001. Other fluoroquinolones make a class of antibacterial drugs and were later synthesized.

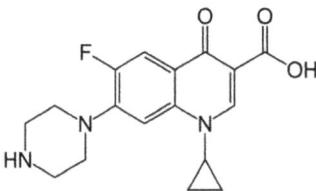

Fig. 1.15: Ciprofloxacin (1-cyclopropyl-6-fluoro-1,4-dihydro-4-oxo-7-(1-piperazinyl)-3-quinolinecarboxylic acid) [40].

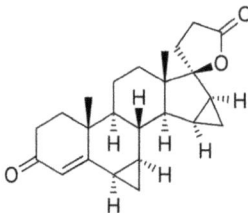

Fig. 1.16: Drospirenone (6R,7R,8R,9S,10R,13S,14S,15S,16S,17S)-1,3',4',6,6a,7,8,9,10,11,12,13,14,15,15a,16-Hexadecahydro-10,13-dimetylspiro-[17H dicyclopropa[6,7:15,16]cyclopenta[a]phenantrene-17,2'(5'H)-furan]-3,5'(2H)-dione [41].

After the war, IG Farben was split into its six constituent companies (BASF, Bayer, Hoechst, Agfa, Chemische Fabrik Griesheim-Elektron, and Chemische Fabrik vorm). In 1951, it was split again into three: BASF, Bayer, and Hoechst.

Processes developed and commercialized at IG Farben and other companies were sophisticated and elaborate (i.e., alternate process chemistries were considered), and principles of good chemistry and chemical engineering were applied. *An example is cited at the end of this chapter in* Figure 1.30.

1.1.9 Eli Lilly

Eli Lilly and Company [42] was founded in 1876 by, and named after, Col. Eli Lilly, a pharmaceutical chemist and veteran of the American Civil War. Lilly's notable achievements include being the first company to mass-produce the polio vaccine developed by Jonas Salk, and insulin.

It was one of the first pharmaceuticals to produce human insulin using recombinant DNA including Humulin (insulin medication), Humalog (insulin lispro), and the first approved biosimilar insulin product Basaglar (insulin glargine) in the United States.

Quinine [43] was one of the first drugs produced by Lilly to treat malaria. It was extracted from cinchona tree bark and first used to treat diarrhea and then malaria in 1631. It was chemically synthesized and perfected over time [44].

Lilly committed himself to producing high-quality prescription drugs, in contrast to the common and often ineffective patent medicines. From its facilities in Indianapolis, the company manufactured and sold "ethical drugs" for use by the medical profession. Lilly's medicine labels disclosed product ingredients.

The company used plants for its raw materials and produced its products by hand. In addition to development of new medicines, the company achieved several technological advances, including automation of its production facilities. Lilly was also an innovator in pill capsule manufacturing. It was among the first manufacturers to insert medications into empty gelatin capsules, which provided a more exact dosage. Lilly manufactured capsules for its own needs and sold its excess capacity to others. Other Lilly's innovations were fruit flavoring for medicines and sugar-coated pills to make their medicines easier to swallow. Over the next few years, the company began to create tens of millions of capsules and pills annually.

Other advances improved plant efficiency and eliminated production errors. Eli Lilly introduced a method for blueprinting manufacturing tickets in 1909. This process, which created multiple copies of a drug formula, helped eliminate manufacturing and transcription errors. In the 1920s, Eli introduced the new concept of straight-line production, where raw materials entered at one end of the facility and the finished product came out the other end, in the company's manufacturing process.

In 1923, Lilly began selling Iletin (Insulin, Lilly), their trade name for the first commercially available insulin product in the United States for the treatment of diabetes. The Second World War brought production at Lilly to a new high with the manufacturing of Merthiolate and penicillin.

Lilly all throughout has continued to develop and introduce new drugs to the marketplace, for example, vancomycin and erythromycin. Lilly was heavily involved in production and distribution of Jonas Salk's polio vaccine.

Lilly is one of the largest manufacturers of psychiatric medications and produces Prozac (fluoxetine) [Figure 1.17], Dolophine (methadone), Cymbalta (duloxetine) [Figure 1.17], and Zyprexa (olanzapine). Prozac is now available as a generic also. The wholesale cost in the developing world is between US $0.01 and US $0.04 per day as of 2014 [45]. In the United States, it costs about US $0.85 per day. Rationale for such price differences is discussed in later chapters.

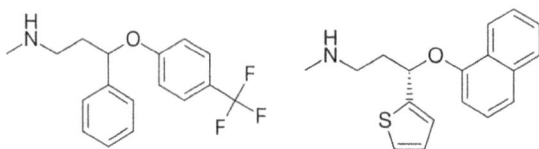

Fig. 1.17: Prozac (fluoxetine) (*N*-methyl-3-phenyl-3-[4-(trifluoromethyl)phenoxy] propan-1-amine) [46]. Duloxetine ((+)-(*S*)-*N*-methyl-3-(naphthalen-1-yloxy)-3-(thiophen-2-yl)propan-1-amine) [47].

1.1.10 Abbott

In 1888, Wallace Abbott started the Abbott Alkaloidal Company [48] in Chicago. He was a practicing physician and drug store owner. His innovation was the use of the active part of a medicinal plant, generally an alkaloid (e.g., morphine, quinine (cinchona), strychnine, and codeine), which he formed into tiny "dosimetric granules" that were essentially the same sized pills. This provided patients more consistent and effective dosages.

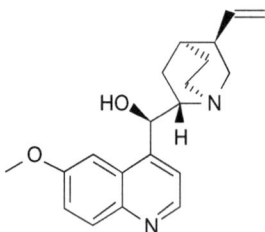

Fig. 1.18: Quinine ((5-ethenyl-1-azabicyclo[2.2.2]octan-2-yl)-(6-methoxyquinolin-4-yl) methanol) [49].

In 2004, it spun off its hospital products division into a new company named Hospira. In February 2010, Abbott acquired a pharmaceutical segment of Solvay S.A., a chemical company. This provided Abbott with a large and complementary portfolio of pharmaceutical products. In 2010, Abbott Laboratories bought Piramal Healthcare Ltd.'s Healthcare Solutions.

In October 2011, the company separated itself into two companies, one research-based pharmaceuticals and the other in medical devices, generic drugs sold internationally, and diagnostics, with the latter retaining the Abbott name. Abbott Nutrition, whose products include Similac, Pedialyte, Glucerna, and Ensure, also retained the Abbott name. The other company was named AbbVie in March 2012.

1.1.11 Imperial Chemical Industries (ICI) and AstraZeneca

Imperial Chemical Company (ICI) [50] was a glorious name in the chemical industry. The company was founded in December 1926 from the merger of four companies: Brunner Mond, Nobel Explosives, the United Alkali Company, and British Dyestuffs Corporation.

In the 1920s and 1930s, the company played a key role in the development of new chemical products, including dyestuff phthalocyanine (1929), acrylic plastic Perspex (1932), Dulux paints (1932, codeveloped with DuPont), polyethylene (1937), and polyethylene terephthalate fiber known as Terylene (1941).

In the 1940s and 1950s, the company established its pharmaceutical business and developed a number of key products, including Paludrine (antimalarial drug), halothane (anesthetic agent), Inderal (beta blocker), and tamoxifen (frequently used drug for breast cancer). In 1993, the company divested its pharmaceutical bioscience businesses: pharmaceuticals, agrochemicals, specialties, seeds, and biological products to a new and independent company called Zeneca. Zeneca subsequently merged with Astra AB to form Astra Zeneca [50]. ICI name disappeared in 2008.

Astra AB is a former international pharmaceutical company headquartered in Södertäljae, Sweden. Astra merged with Zeneca Group in 1999 to form AstraZeneca. Astra's product development was focused on therapeutics for gastrointestinal, cardiovascular, and respiratory disorders and pain control.

One of its very successful drugs Omeprazole was first made in 1979 by Swedish AB Hässle, a division of Astra AB. It was the first of the proton pump inhibitors. It is sold under the Prilosec name. When Prilosec's US patent expired in April 2001, AstraZeneca introduced esomeprazole as a patented replacement drug. Esomeprazole is the "S-isomer" of omeprazole, and its magnesium salt is sold as Nexium (purple pill) Figure 1.19.

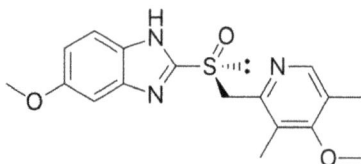

Fig. 1.19: Esomeprazole (6-methoxy-2-[(S)-(4-methoxy-3,5-dimethylpyridin-2-yl)methylsulfinyl]-1H-benzimidazole) [51].

1.1.12 Bristol Myers Squibb

In 1887, William McLaren Bristol and John Ripley Myers purchased the Clinton Pharmaceutical company of Clinton, New York. Later, it was named Bristol, Myers and Company and then changed it to the Bristol-Myers Corporation [52]. The first nationally recognized product was Sal Hepatica, a laxative mineral salt in 1903. Its second national success was Ipana toothpaste, from 1901 through the 1960s. Other divisions were Clairol (hair colors and hair care) and Drackett (household products such as Windex and Drano).

In 1943, Bristol-Myers acquired Cheplin Biological Laboratories, a producer of acidophilus milk in East Syracuse, New York, and converted the plant to produce penicillin for the Second World War. After the war, the company renamed the plant Bristol Laboratories in 1945 and entered the civilian antibiotics market, where it faced competition from Squibb, which had opened the world's largest penicillin plant in New Brunswick, New Jersey. Penicillin production at the East Syracuse plant was stopped in 2005, when it became less expensive to produce overseas, but the facility continues to be used for the manufacturing process development and production of other biologic medicines for clinical trials and commercial use.

It is ironic that the companies who pioneered mass scale production of injectables in the United States are not producing them.

1.1.13 E. R. Squibb

E. R. Squibb, a Navy physician, started his pharmaceutical company in 1858. While at Navy in 1854, he invented an improved distillation method for ether for anesthetic use that he gave away for free. He was a strong advocate of product quality and high purity.

Bristol-Myers and Squibb merged in 1989 to become Bristol Myers Squibb [52]. In 1999, President Clinton awarded Bristol-Myers Squibb the National Medal of Technology, the nation's highest recognition for technological achievement, "for extending and enhancing human life through innovative pharmaceutical research and development and for redefining the science of clinical study through groundbreaking and hugely complex clinical trials that are recognized models in the industry."

1.1.14 Novartis, Ciba Geigy, Sandoz

Roots of Novartis, Ciba Geigy, and Sandoz are very interesting. They pioneered fine/specialty chemical synthesis that served multipurpose global needs that served healthcare needs and changed lifestyle. Over time through reorganization and business needs, they divested and established different entities in fine/specialty chemicals and pharmaceuticals. Formation of Novartis is a complex merger and demerger phenomenon.

1.1.14.1 Ciba

Ciba-Geigy history is an interesting read. Ciba [53] was formed in 1859 by Alexander Clavel when he took up the production of fuchsine, a dye. He sold his factory to the company Bindschedler and Busch. In 1884, Bindschedler and Busch was transformed into a joint-stock company named "Gesellschaft für Chemische Industrie Basel" (Company for Chemical Industry Basel). The acronym, CIBA, was adopted as the company's name in 1945. By 1900, Ciba was the largest chemical company in Switzerland. Ciba, however, started a limited diversification into the pharmaceutical business with the introduction of an antiseptic called Vioform [Figure 1.20].

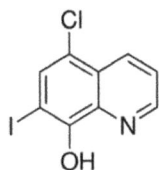

Fig. 1.20: Vioform (5-chloro-7-iodo-8-quinolinol) [54].

The name "Ciba-Geigy" was shortened to Ciba and in 1992 it became part of Sandoz. The chemical producing part was later acquired by BASF in 2008. Ciba in its hay days produced multitude of fine/specialty additive chemicals and dyes. They were ranked number one in their respective product categories.

1.1.14.2 Geigy

Geigy [55] was established in 1857 by Johann Rudolf Geigy-Merian and Johann Muller-Pack. They built a dyewood mill and a dye extraction plant. Two years later, they also began production of synthetic fuchsine. In 1901, they formed the public limited company Geigy, and the name of the company was changed to J. R. Geigy Ltd in 1914.

In 1942, Geigy informed the U.S. Military Attaché in Berne that the Neocid formulation of DDT (dichlorodiphenyltrichloroethane) [Figure 1.21] was very effective against the typhus carrying louse. The U.S. Department of Agriculture verified Geigy's claims. Geigy and Sandoz joint venture started the manufacture of DDT in 1943. This joint venture led to the establishment of the Cincinnati Chemical Works, a subsidiary that gave Basle AG a tariff-free foothold in the American market. DDT was used worldwide as an insecticide. In 1965, author was an apprentice at Hindustan Insecticide plant in India. DDT was eventually banned as an insecticide due to its ill-effects on wildlife.

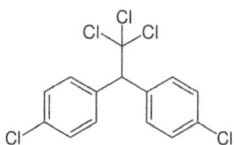

Fig. 1.21: Dichlorodiphenyltrichloroethane (DDT) [56].

Maumee Chemicals, a Toledo base company, bought Cincinnati Chemical Works. It would later go on to become a division of the Sherwin Williams, the paint company. This was later sold to PMC Specialty Group Inc. In 1970s and 1980s, this site was the largest producer of saccharin in the world. Author was associated with saccharin and many products developed and produced at this and other sites.

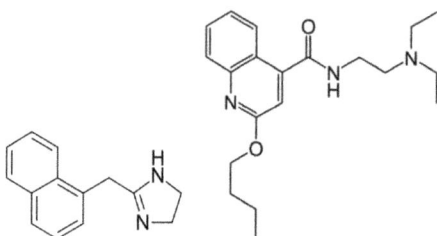

Fig. 1.22: Chemical structures of privine (2-(naphthalen-1-ylmethyl)-4,5-dihydro-1Himidazole) [57] and Nupercaine (2-butoxy-N-[2-(diethylamino)ethyl]quinoline-4-carboxamide) [58].

Research during the war led to the development of several ethical drugs, including Privine Figure 1.22, a treatment for hay fever, and Nupercaine Figure 1.22, a spinal anesthetic used in childbirth. Companies also developed drugs for treatment of high blood pressure and heart disease. CIBA and Geigy merged in 1970 to form Ciba-Geigy Ltd.

Ciba-Geigy in 1960s sold optical brighteners called Tinopal, a specialty chemical, as powder to be added to laundry. Countries that had laundry washers had it mixed in their detergent. After divestitures, BASF owns the business now.

1.1.14.3 Sandoz

The Chemiefirma Kern und Sandoz ("Kern and Sandoz Chemistry Firm") was started in 1886 by Alfred Kern and Edouard Sandoz [55]. The first dyes manufactured by them were alizarinblue and auramine. Company became Chemische Fabrik vormals. It started producing the fever-reducing drug antipyrine [Figure 1.23] in the 1895. In 1899, the company began producing the sugar substitute saccharin [Figure 1.23]. Pharmaceutical research began in 1917. In 1918, ergotamine [Figure 1.23] was isolated from ergot; the substance was eventually used to treat migraine and head-aches and was introduced under the trade name Gynergen in 1921.

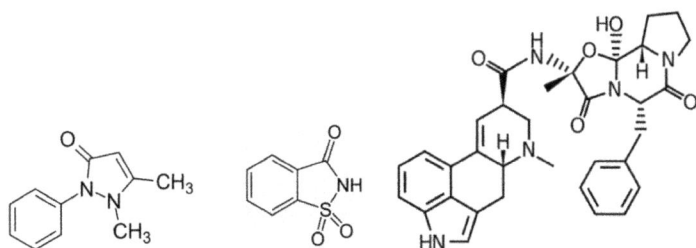

Fig. 1.23: Chemical structures of antipyrine (2-dihydro-1,5-dimethyl-2-phenyl-3H-pyrazol-3-one) [59], saccharin (1,1-dioxo-1,2-benzothiazol-3-one) [60], ergotamine (2'-methyl-5'α-benzyl-12'-hydroxy-3',6',18-trioxoergotaman; 9,10α-dihydro-12'-hydroxy-2'-methyl-5'α-(phenylmethyl) ergotaman-3',6',18-trione) [61].

The psychedelic effects of lysergic acid diethylamide (LSD) were discovered at the Sandoz Laboratories in 1943. Sandoz began clinical trials and marketed the substance, from 1947 through the mid-1960s, under the name Delysid as a psychiatric drug, which is thought useful for treating a wide variety of mental ailments, ranging from alcoholism to sexual deviancy. Sandoz suggested in its marketing literature that psychiatrists take LSD themselves, to gain a better subjective understanding of the schizophrenic experience, and many did exactly that and so did other scientific researchers. In the 1960s, it became a cultural phenomenon.

1.1.14.4 Novartis
In 1995, Sandoz spun off its specialty chemicals and in 1996 merged with Ciba-Geigy to form Novartis. Other Ciba-Geigy and Sandoz businesses were spun off as independent companies. Some parts of Sandoz were spun off also. In 2005, Sandoz became the world leader in generic drugs after acquiring Germany's Hexal AG and US-based Eon Labs. In 2005, Novartis acquired the North American over-the-counter brand portfolio of Bristol-Myers Squibb.

1.1.15 Roche

Hoffman-La Roche [62] started in 1986. It became known for its synthetic production of ascorbic acid [63] Figure 1.24.

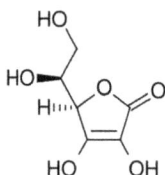

Fig. 1.24: L-ascorbic acid (5R)-[(1S)-1,2-Dihydroxyethyl]-3,4-dihydroxyfuran-2(5H)-one [63].

While looking for a treatment for tuberculosis (isoniazid), it developed an anti-depressant iproniazid. In 2009, Genentech with its focus on biotechnology drugs became part of the Roche Group.

1.1.16 Johnson & Johnson

Robert Wood Johnson and his brothers, James Wood Johnson, and Edward Mead Johnson [64] created ready-to-use surgical dressings in 1885. The company produced its earliest products in 1886. Their initial logo was very similar to the logo used today.

McNeil Consumer Healthcare was founded on March 16, 1879, by Robert McNeil. In 1904, the company focused on the direct marketing of prescription drugs to hospitals, pharmacists, and doctors. The development of acetaminophen began 1959. Johnson & Johnson acquired McNeil Laboratories. Company started selling Tylenol [65] [Figure 1.25] without a prescription.

Janssen Pharmaceuticals was created in 1933. In 1961, the company was acquired by Johnson & Johnson. In 1999, under the name of Janssen Pharmaceutica N.V. became part of Johnson and Johnson.

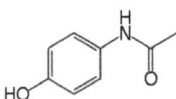

Fig. 1.25: Tylenol (*N*-(4-hydroxyphenyl) acetamide) [65].

1.1.17 Boots

Ibuprofen [Figure 1.26] was derived from propionic acid during the 1960s. Its discovery was the result of research during the 1950s and 1960s to find a safer alternative to aspirin. It was discovered in 1961. Boots [66] is now part of Walgreens.

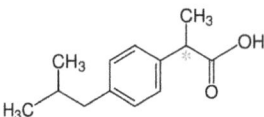

Fig. 1.26: Ibuprofen ((*RS*)-2-(4-(2-methylpropyl)phenyl)propanoic acid) [67].

1.1.18 Boehringer Ingelheim

C. H. Boehringer Sohn AG & Ko. KG is the parent company of the Boehringer Ingelheim [68] group, which was founded in 1885. They started manufacturing tartaric acid [Figure 1.27] for use in the food industry. In 1893, while experimenting with the production of citric acid [Figure 1.28], lactic acid [Figure 1.29] is formed. Albert Boehringer developed the process to produce lactic acid on a larger scale. Commercial-scale production of lactic acid began in 1895.

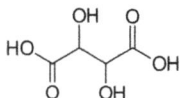

Fig. 1.27: Tartaric acid dihydroxybutanedioic acid [69].

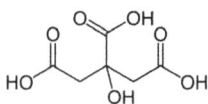

Fig. 1.28: Citric acid 2-hydroxypropane-1,2,3-tricarboxylic acid [70].

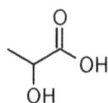

Fig. 1.29: Lactic acid 2-Hydroxypropanoic acid [71].

Boehringer Ingelheim is in human pharmaceuticals, animal health, and biopharmaceuticals. Its focus areas are cardiometabolic, central nervous systems, immunology and respiratory diseases, and oncology and cancer immunology.

1.1.19 Sanofi

Evolution of Sanofi [72] is an interesting pathway for the development of a pharmaceutical company. Sanofi was founded in 1973 as a subsidiary of Elf Aquitaine (a French oil company subsequently acquired by Total), when Elf Aquitaine took control of the Labaz group, a pharmaceutical company formed in 1947, by Societe Belge de l'Azote et des Produits Chimiques du Marly. Labaz developed benziodarone in 1957.

1.1.20 Aventis

Aventis [73] was formed in 1999, when French company Rhône-Poulenc S.A. merged with the German corporation Hoechst Marion Roussel, which itself was formed from the 1995 merger of Hoechst AG with Cassella, Roussel Uclaf, and Marion Merrell Dow.

At the time of the merger, Rhône-Poulenc's business included the pharmaceutical businesses, vaccines, the plant and animal health businesses, and a 67% share in Rhodia, a specialty chemicals company. Hoechst is one of the companies that resulted from the post-Second World War split of IG Farben.

Sanofi-Aventis was formed in 2004, and in 2011 Aventis name was dropped.

1.1.21 Novo-Nordisk

Novo-Nordisk's [74] major business emphasis is diabetes.

1.1.21.1 Novo Industri A/S

August Krogh, a Danish physiologist and Nobel Prize recipient, informed his colleagues of innovative drug research taking place in Toronto. Scientists over there used pancreas extracts as a treatment for diabetes. Inspired by Krogh's enthusiasm, a number of Danes engaged in further investigation of this revolutionary hormone called insulin. Among these early converts were Harald Pedersen, a mechanical engineer, and his brother Thorvald, a pharmacist. Together they established a rudimentary production facility [75] in the basement of Harald's home in Copenhagen. In 1925, just 4 years after the discovery of insulin, the Pedersen brothers were producing a stable, commercially viable solution called "Insulin Novo."

In 1931, production demands required the Pedersen's to leave their cellar and rent space in a former dairy factory. Eventually, the brothers purchased the building along with property surrounding the plant. Growing in just 10 years from a fledgling basement operation into a large-scale enterprise, the company sold insulin in 40 countries. Pancreas from oxen, calves, and swine were procured from slaughterhouses across Europe and transported to Novo first by refrigerated car, then by railway van, and finally by trucks.

1.1.21.2 Nordisk Gentofte A/S

In 1966, the Nordic Insulin Laboratory became involved in the production of growth hormone and was among the first companies in the world to research growth disorders. In 1973, the company's first growth hormone preparation Nanormon *was* approved by the Danish health authorities. Nanormon was based on growth hormone extracted from pituitary glands. In 1982, Nordisk Gentofte began the development of biosynthetic growth hormone and commercialized its first biosynthetic human growth hormone. The product was named Norditropin.

In 1989, Novo Industri A/S (Novo Terapeutisk Laboratorium) and Nordisk Gentofte [74, 75] A/S (Nordisk Insulin laboratorium) merged to become Novo Nordisk A/S, the world's largest producer of insulin.

1.1.22 Other companies

Sterling Drug [76], an American pharmaceutical company, also known as Sterling Winthrop Inc., started in 1901, and purchased US assets of Bayer AG at the end of the First World War in 1918. It had the rights to sell Bayer aspirin in the United States, the United Kingdom, and Commonwealth countries. In 1940, a cross-contamination from equipment sharing resulted in Winthrop Chemical producing contaminated sulfathiazole tablets contaminated with phenobarbital. This led to the cGMP [77] practices for the manufacture of drugs. Sterling was sold to Eastman Kodak in 1988 and in 1994 it sold its drug subsidiary to what is Sanofi today [72].

Gilead, Amgen, Shire, Allergan, and Vertex to name a few are some companies of recent vintage. Some of these have developed their own drugs, and others have acquired companies to serve specific patient needs.

1.2 Second half of twentieth century and generic pharma companies

Most of the companies mentioned above patented their products sold them through prescriptions and were called ethical drugs or brand drugs. They produced disease-curing chemicals that were formulated and sold and were sold under their trademarks. They did not use the patent system as used in the second half of the twentieth century and the current times.

Around the Second World War, the use of the patent system became increasingly important. In addition, proof of drug safety and usefulness became progressively important. Patent attorneys understood the value of patent system and validity of patent length. Ethical/brand pharma companies used the length of patents to capitalize on opportunity offered to keep the competition off the landscape.

In the United States, once the patents expired, the drugs were called "Generic," and Kefauver Harris Amendment or "Drug Efficacy Amendment" of 1962 [78, 79] led to the following changes to the drug/pharmaceutical landscape:
1. Required that manufacturers prove the effectiveness of drug products before they go on the market, and afterward report any serious side effects.
2. Required that evidence of effectiveness be based on adequate and well-controlled clinical studies conducted by qualified experts. Study subjects would be required to give their informed consent.
3. Gave FDA 180 days to approve a new drug application, and required FDA approval before the drug could be marketed in the United States.
4. Mandated that FDA conduct a retrospective evaluation of the effectiveness of drugs approved for safety – but not for effectiveness – between 1938 and 1962.
5. Allowed FDA to set good manufacturing practices for industry and mandated regular inspections of production facilities.

6. Transferred to FDA control of prescription drug advertising, which would have to include accurate information about side effects.
7. Controlled the marketing of generic drugs to keep them from being sold as expensive medications under new trade names.

FDA's drug development process became the gold standard to which other countries aspired.

This Act reduced the number of new drugs approved and increased the cost of approval of drugs. Eventually this Act also has resulted in significant reduction in number of new drug developments and their development costs have gone up [80].

In 1983 [81], Orphan Drug Act was passed for rare diseases. This law is designed for less than 200,000 patients. Pharmaceutical companies did not focus on most rare diseases till recently. Most of the recent drugs for rare disease run in six figures. There is significant bruhaha about their prices when fundamental research is funded by the US universities and government.

In 1984, Hatch-Waxman Act [82] was adopted to facilitate introduction of generic drugs through abbreviated new drug application (ANDA) process. These drugs do not require preclinical trials and clinical data for safety and efficacy but are required proof of bioequivalence. Flood gates for generic competition opened. There were many issues at FDA and still to date the ANDA filing and approval process is mired with bureaucracy and delays [79]. This will be discussed in later chapters.

Hatch-Waxman Act [82] encouraged generics to challenge innovator companies. Brand Companies used every loop hole, for example, "pay for delay [82, 83]." Excellent examples are AstraZeneca's sale of Nexium and AbbVie's sale of Humira. AbbVie has over 130 patents to extend the life of its patent on Humira. There are too many examples to enumerate here. AbbVie's pricing of Humira has been a cause of US Congressional hearings [84].

Generic drugs make up better than 50% of the drugs used globally. Their manufacturing technologies, affordability, and regulatory compliance are important. These are discussed in the following chapters.

1.3 IG Farben's contribution

IG Farben [39] affiliates and other European companies did considerable contribution for the well-being of humanity during the first half of the twentieth century. The United States and Great Britain obviously must have been extremely impressed with the state of chemical technology developed and practiced in Germany and as it applied to human health and lifestyle. After the Second World War, US and British Intelligence Offices catalogued these chemistries and manufacturing methods of many drugs and chemicals.

It is worth reviewing the synthesis of some of drugs that were produced at IG Farben. They illustrate the chemical synthesis and process innovation during the first half of the twentieth century. Following are illustration of two important drugs that were produced at IG Farben.

1.3.1 Sontochin

During the Second World War, malaria was very prevalent. A cure was needed. Sontochin [85, 86] was the original chloroquine replacement drug, arising from research by Hans Andersag 2 years after chloroquine (known as "resochin" at the time) had been shelved due to the mistaken perception that it was too toxic for human use. Sontochin, that is, 3-methylchloroquine, had significant activity against chloroquine-resistant strains of *Plasmodium falciparum*.

1.3.1.1 Preparation of sontochin, resochin, and brachysan

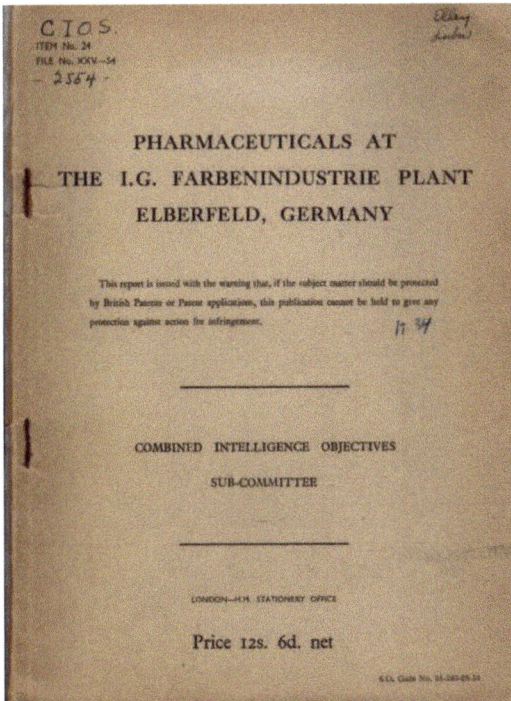

Fig. 1.30: Cover Page "Pharmaceuticals at the I.G. Farbenindustrie plant Elberfeld, Germany" [87].

Sontochin has the following structure:

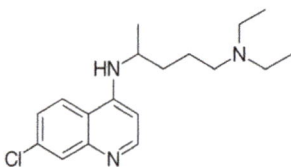

Fig. 1.31: Synthesized structures.

All of these compounds are made in a similar manner. A description of the preparation of sontochin will suffice for this class of compounds. Resochin and sontochin are patented in D. R.P. 683692 (Nov. 13, 1939).

1. *Preparation of sontochin:* Preparation of 3-methyl-4-hydroxy-7[5]-chlorquinoline-carboxyl acid ester.

In an apparatus for continuous azeotropic distillation, 4,450 g of oxalylpropionic acid ester, 2,800 g of *m*-chloroaniline, 4 L of chloroform, and 7 cc of concentrated hydrochloric acid are refluxed until the calculated amount of water is distilled off (22 mol or 396 cc). This requires about 8–10 h. The chloroform is then removed by distillation *in vacuo*. The Schiff's base remains behind as a red oil.

To this raw product, 70 L of paraffin oil is added and heated to 255 °C in an iron stirring vessel with direct heating. The alcohol is removed by distillation as formed and after this has occurred, the lot is cooled. The reaction product crystallizes at about 160 °C, is filtered and washed twice with ligroin, and dried at 100 °C. In this reaction, a mixture of isomers results and melts at 190 °C.

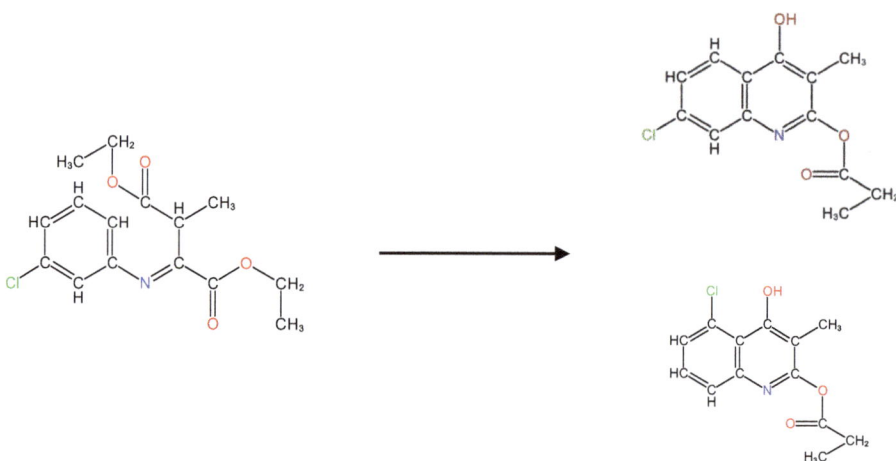

Fig. 1.32: Preparation of 3-methyl-4-hydroxy-7 (5) chloroquinoline-carboxyl acid ester.

The isomeric mixture is separated by dissolving in 18 L of alcohol and heated to boiling. While being held at this temperature, 1 kg of dry HCl gas is led in. The solution is gradually cooled in 2 h. During cooling, 2 kg more of HCl is added causing the 7-chloro compound to crystallize out. This is filtered off and washed once with 2.5 L of alcohol containing HCl and finally washed with alcohol. The product is dried.

The yield is 3,110 g (47% calcined on *m*-chloroaniline).

The melting point (MP) of HCl salt is 196 °C and of free ester is 226 °C. The alcoholic mother liquor contains the 5-chloroisomer and a little of the 7-chloroisomer.

2. *Preparation of 3-methyl-4-hydroxy-7-chloroquinoline-2-carboxylic acid*

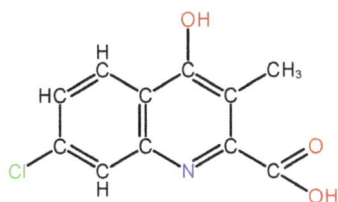

About 3,110 g of the ester-HCl is heated for 2 h under a reflux with 15.5 L of 2 N sodium hydroxide.

After the reaction is complete, the solution is diluted to 40 L with water and acidified with dilute hydrochloric acid. The product is filtered, washed with water, and dried at 100 °C.

The MP = 264 °C (CO_2 evolution).

The yield is 2,410 g (100%).

3. *Preparation of 3-methyl-4-hydroxy-7-chlorquinoline*

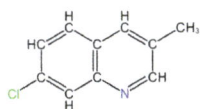

About 10 L of paraffin oil is heated at 275 °C. To this 2.5 kg of the acid is added in small portions during 20–30 min. Carbon dioxide is split out of the molecule and is led away with the help of a water pump. After the evolution of CO_2 has ceased, the mixture is cooled and the product crystallizes out at 80 °C. It is filtered off and washed three times with ligroin and dried at 100 °C.

It has an MP of 334 °C.

Yield is 1,970 g (97%).

The oil may be used for four to five runs and then purified with sulfuric acid and sodium hydroxide.

4. *Preparation of 3-methyl-4,7-dichloroquinoline*

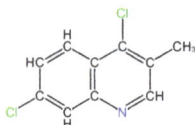

About 5 kg of the free hydroxy compound is added to 15 L of dry chlorbenzol at 170 °C. To this 3.4 kg of $POCl_3$ is added during 2 h in small quantities, beginning with portions of about 50 cc which end later for the last 800 cc, and 100 cc at a time. After each addition, the course of the reaction is watched (lively boiling). After adding all $POCl_3$, the mixture is heated for 3 h and allowed to cool with stirring.

The cooled chlorobenzene solution is added to 25 L of water and stirred for ½ h. Then about 10 kg of ice and 5 L of chlorbenzol are added, and when the temperature has reached 20 °C, ammonium hydroxide is added to alkalinize the solution (about 6 L). The lot is stirred for 15 min and allowed to stand for 1 h for the product to separate into layers. The lower layer contains the dichlor compound dissolved in chlorbenzol. This layer is drawn off and washed with equal volume of water.

To determine the MP of the product dissolved in the chlorbenzol, a small amount of the solution is distilled and the MP of the product determined:

$BP_2 = 140 °C$
$MP = 90 °C$
Yield = 92%

5. *Sontochin base*

The solution of the dichlor compound from (d) above is heated to distill off enough chlorbenzol to remove water.

Then 2 kg of pyridine, 2.5 kg of phenol, and 2.5 kg of sodium iodide are added and the temperature is raised to 150–160 °C. Over a period of 3 h, 5.6 kg (1.5 mol) of novoldiamine is added to this mixture. The temperature is then raised to 160–170 °C and held at this point for 15 h.

After cooling, the melt is diluted with 15 L of methylene chloride and 6.5 L of glacial acetic acid, and 3.0 kg crystalline sodium acetate in 15 L of water are added. After 5 min stirring, the lot is allowed to stand for ½ h to separate. The methylene chloride layer which contains phenol and weakly basic portions is filtered. The methylene chloride is washed with 1 L of 10% acetic acid which is added to the chief portion of the lot.

The combined acetic acid contains sontochin and excess novoldiamine. About 10 L of benzol is added to it, cooled with ice, and made strongly alkaline with 150 mol of sodium hydroxide solution. The layers are allowed to separate and the benzol layer containing the sontochin is removed. This is washed with 5% sodium hydroxide solution and twice with water.

The benzol is removed *in vacuo* and the residue is distilled *in vacuo*:

$PP_2 = 210-220\_°C$

Yield = 6.1 kg (77.6%)

In Kikuth's opinion, sontochin is a promising antimalarial, but is rather diffi-cult to prepare. Brachysan, on the other hand, is just as effective according to Sioli, but is much easier to prepare. The basic side chain is prepared at Ludwigshafen by the following reaction:

$$CH_2-CH-C \equiv N + HN(C_2H_5)_2 \; -> N \equiv -CH_2CH_2-N\;(C_2H_5)_2 \; -> H_2N(CH_2)_3N(C_2H_5)_2$$

1.3.2 Salicylic acid

Salicylic acid [88] chemistry is simple. The process is capitalized on the sublimation property of the product to produce a high purity product. This indicates that the process developers were cognizant of product quality and purity of the commercial product.

1.3.2.1 Introduction

Objectives
The salicylic acid plant of the I. G. Farbenindustrie at Leverkusen was investigated on November 5, 1945. Dr. Wingler (acting agent for contacting American and British investigators) and Dr. Böhme (in charge of intermediates) were interviewed. The purpose of the investigation was to obtain more information in regard to details of the construction of the sublime and the main reactor.

Evaluation
The design of the sublime and the reactors appeared to be of interest and are described in detail. Salicylic acid was also produced by the Chemische Fabrik Hayden near Dresden and the Alpine Chemische Industry at Kufastein. Since these plants were not covered by the investigators, a comparison of the processes cannot be made.

Process details

Process summary
Phenol reacted with caustic and the resulting phenolate is treated with CO_2 to form sodium salt of salicylic acid. This solution is acidified and the precipitated acid is filtered off, washed, dried, and sublimed.

OFFICE OF MILITARY GOVERNMENT FOR GERMANY (US)

FIAT FINAL REPORT NO. 744   28 February, 1946

SALICYLIC ACID

BY

W.H. VILCOXEN

Joint Intelligence Objectives Agency

THIS REPORT IS ISSUED WITH THE WARNING THAT, IF THE SUB-
JECT MATTER SHOULD BE PROTECTED BY U.S. PATENTS OR PATENT
APPLICATIONS, THIS PUBLICATION CANNOT BE HELD TO GIVE ANY
PROTECTION AGAINST ACTION FOR INFRINGEMENT.

FIELD INFORMATION AGENCY, TECHNICAL

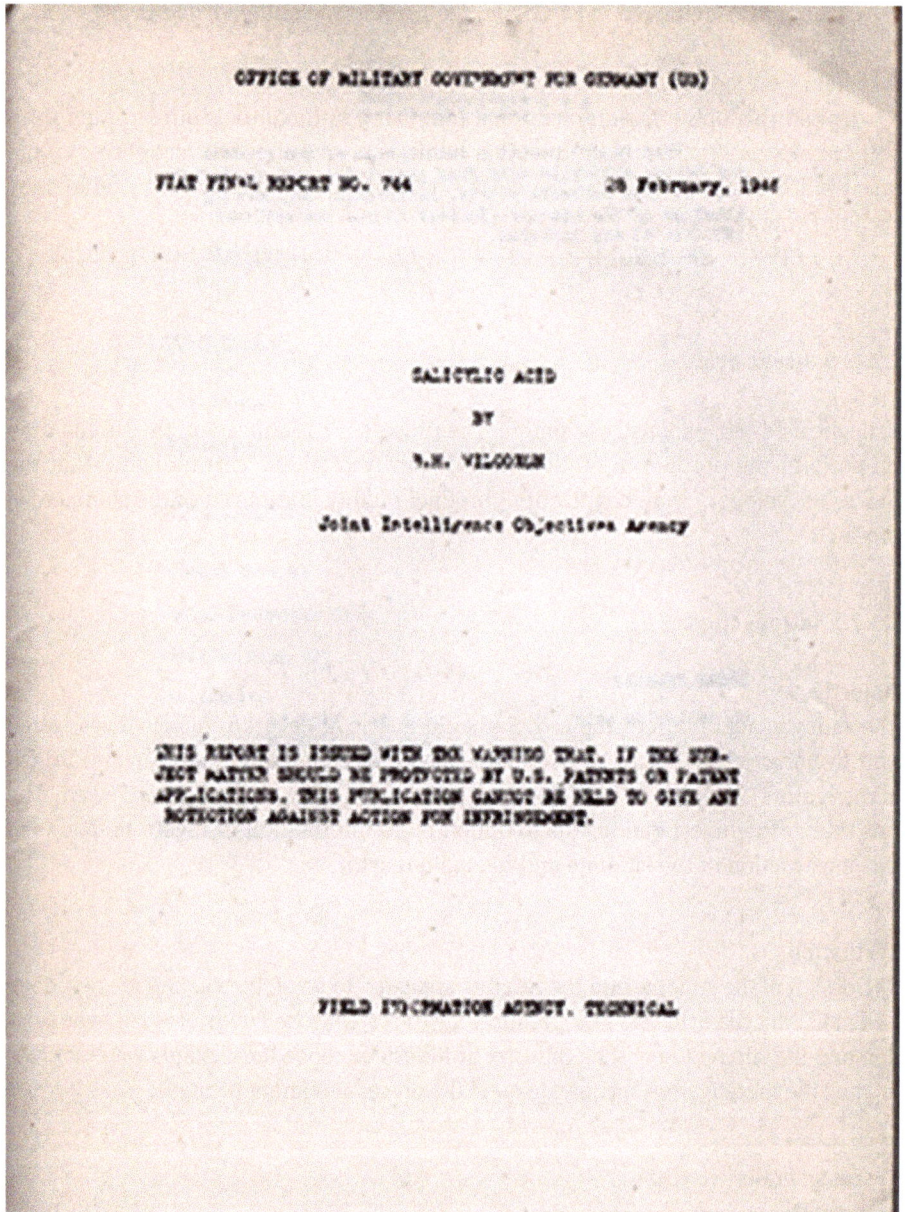

Fig. 1.33: Title page "Salicylic Acid," Joint Intelligence Objective Agency, February 1946.

Sodium phenolate step

About 40.6 °C phenol and 40% sodium hydroxide solution are fed from separate weigh tanks into a 10,000 m steel, agitated, closed mixer. This cycle requires 2–3 h and sodium phenolate solution is blown from a storage tank and then into a weigh tank before charging the main reactor. Approximately 1% excess sodium hydroxide is used for the reaction.

Reaction with CO_2

The main reactor consists of a 3% nickel steel, agitated, closed container, equipped for pressure operation. The vessel is wound on the outside with 1-in steel pipe coils which are welded to the vessel with a full fillet weld on both the top and bottom sides in order to fill in the space between the coil and the vessel. This kettle was fabricated by Samesreuther at Butzbach Hessen. The iron coils are spaced on approximately 2½ in centers, and the vessel should be insulated.

Refer to the attached (Exhibit A) for details of the agitator and vessel construction. Also refer to FIAT Report No. 725 entitled "High Pressure High Temperature Heating to 250 ATM," by Ernst W. Halbach, which describes the method of high-temperature hot water heating used here.

The main reactor is charged with sodium phenolate solution equivalent to 2,500 kg of phenol.

With agitation, the mixture is first heated with 5 atm hot water. Heat is continued, using 30 atm. At a temperature of 180 °C, water pressure is reached. A vacuum of 20–30 mm is then applied with continued stirring. Approximately 24 h is required to remove all the water.

After the water is removed the vessel is cooled, using 10 atm hot water, until a temperature of 150 °C is obtained. This requires 6–7 h. Stirring is maintained during the entire previous and following procedures.

The CO_2 which is now to be added must be free from air. Other gases must be less than 0.4%. CO_2 is added and a pressure of 4–4½ atm is maintained for 36 h. The heat generated is removed by circulating hot water in the jacket coils. The temperature of the mixture must be kept between 150 and 160 °C.

After 36 h, the CO_2 is released, and vacuum is applied in order to remove any remainder phenol. After all phenol is removed, the vacuum is turned off and CO_2 is again added to the vessel. A pressure between 4 and 5 atm is applied. And the reactants are heated to 180 °C. If the CO_2 pressure holds substantially constant with the addition of no further CO_2, the reaction is then considered finished. A vacuum of 10–15 mm absolute is again applied in order to remove final traces of phenol.

Reactants are cooled to room temperature and approximately 600 L of water is added, or enough to fill the kettle to its working level. This gives approximately a 50% solution of sodium salt.

Approximately 4 days is required for a cycle on this main reactor. The solution is blown to the storage tank where it is boiled together with decoloring carbons, called carboraffin (made at Leverkusen from peat or sawdust, and activated with zinc chloride). Approximately 75 kg of activated carbon is used per batch. The solution is then blown through a ceramic filter into a rubber-lined tank to which is added 40% sulfuric acid. A temperature of approximately 60 °C is maintained during this addition. Sufficient sulfuric acid is added until the mixture is blue to Congo red. After the main reactor, everything is rubber-lined. The precipitation by addition of acid is accomplished in a closed vertical tank of 16,000–20,000 L, approximately 2.4 m in diameter, equipped with a stirrer consisting of three long paddles, approximately equally spaced, driven by 17–20 RPM by means of 4–5 HP motor.

The material is removed from the precipitator into a centrifuge located underneath the vessel and from which the crystals are blown by air to a hopper which feeds the dryer. Salicylic acid leaving water. Drying is accomplished in a two-pass porcelain-lined screw-conveyor-type dryer, using stainless steel paddles to move the material along. Aluminum is not good for this but V2A stainless steel is quite satisfactory. Waste steam is used for heating and approximately 60 °C temperature is used. Porcelain was used because of shortage of stainless steel and Dr. Böhme stated that stainless steel would be used if available.

The wastewater in the centrifuge is precipitated a second time using a lower temperature. The water from the filtration is sent to sewer. It contains no phenol and, hence, presents no waste problem.

Sublimation

The crude salicylic acid which is quite white in appearance and already has high degree of purity is further purified by sublimation. The crude salicylic acid is loaded on flat trays 10–12 kg per tray. The sublime is operated on a 24 h cycle, using 150 °C and 15 mm vacuum. Approximately 20 h is required for heating and 2 h for emptying. About 35–400 kg is charged per sublime and 320–330 of sublime product is obtained (refer to Exhibit B for details of sublimer).

The salicylic acid, which is being sublimed, becomes covered with a hard surface layer, and a film of brown impurity restricts further sublimation. Accordingly, there is an appreciable amount of salicylic acid left at the end of sublimation. This is fed into an apparatus that recrushes the salicylic acid. The apparatus is subjected to a slight draft by means of an exhaust fan. This fan removes a certain amount of the dust which would be detrimental to the operator and also to the subsequent sublimation. Approximately 15–20% of the salicylic acid is recycled, but the yield on sublimation, considering the recycling, is approximately 98%.

Yield

An overall yield of 90–92% based on phenol and the refined salicylic acid is obtained. Approximately 2,500 kg of phenol is required for 3,000 kg of salicylic acid.

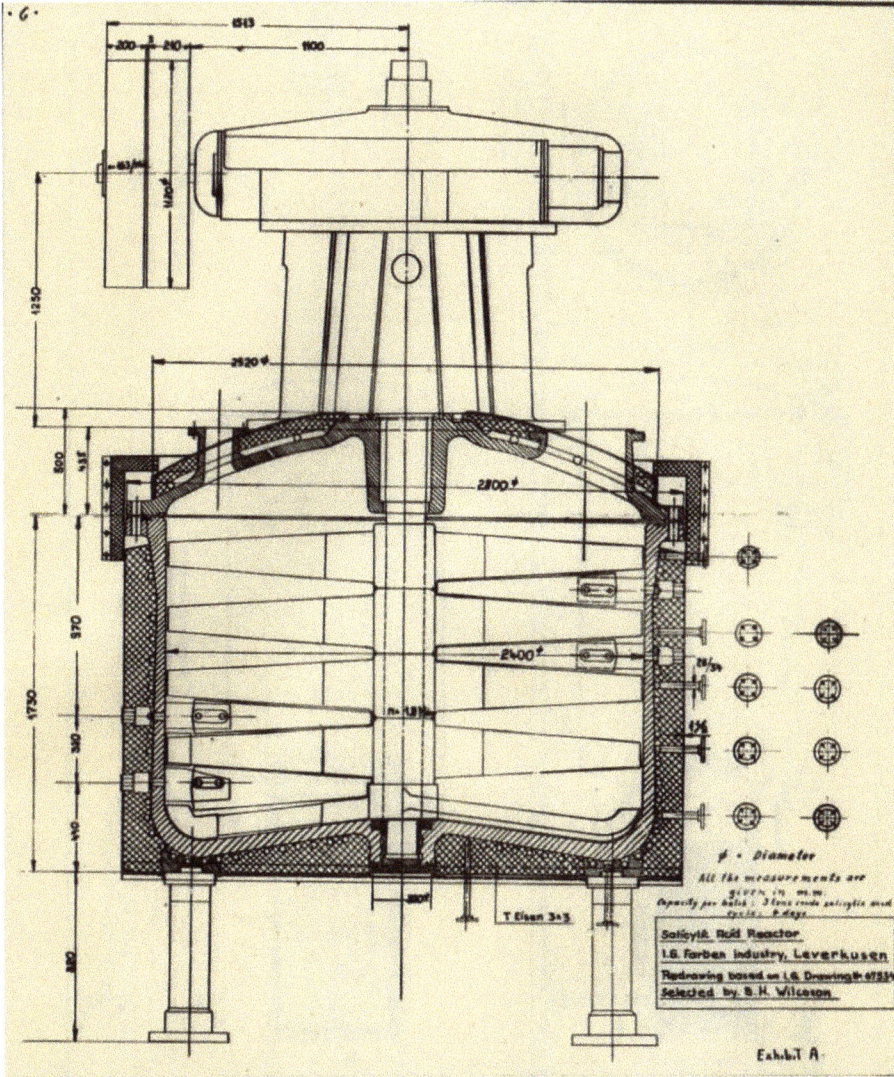

Fig. 1.34: Salicylic Acid Reactor.

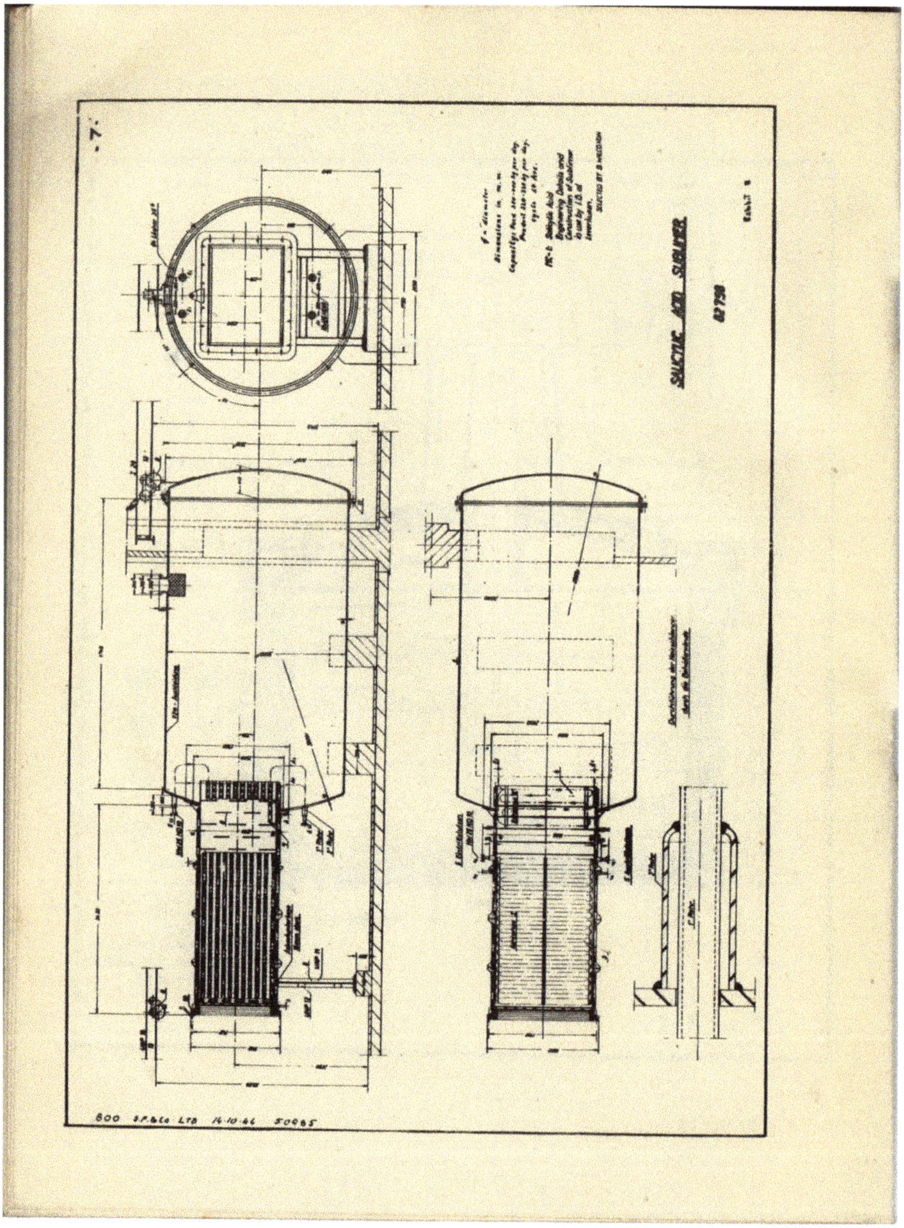

Fig. 1.35: Salicylic Acid Sublimer details.

The plant is now operating, and approximately 20 men are required in the production of 60 tons per month. The plant has a maximum capacity of 90–100 tons, and that rate would require about 30 men.

Size of agitators

The reaction is accomplished in two sizes of reactors. One size of which can produce 5 tons in 6 days and another one can produce 3 tons in 4 days.

Dr. Böhme preferred the smaller sized reactor and would build any additions on new plants on the smaller basis.

The following is a summary of a comparison of the two types of agitators:

	Small reactor (recommended size)	Large reactor
Capacity per batch Crude salicylic acid, tons	3	5
Cycle days	4	6
Agitator H.P.	10	15
Capacity: volumetric, m^3	6	10
RPM of agitator	9–10	9–10

A review of the history of each of the enterprises, which are now called the pharmaceutical companies, clearly indicates that most started as fine/specialty chemical companies and over a period of time have moved to become world's famous drug innovator companies. Their process designs were novel and creative. However, the companies on an ongoing basis did not explore alternate manufacturing methods and technologies. As a result, the methods that prevailed then for manufacturing of fine/specialty chemicals have continued. Innovation of manufacturing practices seems to have bypassed pharmaceuticals. It may be the best time to consider and practice methods that are based on fundamentals of chemistry and engineering and significantly lower their environmental impact.

Abbreviations

API	Active pharmaceutical ingredient
PLC	Public limited company
R&D	Research and development
MSD	Merck Sharpe & Dohme
WHO	World Health Organization
LSD	Lysergic acid diethylamide

Bibliography

[1] Merck George, W.: Quote What the Doctor Ordered, Time August 18, 1952, pg. 38–44 https://www.azquotes.com/quote/825854, Accessed February 27, 2020.

[2] Walsh, R.: A History of the Pharmaceutical Industry, Pharmaphorum, October 1, 2010, https://pharmaphorum.com/articles/a_history_of_the_pharmaceutical_industry/ Accessed February 21, 2020.

[3] Wikipedia https://en.wikipedia.org.

[4] Daemmrich, A., Bowden, M. E. Chemical Heritage Foundation, Chemical and Engineering News, https://pubsapp.acs.org/subscribe/journals/cen/83/i25/toc/toc_i25.html, June 2005, Accessed March 4, 2020.

[5] GlaxoSmithKline, https://en.wikipedia.org/wiki/GlaxoSmithKline) Accessed February 21, 2020.

[6] Allen & Hanburys https://en.wikipedia.org/wiki/Allen_%26_Hanburys Accessed February 25, 2020.

[7] Wellcome Trust, https://wellcome.org/who-we-are/history-wellcome Accessed February 28, 2020.

[8] Mercaptopurine, https://commons.wikimedia.org/wiki/File:Mercaptopurine.svg Accessed February 21, 2020.

[9] Horlicks, https://www.horlicks.co.uk Accessed March 1, 2020.

[10] Merck and Merck KGaA, https://en.wikipedia.org/wiki/Merck_Group Accessed February 10, 2020.

[11] What the Doctor Ordered, TIME Magazine, August 18, 1952, 60, 7, 38–44. http://content.time.com/time/subscriber/article/0,33009,816710,00.html. February 27, 2020.

[12] Morphine https://en.wikipedia.org/wiki/Morphine#/media/File:Morphin_-_Morphine.svg, Accessed February 20, 2020.

[13] Dexamethasone (https://commons.wikimedia.org/wiki/File:Dexamethasone_structure.png) Accessed February 20, 2020.

[14] Nasivin (https://commons.wikimedia.org/wiki/File:Oxymetazoline.svg) Accessed February 20, 2020.

[15] Gestafortin (https://commons.wikimedia.org/wiki/File:Chlormadinone_acetate.svg) February 20, 2020.

[16] Regulatory history: Elixir Sulfanilamide https://www.ivtnetwork.com/sites/default/files/IVTGXPxxxx_CoverStory-2%20pr1.pdf Accessed February 22, 2020.

[17] Sulfanilamide https://commons.wikimedia.org/wiki/File:Sulfanilamide-skeletal.svg Accessed February 2, 2020.

[18] Prontosil https://commons.wikimedia.org/wiki/File:Prontosil.svg February 2, 2020.

[19] Cortisone (https://en.wikipedia.org/wiki/Cortisone) Accessed February 25, 2020.

[20] Schering-Plough Corporation: https://en.wikipedia.org/wiki/Schering-Plough Accessed February 10, 2020.

[21] Pfizer https://en.wikipedia.org/wiki/Pfizer Accessed February 19, 2020.

[22] Santonin: https://en.wikipedia.org/wiki/Santonin#/media/File:Santonin-2D-skeletal.png Accessed March 4, 2020.

[23] Phenergan: https://commons.wikimedia.org/wiki/File:Promethazine.svg Accessed March 2, 2020.

[24] Ibuprofen (Advil): https://commons.wikimedia.org/wiki/File:Ibuprofen3.svg.

[25] Warner Lambert: https://en.wikipedia.org/wiki/Warner–Lambert Accessed February 15, 2020.

[26] Lipitor (Atorvastatin): https://commons.wikimedia.org/wiki/File:Atorvastatin.svg Accessed February 15, 2020.

[27] Parke, D.: https://en.wikipedia.org/wiki/Parke-Davis Accessed February 10, 2020.

[28] Chloramphenicol Synthesis: https://commons.wikimedia.org/wiki/File:Chloramphenicol_syn thesis.svg Accessed February 10, 2020.

[29] WHO model list of essential medicines – 22nd list, 2021 https://www.who.int/publications/ i/item/WHO-MHP-HPS-EML-2021.02 Accessed November 30, 2021.

[30] Upjohn: https://en.wikipedia.org/wiki/Upjohn Accessed February 15, 2020.

[31] Injectable vs. Oral Drugs: What's the Best for Human Health? https://www.aplususapharma. com/blog/injectable-vs-oral-drugs-whats-the-best-for-human-health/ Accessed October 20, 2021.

[32] Bayer, A. G.: https://en.wikipedia.org/wiki/Bayer Accessed February 29, 2020.

[33] Fuchsine: https://commons.wikimedia.org/wiki/File:Rosaniline_hydrochloride.svg Accessed February 10, 2020.

[34] Aniline: https://en.wikipedia.org/wiki/Aniline#/media/File:Structural_formula_of_aniline. svg Accessed February 10, 2020.

[35] A Look Back: The Wonder Drug: https://www.intljourtranur.com/article/S0099-1767(00) 70022-5/fulltext Accessed March 10, 2020.

[36] Acetyl Salicylic Acid (Aspirin) https://en.wikipedia.org/wiki/Aspirin#/media/File:Aspirin-skeletal.svg Accessed February 28, 2020.

[37] Heroin (Diacetylmorphine): https://commons.wikimedia.org/wiki/File:Heroin_-_Heroine.svg Accessed February 16, 2020.

[38] Suramin (8,8'-[ureylenebis[m-phenylenecarbonylimino(4-methyl-m-phenylene) Carbonylimino]]di-1,3,5-naphthalenetrisulfonic Acid) https://en.wikipedia.org/wiki/Suramin Accessed February 22, 2020.

[39] Glaser-Schmidt, E.: Foreign Trade Strategies of I. G. Farben after World War I, https://web. archive.org/web/20050429021019/http://www.h-net.org/~business/bhcweb/publications/ BEHprint/v023n1/p0201-p0211.pdf, Accessed August 10, 2020.

[40] Ciprofloxacin (1-cyclopropyl-6-fluoro-1,4-dihydro-4-oxo-7-(1-piperazinyl)-3-quinolinecarboxylic acid): https://en.wikipedia.org/wiki/Ciprofloxacin#/media/File:Ciproflox acin.svg Accessed February 22, 2020.

[41] Drospirenone https://en.wikipedia.org/wiki/Drospirenone#/media/File:Drospirenone.svg Accessed February 22, 2020.

[42] Eli Lilly: https://en.wikipedia.org/wiki/Eli_Lilly_and_Company Accessed February 17, 2020.

[43] Quinine https://en.wikipedia.org/wiki/Quinine Accessed February 29, 2020.

[44] Quinine synthesis: https://en.wikipedia.org/wiki/Quinine_total_synthesis Accessed February 29, 2020.

[45] Prozac prices https://www.nootropics-information.com/pharmaceuticals/prozac/.

[46] Prozac (Fluoxetine): https://en.wikipedia.org/wiki/Fluoxetine#/media/File:Fluoxetine.svg Accessed February 29, 2020.

[47] Duloxetine: https://en.wikipedia.org/wiki/Duloxetine#/media/File:Duloxetine.svg Accessed February 29, 2020.

[48] Abbott: https://en.wikipedia.org/wiki/Abbott_Laboratories Accessed February 18, 2020.

[49] Quinine (5-ethenyl-1-azabicyclo[2.2.2]octan-2-yl)-(6-methoxyquinolin-4-yl)methanol https:// en.wikipedia.org/wiki/File:Quinine_structure.svg Accessed February 18, 2020.

[50] Imperial Chemical Industries (ICI) and AstraZeneca https://en.wikipedia.org/wiki/Imperial_ Chemical_Industries Accessed February 24, 2020.

[51] Esomeprazole [6-methoxy-2-[(S)-(4-methoxy-3,5-dimethylpyridin-2-yl)methylsulfinyl]-1H-benzimidazole]: https://en.wikipedia.org/wiki/File:Esomeprazole.svg Accessed February 25, 2020.

[52] Bristol Myers Squibb: https://en.wikipedia.org/wiki/Bristol_Myers_Squibb Accessed February 2, 2020.

[53] Ciba-Geigy History: http://www.fundinguniverse.com/company-histories/ciba-geigy-ltd-history/, Accessed March 12, 2020.

[54] Vioform: 5-Chloro-7-iodo-8-quinolinol https://en.wikipedia.org/wiki/File:Clioquinol.png Accessed February 16, 2020.

[55] Geigy: https://en.wikipedia.org/wiki/Novartis Accessed March 10, 2020.

[56] Dichlorodiphenyltrichloroethane (DDT): https://en.wikipedia.org/wiki/File:P,p%27-dichlorodiphenyltrichloroethane.svg Accessed February 16, 2020.

[57] Privine: https://en.wikipedia.org/wiki/Naphazoline#/media/File:Naphazoline.svg Accessed February 28, 2020.

[58] Nupercaine: https://en.wikipedia.org/wiki/Cinchocaine#/media/File:Cinchocaine.svg Accessed February 28, 2020.

[59] Antipyrine: https://en.wikipedia.org/wiki/File:Phenazone2DCSD.svg Accessed February 15, 2020.

[60] Saccharin: https://en.wikipedia.org/wiki/File:Saccharin.svg Accessed February 15, 2020.

[61] Ergotamine: https://en.wikipedia.org/wiki/File:Ergotamine-skeletal.svg Accessed February 15, 2020.

[62] Hoffman-LaRoche: https://en.wikipedia.org/wiki/Roche Accessed February 29, 2020.

[63] L-Ascorbic Acid: https://en.wikipedia.org/wiki/Vitamin_C#/media/File:L-Ascorbic_acid.svg Accessed February 10, 2020.

[64] Johnson & Johnson: https://en.wikipedia.org/wiki/Johnson_%26_Johnson Accessed February 2, 2020.

[65] Tylenol: https://en.wikipedia.org/wiki/Paracetamol#/media/File:Paracetamol-skeletal.svg Accessed February 10, 2020.

[66] Boots: https://en.wikipedia.org/wiki/Boots_(company) Accessed March 1, 2020.

[67] Ibuprofen: https://en.wikipedia.org/wiki/File:(RS)-Ibuprofen_Structural_Formula_V1.svg Accessed February 10, 2020.

[68] Boehringer Ingelheim (https://en.wikipedia.org/wiki/Boehringer_Ingelheim) Accessed February 10, 2020.

[69] Tartaric Acid: https://en.wikipedia.org/wiki/File:Tartaric_acid.svg Accessed Faeb17, 2020.

[70] Citric Acid: https://en.wikipedia.org/wiki/File:Zitronensäure_-_Citric_acid.svg Accessed Faeb17, 2020.

[71] Lactic Acid https://en.wikipedia.org/wiki/Lactic_acid#/media/File:7_Milchsäure.svg Accessed Feb17, 2020.

[72] Sanofi: https://en.wikipedia.org/wiki/Sanofi Accessed February 18, 2020.

[73] Aventis: https://en.wikipedia.org/wiki/Sanofi Accessed February 18, 2020.

[74] Novo-Nordisk: https://en.wikipedia.org/wiki/Novo_Nordisk Accessed March 13, 2020.

[75] Novo Industri A/S: https://www.referenceforbusiness.com/history2/30/NOVO-INDUSTRI-A-S.html#ixzz6FrCJy84o Accessed March 13, 2020.

[76] Sterling Drug: https://en.wikipedia.org/wiki/Sterling_Drug Accessed February 28, 2020.

[77] Current Good Manufacturing Practice (CGMP) Regulations: https://www.fda.gov/drugs/pharmaceutical-quality-resources/current-good-manufacturing-practice-cgmp-regulations Accessed September 20, 2020.

[78] Promoting Safe & Effective Drugs for 100 Years, https://www.fda.gov/about-fda/histories-product-regulation/promoting-safe-effective-drugs-100-years Accessed March 27, 2020.

[79] Kefauver-Harris Amendments Revolutionized Drug Development, GVSU.EDU, https://www.gvsu.edu/cms4/asset/F51281F0-00AF-E25A-5BF632E8D4A243C7/kefauver-harris_amendments.fda.thalidomide.pdf, March 27, 2020.

[80] Peltzman, S., Pharmaceutical Regulation: A Matter of Life and Death, https://slideplayer. com/slide/6230805/ Accessed March 27, 2020.

[81] Orphan Drug Act, FDA.gov, https://www.fda.gov/industry/designating-orphan-product-drugs -and-biological-products/orphan-drug-act-relevant-excerpts.

[82] Boehm, G., Yao, L., Han, L., Zheng, Q. Development of the Generic Drug Industry in the US after the Hatch-Waxman Act of 1984, Acta Pharmacologica Sinica B, September 2013, 3, 5, 297–311. Accessed March 20, 2020.

[83] Pay for Delay, Federal Trade Commission.gov, https://www.ftc.gov/news-events/media-resources/mergers-competition/pay-delay Accessed March 10, 2021.

[84] Oversight Committee Held Hearing with AbbVie CEO and Experts on Pricing Practices for Humira and Imbruvica, https://oversight.house.gov/news/press-releases/oversight-committee-held-hearing-with-abbvie-ceo-and-experts-on-pricing, May 18, 2021 Accessed October.

[85] Pou, S., Winter, R. W., Nilsen, A., Kelly, J. X., Li, Y., Doggett, J. S., Riscoe, E. W., Wegmann, K. W., Hinrichs, D. J., Riscoe, M. K. Antimicrobial Agents and Chemotherapy, 2012 Jul, 56, 7, 3475–3480. Accessed May 12, 2020.

[86] History of antimalarials, Medicines of Malaria Venture, https://www.mmv.org/malaria-medicines/history-antimalarials, Accessed May 09, 2020.

[87] Preparation of Sontochin, Rosochin and Brachysan, Combined Intelligence Objectives Sub-Committee, Item No. 24, File No. XXV-54, pgs. 33–37.

[88] Salicylic Acid, F. I. A. T- Final Report No. 744, United States Group Control Council for Germany Accessed April 25, 2020.

Chapter 2
Business model: brand versus generics, product demand, process selection, and economics

2.1 What is a drug?

A drug is a chemical (active pharmaceutical ingredient, API) which is given to people in order to treat or prevent an illness or disease. Consumption of drugs can be via inhalation, injection, smoking, ingestion, absorption via a patch on the skin, or dissolution under the tongue. Traditionally, drugs were obtained through extraction from medicinal plants, but more recently also by organic synthesis. Pharmaceutical drugs are generally used for a limited duration, or on a regular basis for chronic illnesses. APIs in unformulated form are difficult to dispense. APIs are used in minute quantities (micrograms to milligrams). For easier and safer dispensing, they are formulated with inert excipients. Consumption of unformulated APIs can be difficult and more than needed dose can be toxic.

Most of the API are also called small-molecule drugs. They are fine/specialty chemicals and are manufactured by chemical synthesis. Besides small-molecule drugs, there are biologic drugs, a product that is produced from living organisms or contain components of living organisms. Biologic or biosimilar drugs are produced using a living or attenuated microorganism [1]. They are not discussed in detail here but, in most cases, chemical engineering principles apply for their process commercialization and optimization.

2.2 Brand and generic drugs

Most of the drugs are small molecules that fall into two categories: brand or generic. How they came about has been discussed in Chapter 1. The business model for the brand and the generic drugs is very similar. Each drug (injectable, solid tablet, an ointment, spray, or drops) in its finished dose form (FDF) has two components. One is an API that has the effect to cure the disease [2]. Generally, very small quantities (micrograms to milligrams) are needed to cure ailments. Since such small quantities cannot be picked by a pinch, they are admixed with a mixture of inert components, the second component, known as the excipients. This mix is converted to FDF to facilitate dispensing. Focus here is on solid dosage. Discussion and concepts can be extended to other dispensable forms.

https://doi.org/10.1515/9783110702842-002

2.2.1 Brand drugs

Brand drugs generally have a trade name and are protected by a patent. A patent protects the investment of the drug company that developed the drug and gives the company the sole right to sell the drug while the patent is in effect. When the patent expires, the same drug is sold as a generic drug.

Brand drug development is expensive. Their costs have escalated from about $802 million in 2003 to approximately $2.6 billion dollars in 2019 [3]. They go through a rigorous discovery, development, clinical research/trials, review, and approval process. Their manufacturing processes have to follow cGMP practices [4]. Once approved, a brand drug is sold by the drug molecule pioneering company exclusively.

For decades, the regulation and control of new drugs in the United States has been based on the new drug application (NDA). Since 1938, every new drug has been the subject of an approved NDA [5] before US commercialization. The NDA application is the vehicle through which drug sponsors formally propose that the FDA approve a new pharmaceutical for sale and marketing in the United States. The data gathered during the animal studies and human clinical trials of an investigational new drug (IND) becomes part of the NDA.

The goals of the NDA are to provide enough information to permit an FDA reviewer to reach the following key decisions:

– Whether the drug is safe and effective in its proposed use(s), and whether the benefits of the drug outweigh the risks
– Whether the drug's proposed labeling (package insert) is appropriate, and what it should contain
– Whether the methods used in manufacturing the drug and the controls used to maintain the drug's quality are adequate to preserve the drug's identity, strength, quality, and purity

The documentation required in an NDA is supposed to tell the drug's whole story, including what happened during the clinical tests, what the ingredients of the drug are, the results of the animal studies, how the drug behaves in the body, and how it is manufactured, processed, and packaged. The following provides summaries on NDA content, format, and classification, plus the NDA review process.

2.2.2 Brand drug development process

New drug discovery and development process is complex. It has multiple steps [6].

2.2.2.1 Step 1: discovery and development
Discovery
Typically, researchers discover new drugs through:
- new insights into a disease process that allows researchers to design a product to stop or reverse the effects of the disease;
- many tests of molecular compounds to find possible beneficial effects against any of a large number of diseases;
- existing treatments that have unanticipated effects;
- new technologies such as those that provide new ways to target medical products to specific sites within the body or to manipulate a genetic material.

At this stage in the process, thousands of compounds may be potential candidates for development as a medical treatment. After early testing, however, only a small number of compounds look promising and call for further study.

Development
Once researchers identify a promising compound for development, they conduct experiments to gather information on:
- how it is absorbed, distributed, metabolized, and excreted;
- its potential benefits and mechanisms of action;
- the best dosage;
- the best way to give the drug (such as by mouth or injection);
- side effects or adverse events that can often be referred to as toxicity;
- how it affects different groups of people (such as by gender, race, or ethnicity) differently;
- how it interacts with other drugs and treatments; and
- its effectiveness as compared with similar drugs.

2.2.2.2 Step 2: preclinical research
Before testing a drug in people, researchers must find out toxicity of the compound. The two types [7] of preclinical research are:
- in vitro (Petri dish studies) and
- in vivo (studies performed in living organism).

FDA requires researchers to use good laboratory practices (GLP), defined in medical product development regulations, for preclinical laboratory studies. The GLP regulations are found in CFR – Code of Federal Regulations Title 21 [8]. These regulations set the minimum basic requirements for:
- study conduct
- personnel

- facilities
- equipment
- written protocols
- operating procedures
- study reports
- system of quality assurance oversight for each study to help assure the safety of FDA-regulated product

Usually, preclinical studies are not very large. However, these studies must provide detailed information on dosing and toxicity levels. After preclinical testing, researchers review their findings and decide whether the drug should be tested in people.

2.2.2.3 Step 3: clinical research

While preclinical research answers basic questions about a drug's safety, it is not a substitute for studies of ways the drug will interact with the human body. "Clinical research" refers to studies, or trials, that are done in people. As the developers design the clinical study, they will consider what they want to accomplish for each of the different clinical research phases and begin the IND process, a process they must go through before clinical research begins.

Designing clinical trials

Researchers design clinical trials to answer specific research questions related to a medical product. These trials follow a specific study plan, called a protocol, that is developed by the researcher or manufacturer. Before a clinical trial begins, researchers review prior information about the drug to develop research questions and objectives. Then, they decide:

- Who qualifies to participate (selection criteria)?
- How many people will be part of the study?
- How long the study will last?
- Whether there will be a control group and other ways to limit research bias.
- How the drug will be given to patients and at what dosage?
- What assessments will be conducted, when, and what data will be collected?
- How the data will be reviewed and analyzed?

Clinical trials follow a typical series from early, small-scale, phase 1 studies to late-stage, large-scale, phase 3 studies.

The investigational new drug process
Drug developers, or sponsors, must submit an IND application to FDA before beginning clinical research.

In the IND application, developers must include:
- Animal study data and toxicity (side effects that cause great harm) data
- Manufacturing information
- Clinical protocols (study plans) for studies to be conducted
- Data from any prior human research
- Information about the investigator

Drug developers are free to ask for help from the FDA at any point in the drug development process, including:
- Pre-IND application, to review FDA guidance documents and get answers to questions that may help enhance their research
- After phase 2, to obtain guidance on the design of large phase 3 studies
- Any time during the process, to obtain an assessment of the IND application

Even though FDA offers extensive technical assistance, drug developers are not required to take FDA's suggestions. As long as clinical trials are thoughtfully designed, reflect what developers know about a product, safeguard participants, and otherwise meet Federal standards, FDA allows wide latitude in a clinical trial design.

Approval
The FDA review team has 30 days to review the original IND submission. The process protects volunteers who participate in clinical trials from unreasonable and significant risk in clinical trials. FDA responds to IND applications in one of two ways:
- Approval to begin clinical trials.
- Clinical hold to delay or stop the investigation. FDA can place a clinical hold for specific reasons, including:
  - Participants are exposed to unreasonable or significant risk.
  - Investigators are not qualified.
  - Materials for the volunteer participants are misleading.
  - The IND application does not include enough information about the trial's risks.

A clinical hold is rare; instead, FDA often provides comments intended to improve the quality of a clinical trial. In most cases, if FDA is satisfied that the trial meets Federal standards, the applicant is allowed to proceed with the proposed study.

The developer is responsible for informing the review team about new protocols, as well as serious side effects seen during the trial. This information ensures that the team can monitor the trials carefully for signs of any problems. After the trial ends, researchers must submit study reports.

This process continues until the developer decides to end clinical trials or files a marketing application. Before filing a marketing application, a developer must have adequate data from two large, controlled clinical trials.

2.2.2.4 Step 4: FDA drug review

If a drug developer has evidence from its early tests and preclinical and clinical research that a drug is safe and effective for its intended use, the company can file an application to market the drug. The FDA review team thoroughly examines all submitted data on the drug and makes a decision to approve or not to approve it.

New drug application

An NDA tells the full story of a drug. Its purpose is to demonstrate that a drug is safe and effective for its intended use in the population studied.

A drug developer must include everything about a drug – from preclinical data to phase 3 trial data – in an NDA. Developers must include reports on all studies, data, and analyses. Along with clinical results, developers must include:

- Proposed labeling
- Safety updates
- Drug abuse information
- Patent information
- Any data from studies that may have been conducted outside the United States
- Institutional review board compliance information
- Directions for use

FDA review

Once FDA receives an NDA, the review team decides if it is complete. If it is not complete, the review team can refuse to file the NDA. If it is complete, the review team has 6–10 months to make a decision on whether to approve the drug. The process includes the following:

- Each member of the review team conducts a full review of his or her section of the application. For example, the medical officer and the statistician review clinical data, while a pharmacologist reviews the data from animal studies. Within each technical discipline represented on the team, there is also a supervisory review.
- FDA inspectors travel to clinical study sites to conduct a routine inspection. The Agency looks for evidence of fabrication, manipulation, or withholding of data.
- The project manager assembles all individual reviews and other documents, such as the inspection report, into an "action package." This document becomes the record for FDA review. The review team issues a recommendation, and a senior FDA official makes a decision.

FDA approval

In cases where FDA determines that a drug has been shown to be safe and effective for its intended use, it is then necessary to work with the applicant to develop and refine prescribing information. This is referred to as "labeling." Labeling accurately and objectively describes the basis for approval and how best to use the drug.

Often, though, remaining issues need to be resolved before the drug can be approved for marketing. Sometimes FDA requires the developer to address questions based on the existing data. In other cases, FDA requires additional studies. At this point, the developer can decide whether or not to continue further development. If a developer disagrees with an FDA decision, there are mechanisms for formal appeal.

FDA advisory committees

Often, the NDA contains sufficient data for FDA to determine the safety and effectiveness of a drug. Sometimes, though, questions arise that require additional consideration. In these cases, FDA may organize a meeting of one of its Advisory Committees to get independent, expert advice and to permit the public to make comments. These Advisory Committees include a Patient Representative that provides input from the patient perspective.

2.2.2.5 Step 5: FDA post-market drug safety monitoring

Even though clinical trials provide important information on a drug's efficacy and safety, it is impossible to have complete information about the safety of a drug at the time of approval. Despite the rigorous steps in the process of drug development, limitations exist. Therefore, the true picture of a product's safety actually evolves over the months and even years that make up a product's lifetime in the marketplace. FDA reviews reports of problems with prescription and over-the-counter drugs, and can decide to add cautions to the dosage or usage information, as well as other measures for more serious issues.

2.2.3 Manufacturer inspections

FDA officials conduct routine inspections of drug manufacturing facilities across the United States and abroad if approved products are manufactured overseas. Manufacturers may be informed of inspections in advance, or the inspections may be unannounced. Inspections may be routine or caused by a particular problem or concern. The purpose of these inspections is to make sure that developers are following good manufacturer practice. FDA can shut down a facility if minimum standards are not met.

2.2.4 Generic drugs

For the drugs, coming off patent, generic companies under the Hatch-Waxman Act [9, 10] apply for ANDA (abbreviated new drug application). Paragraph IV matters if many companies file ANDA application on the same day. The company with IV certification gets the 180-day exclusivity [11]. After 180-day exclusivity, many other companies join the landscape. As explained later, this generally results in companies having less than optimum processes.

Since generic drugs are comparable to the brand drugs already on the market, their drug manufacturers do not have to conduct clinical trials to demonstrate that their product is safe and effective. Instead, they conduct bioequivalence studies and file an ANDA [12]. Companies selling generic drugs have to prove their product's bioequivalence to the brand drug [13–15]. Since generic drugs are comparable to brand name drugs they must have the same:

– Dosage form
– Strength
– Safety
– Quality
– Performance characteristics
– Intended use

Current regulatory landscape

The process of filing and approval of NDA [5] and ANDA [12] are long and cumbersome. No one in the public domain knows the real time it takes to get respective approvals. The best guess for ANDA approval is about 36–48 months. FDA publishes how many NDA and ANDA approvals happen in the fiscal year. However, these numbers are meaningless as they do not tell the reality. FDA most likely knows the actual number or the time it takes for every application.

FDA has not made any attempt to simplify the existing filling processes. It would be helpful if they did that. They could use templates for illustration for different drug categories. If FDA personnel did a mock filling and approvals, such exercises would facilitate and simplify filing and approval processes [16, 17]. They will improve time availability and improve profitability of the companies.

Regulators recently introduced ICHQ12 [18]. This guidance reads like a legal document rather than a simple instruction of what is expected. It will essentially stop any innovation and improvement of the existing processes if the process developers and designers have to get approval prior to any change or improvement. It is most likely that many companies will not spend any time, effort, and money to improve their existing processes. In addition, this guidance talks about marketing authorization holder (MAH). If these MAH are same as the distributors and sellers of the drugs, then they would be the pharmacy benefit managers (PBMs) and members of the supply chain who distribute and sell drugs in the United States. Would they be included in the process and will be held accountable for less than quality drugs? It is possible that the companies have not realized this inclusion.

The following excerpt basically eludes that the process developers and designer's knowledge to manage their processes will increase as if they did not have that knowledge. It has to be recognized that these are the same people who developed, designed, and commercialized the sold products.

> This guideline is also intended to demonstrate how increased product and process knowledge can contribute to a more precise and accurate understanding of which post-approval changes require a regulatory submission as well as the definition of the level of reporting categories for such changes (i.e., a better understanding of risk to product quality). Increased knowledge and effective implementation of the tools and enablers described in this guideline should enhance industry's ability to manage many CMC (Chemistry, Manufacturing, and Controls) changes effectively under the company's Pharmaceutical Quality System (PQS) with less need for extensive regulatory oversight prior to implementation.

Latently, this guideline suggests that the process developers and designers should not practice any continued improvement and innovation as they do not understand the impact of changes. It is ironic that these are the same people who developed, designed, and commercialized the process and product. Regulators have to recognize that no producer will change or commercialize a product if the changed processing conditions and raw materials will alter the established product quality. In addition, regulators can stop the sale of the drug if process change impacts drug performance.

It is interesting that the regulators are pushing continuous manufacturing [19, 20] when they do not understand the economics and the fundamental basis of such process designs. Regulators have never developed, designed, or commercialized such processes and have even altered the established definition to call a batch process a continuous process. Basics of such processes are very different from batch processes. They have established definitions [21, 22] and are discussed later in this chapter and in different chapters. Regulators have shied away from giving their definition for a continuous API and formulation process.

2.2.5 Process selection

Product demand is the basic building block of every process in any manufacturing industry. In pharmaceuticals and fine/specialty chemicals, process chemistry and its method of execution sway the type of process and technology used. Economics is an important part of manufacturing process selection. These guide designers to make sure that an optimum process is selected for the quality product and it can fulfill the needed demand. This is not a new revelation.

Selection processes in the chemical and petrochemical industry have been discussed and taught in chemical engineering curriculums as early as 1926 (just a few are cited [23–26]). Since pharma is a subset of the specialty chemical industry, the same selection criterion is applied. However, as explained later in this chapter, pharma has its own process selection and business criterion model.

Processes used can be batch [27, 28] or continuous [21, 22] and they have established definitions. It is beneficial to review these as they assist in explaining the rationale for pharma's preferred process selection. Batch manufacturing relates to specified amount of product in a limited time. Continuous process relates to manufacturing a single product independent of time. Schematically, they are illustrated in Figure 2.1 and Figure 2.2 respectively. Such processes will be discussed in other chapters.

2.2.5.1 Batch process

Figure 2.1 is a simple schematic of a batch process. In batch processes, the product is held after each process step and then processed. Thus, it is critical and necessary to have storage space/tanks/vessels where intermediate products can be stored before the next processing step. They are tested to assure that they meet the established specifications. Intermediates, if need be, are reworked for a quality product or disposed of if they cannot be reworked. This can cost as much as much as 40% of sales of the final product [29–35].

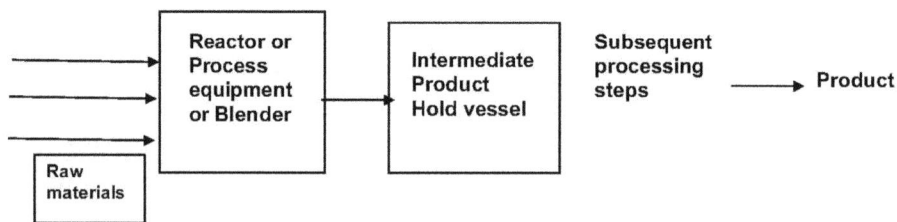

Fig. 2.1: Schematic of a batch process: *Intermediates held for further processing.*

In batch manufacturing, laboratory processes are modified to fit in the available equipment [36, 37]. This minimizes new capital investment and provides operational flexibility to companies to produce different products but results in asset utilization of about 22–40% [38–42] for the pharmaceutical industry. Compared to the chemical industry, these numbers are low. Any other industry would have difficulty sustaining itself with such low asset utilization numbers. However, even with low numbers, pharmaceuticals are profitable and it is due to its product pricing, discussed later.

Quality level of about 2–3 sigma [29–35], low asset utilization [39–43], and un-optimized processes present pharmaceutical industry with tremendous opportunity to improve profitability. These are discussed in later chapters.

2.2.5.2 Continuous process

Processes are selected on the basis of product demand, process chemistry, formulation method, and financial justification process. Unlike batch process [27, 28], continuous process [21, 22] is designed to produce a single product. This is the most critical aspect of the process. Seldom an exactly similar product processing can be fitted in the same equipment. This is another distinct difference between a batch process and a continuous process. In continuous processes, flow of materials does not stop during the operating year [24 × 7 × 50 = 8,400 hr.] except for the necessary downtime for planned maintenance or unexpected shutdown, which generally is as short as possible to minimize financial loss. Downtime is recognized in product cost structure. It is ironic that the regulators [19, 20] have been promoting and pushing continuous manufacturing for API manufacturing and their formulations, realm of operating companies who decide the type of processes are used, when no regulator has ever developed, designed, justified, or commercialized such processes. Figure 2.2 is a generic schematic of a continuous chemical synthesis process.

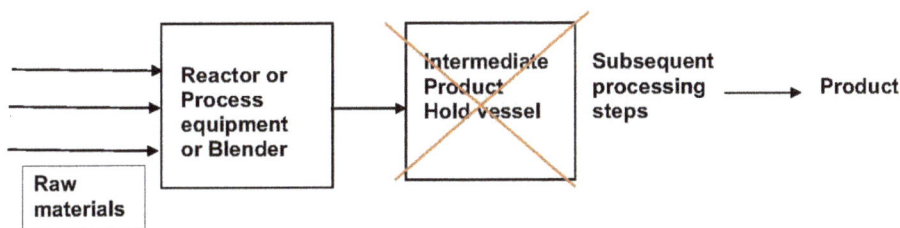

Fig. 2.2: Schematic of a continuous process: *No intermediate product hold.*

Figure 2.3 is a simple schematic of a solid dose continuous formulation process. It is not different from batch formulation processes. The only difference is batch formulation processes are replicate of batch API manufacturing processes, where in the process material is processed, sampled, and tested after every process step to assure quality. Necessary equipment and technology for continuous formulation like continuous reactive process equipment and process control technologies have been available for more than 60 years and used in many chemical manufacturing and their formulation processes.

Pharma process design engineers, who are basically chemical engineers, have also shied away from using them to design and commercialize automated processes that can deliver quality. These have been successfully applied in other chemical

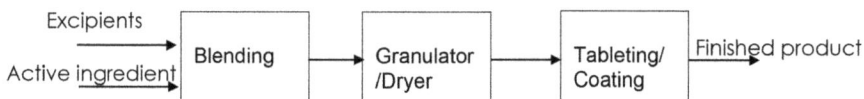

Fig. 2.3: Continuous formulation process schematic.

operations. "Is it the fear of not being able to design a process which will not produce specification quality product on an ongoing basis for the ~ 7,140 h per year production coming in the way?" Equipment for continuous formulations has existed but is not used. Could it be due to the same reasons as chemical synthesis or due to traditions of mortar and pestle used in the early stages of dose dispensation?

Figure 2.4 is a schematic of a real continuous process. It operates about 7,500 h per year producing same fine/specialty chemical. Again, we have to recognize that chemistries of API are similar to their older cousin fine/specialty chemicals. The only differentiation is: API have a disease-curing value, whereas fine/specialty chemicals do not have a disease-curing value. They are used as additives to facilitate and improve life.

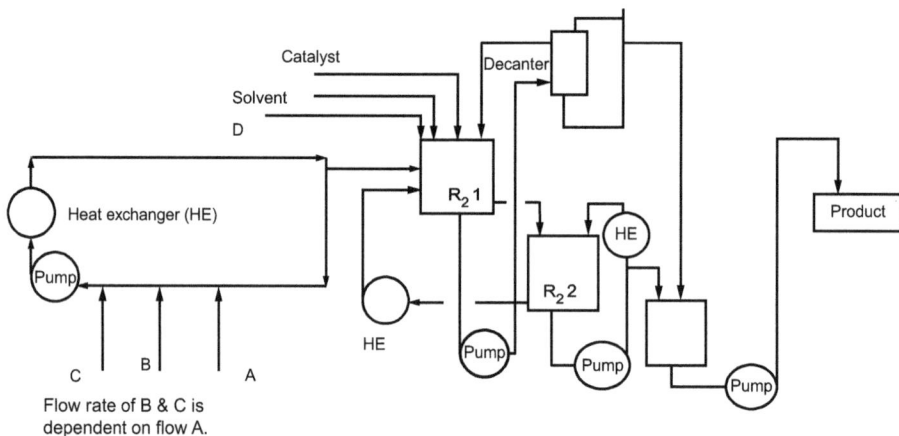

Fig. 2.4: Continuous synthesis process schematic.

As shown in Figure 2.4, process chemicals A, B, and C are introduced at a predefined rate to produce a product in a pipe flow reactor, reacted, and continuously pumped to reactors 1 and 2, where chemical D is introduced and reacted to produce the product. Stoichiometry was precisely controlled using the existing commercially available process controllers. Product D was withdrawn from the reaction system continuously. Thus, continuous processing/manufacturing is not a new technology but is being touted by many (FDA and others who are equipment suppliers rather than actual practitioners) as a NEW technology.

Unlike in batch processes, there is no intermediate product hold tank/space in any continuous processes. Since the process is producing a product in every operating moment, the product quality cannot deviate outside the designed and established upper and lower control limits at any instant, sine curve [44, 45] (Figure 2.5). Deviation outside the control limits means the process is out of control and the product is not meeting specs and will result in significant financial loss.

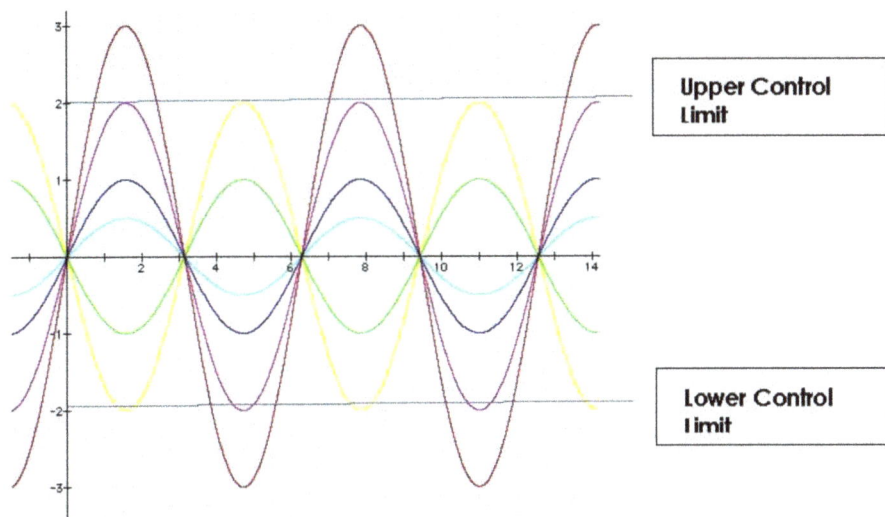

┌──────────────────┐
│ Upper Control   │
│ Limit   │
└──────────────────┘

┌──────────────────┐
│ Lower Control   │
│ l imit   │
└──────────────────┘

Fig. 2.5: Product quality range.

Process design of a continuous process is much more stringent than a batch process. Process controls to manage every process parameter are well established and have been used for over 50 years for respective processes. They are used to maintain quality regiment and have been improved. Since the process is time independent, in-line testing gives an instant image of the process to assure that the product quality is within the design parameters. Any properly designed feedback controls assure that the process does not deviate out of the established control limits. This is extremely critical in the API manufacturing and formulations.

2.2.5.3 Batch versus continuous process in pharmaceuticals

Like in any chemical and petrochemical manufacturing, pharmaceuticals also have a choice of manufacturing processes. They are selected on the basis of economics and product demand. Pharmaceuticals, even with low-quality performance (2–3 sigma) and less than 50% asset utilization, have, as discussed earlier, opted to stay with batch processes. It is necessary to understand why batch manufacturing is the preferred route for the manufacture of APIs and their formulations. An alternate question could be "why pharma has shied away from better technologies and methods, for example, process of continuous improvement and adopting and commercializing continuous processes in the manufacture of API or their formulations?" As stated earlier, these technologies and equipment are commercially available and are being used in the manufacture of the same family of chemicals, fine/specialty included, and their formulations.

There are plausible explanations and answers and they are reviewed. Could either or combination of the following be contributing to the current decision making?

1. Profitability exceeding expectation with the current methods and technologies even with known quality and asset utilization issues.
2. FDA/regulator's practice to question and have the companies prove every manufacturing technology innovation before they can be commercialized.
3. Fear of failure to achieve quality instilled by equipment vendors.
4. Lack of confidence and/or capabilities of engineers and chemists in their own process development and design capabilities and need to prove to the regulators every process technology innovation.
5. Excess idle capacity at operating sites has negated use of better technologies.
6. Lack of differential profitability analysis between existing operating mortar/ pestle model versus alternate model that could be combination of better batch/ continuous processes.

If some or all of the above questions can be addressed, it is possible that current operating model could see a shift to higher profitability through larger customer base that comes from use of better technologies [33].

2.3 Review of brand and generic drug API production

The current operating landscape, product demand, production methodology, and profitability are reviewed. Small-molecule brand and generic drugs are fine/specialty chemicals that have disease-curing value. Due to disease-curing capabilities, manufacturing and selling strategies of these chemicals are different from other chemicals. Analysis presented in Table 2.1 applies to API manufacturing and formulations.

When a drug is under patent, brand companies generally produce API and its formulations at limited sites. This is logical as the companies do not know a drug's potential need from the start. They cope up with the incremental demand through batch API and formulation processes. These processes are designed using quality by design principles to produce quality products, a fundamental trait every chemist and chemical engineer is taught and practices. Since companies do not invest in new equipment for these products, they are fitted in the available equipment [36, 37]. However, such product accommodation results in a quality by analysis behavior of testing materials, an age-old practice of laboratory bench where after every process step samples are taken and analyzed, a hallmark of poor asset utilization and quality by analysis even when the process is a quality by design process. This practice continues during a drug's patent life (brand) and generic's life.

Generics use their own or processes that are similar to brand processes to produce the API and their formulations once they have proven bioequivalence. They have an opportunity to capitalize on economies of scale but generally do not. It could be due to not having the right equipment.

Table 2.1 illustrates API production and formulation needs for 50,000,000 (0.06% of global population) patients requiring different dose tablets. Numbers in the table can be factored to determine the API and tablet needs per year for different population size. Companies can achieve certain economies of scale benefits if they have large production volume. Very limited number of API can be produced using continuous process per established definition [21, 22]. Even if some APIs could be produced using continuous process, pharma companies due to imbedded tradition and many others producing the same API, all opt for batch production. Many APIs can be definitely formulated using continuous process but they are not. This is due to some of the issues discussed earlier.

Tab. 2.1: Population, tablets/day, dose/day, API/year, and tablet rate, year and hour.

Case	Population	Tablets /day	Milligram /tablet	API, kg/year	Tablets needed/year	Tablets needed/hour
1	50,000,000	1	10	182,500	18,250,000,000	2,556,022
2	50,000,000	2	10	365,000	36,500,000,000	5,112,045
3	50,000,000	1	50	912,500	18,250,000,000	2,556,022
4	50,000,000	2	50	1,825,000	36,500,000,000	5,112,045
5	50,000,000	1	200	3,650,000	18,250,000,000	2,556,022
6	50,000,000	2	200	7,300,000	36,500,000,000	5,112,045

Brand or generic APIs (cases 1 and 2, Tab. 2.1) could be produced at a single plant. However, brand companies expand manufacturing to produce at many plants to meet the increasing demand. Brand companies do not consolidate production of these sites mostly due to their focus on new patented drugs. They let the generics take over after patent expiration. Since many generics apply for ANDA, they fit these products in their existing equipment [36, 37]. These operations, due to the lack of economies of scale, do not have optimum processes. Due to their profitability, 2–3 sigma quality level and less than 50% asset utilization are acceptable. Quality by analysis is acceptable. API demand of more than 1,500 metric ton per year could be produced using continuous process at a single plant but most likely will be produced at multiple plants using batch processes.

Since many companies have approved ANDA, they produce their API at their own sites or use contract manufacturers. Annual API need is divided into multiple plants. Table 2.2 is an illustration of division of numbers in Tab. 2.1.

Tab. 2.2: API/year, number of API plants, and number of formulation plants for 50 million population.

Case	Milligram	Kilogram needed/year	Kilogram/plant	No. of API plants	Tablets/API plant
A	10	182,500	18,250	10	1,825,000,000
B	10	365,000	18,250	20	1,825,000,000
C	50	912,500	22,813	40	456,250,000
D	50	1,825,000	45,625	40	912,500,000
E	100	1,825,000	45,625	40	456,250,000
F	100	3,650,000	91,250	40	912,500,000
G	200	3,650,000	91,250	40	456,250,000
H	200	7,300,000	146,000	50	730,000,000

2.4 Review of formulations

Due to their volume (Tab. 2.2) (tablets needed per year for different doses), formulations could be produced using continuous processes. Technologies to formulate continuously [46, 47] have existed but have not been commercialized. Current commercially available tableting machines can operate to about ~230,000 tablets per hour. Tablet production at this rate would be a continuous process. However, pharma has not taken that route. Lack of confidence in engineering capabilities and easy mortar/pestle philosophy, pharma has consistently used batch formulation methods. Costs due to low (2–3) sigma quality [29–35] and low asset utilization [39–43] are built in product costs. Mortar/pestle thinking and past practices prevail. Adoption and implementation of continuous formulation will change the landscape. This is briefly reviewed but is not part of the detailed discussion.

Table 2.3 is an extension of Table 2.2 to illustrate how many batch tableting plants would be needed per API plant at two different tableting rates and at two different dosages. Metformin hydrochloride, a common diabetes drug, has about 158 ANDA companies listed in the FDA's *Orange Book* and is just an illustration [48] of such fragmentation. In addition, if a drug dose is a combination of multiple APIs

Tab. 2.3: Number of formulation plants needed at different tablet rates.

Case	Milligram	Kilogram needed/year	Kilogram/plant	No. of API plants	Tablets/API plant	No. of formulation plants needed		Total number of FDF needed for API in column no. 4 at 75,000 tablets/hour
						Tablet rate 230,000 per hour operating 8,000 h per year	Tablet rate 75,000 per hour operating 3,000 h per year	
A	10	182,500	18,250	10	1,825,000,000	~1	~9	~82
B	10	365,000	18,250	20	1,825,000,000	~1	~9	~163
C	50	912,500	22,813	40	456,250,000	~1	~3	~82
D	50	1,825,000	45,625	40	912,500,000	~1	~5	~163
E	100	1,825,000	45,625	40	456,250,000	~1	~3	~82
F	100	3,650,000	91,250	40	912,500,000	~1	~5	~163
G	200	3,650,000	91,250	40	456,250,000	~1	~3	~82
H	200	7,300,000	146,000	50	730,000,000	~1	~4	~163

(e.g., lisinopril/HCTZ, aspirin/paracetamol/caffeine, lisinopril/metformin, and atorvastatin/metformin), they add complexity to formulations [49] and batch manufacturing becomes a necessity.

Table 2.3 illustrates the number of dedicated formulation lines needed if tablet rate of 230,000 per hour operated at 7,300 h per year or 75,000 per hour and 3,000 h per year (<50% asset utilization), respectively. Number of formulation facilities for continuous and batch production to fulfill the needs of 50 million population are illustrated. Most batch pharma companies produce tablets at variable hourly (anywhere from 30,000 to 75,000 tablets per hour) rates.

Continuous formulation and tableting technologies (230,000 tablets per hour operating at about 7,300 h per year) have existed [46, 47] but require absolute command of the process stoichiometry, process controls, and ultimate understanding of involved unit processes and unit operations [50, 51].

Profitability achieved through the current technologies and the challenges posed by the regulators have discouraged and intervened incorporation of efficient technologies for API synthesis and formulations. Lack of confidence in design capabilities and equipment performance should not be an issue in light of our capabilities to send a man to the Moon and bring him back and of sending the Cassini rocket around Saturn. Regulations are most likely the cause.

2.5 Manufacturing process technologies and their impact on profitability

As indicated earlier, batch processes are the preferred processes for the manufacture of APIs and their formulations. Sustained profitability from such processes is the reason. If a fine/specialty chemical finds its use as a drug, the selling price of the API goes up and profits go up. This prevents or comes in the way of incorporating any improved manufacturing technology in existing processes. This can best be illustrated through an example.

Dimethyl fumarate is used as one of the examples. Its initial use was as a biocide for the furniture and shoe industry to prevent the growth of mold during shipments in humid climate. Selling of dimethyl fumarate was low at this stage. Due to its allergic reaction to the skin [52], its use was curtailed. Its use for multiple sclerosis was discovered and approved by the US FDA. Selling price of the biocide increased as it became an active ingredient (Table 2.4).

Review of Table 2.4 suggests that profits realized as a drug do not warrant significant investment to improve the process, its yield, or reduce the effluent to have an increase in profits at API level. Such situations apply across the board on pharma landscape.

With US patent, expiring multiple companies will file ANDAs [45]. When the brand drug becomes generic, API selling price will be lower than the current API

Tab. 2.4: Comparison of selling price as a fine chemical versus an API and number formulation plants needed.

Dimethyl fumarate (Tecfidera)		
	Fine chemical	Active pharmaceutical ingredient
Sale price, $/kg	$55.00 [53]	$402.00 [54]
	Dose 240 mg/day	
API cost per day per 240 mg tablet	=(55 × 240)/1,000,000 = $0.013	=(402 × 240)/1,000,000 = $0.097
Formulated drug price, $/tablet	It is assumed that the factory cost of the formulated drug will be ten times the cost of active.	
	Factory cost of formulated tablet = $0.13 per 240 mg	Factory cost of formulated tablet = $0.97 per 240 mg
Selling price [55] of dimethyl fumarate as brand drug (TECFIDERA = $217.00 per day)		
Global multiple sclerosis population [56] ~2.3 million		
API	Total need = 205,000 kg/year	
Dose	120 mg; twice a day	
Total tablet plants needed at 50,000 tablets per hour for 3,000 h per year = ~12		

price but will still be significantly higher compared to the selling price of the specialty chemical when it is being used as a biocide. This happened, as explained later.

Dimethyl fumarate's US patent 8399514 has been held invalid [57] and as a result its API price has dropped from $402.00/kg to about $210.00/kg [58]. Many other API molecules fit similar price drop criterion.

With brand company's profitability assured in a product's short patent life, they would not want to make changes in their approved manufacturing processes as they might have to go to FDA for additional review and associated but necessary paperwork and approvals. This is an expense (time and effort) that companies do not want to have. With many companies filing for ANDA for the same drug, brand companies let the generic companies use and commercialize their technologies to manufacture the API and its formulation. A brand company move to new products is the tradition.

No one knows the real price of the API or their formulation. They can be generated through cost analysis using raw material prices, conversion costs, and profit margins. This is standard cost analysis practice used in every industry. They are reviewed in later chapters.

Tables 2.5 and 2.6 are snapshot of prices of a few generic drugs in India and the United States. These prices were collected by contacting pharmacies in these countries. Using API selling prices and guesstimating formulation costs and profits along the supply chain, one can conjecture that companies (formulators and supply chain facilitators, and PBMs) have their profitability assured. Generics like brand companies have little or no incentive to make any continual improvements in their processes. Paperwork necessary to update the regulators about improvements can be an unnecessary encumbrance and an expense with not much financial return.

Tab. 2.5: Drug dose, API cost, and sell price per tablet [59].

Drug	API selling price, rupees per kg	Dose, mg	API content, rupees per tablet	Sell price, rupees per tablet
Lasix	3,375.00	40.00	0.14	0.50
Metformin	375.00	500.00	0.19	4.00
Atorvastatin	24,000.00	10.00	0.24	5.50
Levothyroxine	322,950.00	0.11	0.04	1.65
Nexium	6,150.00	40.00	0.25	9.90
Ciprofloxacin	1,800.00	500.00	0.90	3.47

Tab. 2.6: Sample of drug prices in the United States [60].

Drug	API, $/kg	Milligram/ tablet	Tablets/ kg	API, $/tablet	$ per tablet, 10× API	Price per tablet in the United States, $	
						With health coverage	With no coverage
Ciprofloxacin	24	500	2,000	0.012	0.12	0.2	3.58
Atorvastatin calcium	302	10	100,000	0.00302	0.0302	0.02	3.5
Omeprazole	2	20	5,000	0.00004	0.0004	0.05	0.3
Metformin HCl	4	500	2,000	0.002	0.02	0.07	0.7

Actual selling prices of these drugs in the developed countries are not known as they get camouflaged in mutually subsidized healthcare systems. However, they can be estimated at different levels. Table 2.7 is an illustration [61].

Table 2.8 [45] is a snapshot of a few drugs illustrating the number of site where the API is produced and formulated.

Most of these API can be produced at less than five sites using batch or continuous processes to meet the global demand. Metformin hydrochloride and omeprazole API could be produced using continuous process [62, 63]. The process for each API and its formulation has to be on its own merit and financial justification. Process of

Tab. 2.7: Formulator drug selling price $/kg and patient purchase price.

Drug	Metformin HCl	Ciprofloxacin	Levothyroxine	Atorvastatin
API cost, $/kg [22]	4.00	25.00	4,400.00	310.00
Inert excipients, $/kg (@ 40% API cost)	1.60	10.00	1,760.00	124.00
Conversion cost, $/kg (@ 40% API cost)	1.60	10.00	1,760.00	124.00
Profit (@ 40% above)	2.88	18.00	3,168.00	223.20
Total. $/kg	10.08	63.00	11,088.00	781.20
Average dose	**500 mg**	**500 mg**	**0.112 µg**	**20 mg**
Formulator sale price per tablet, $	0.005	0.032	0.001	0.008
Patient purchase price, $/tablet				
Walmart	0.07	1.04	0.11	0.30
Rite-Aid With insurance	0.07	0.2	0.17	0.31
Rite-Aid Without insurance	0.7	4.77	0.82	3.97

Tab. 2.8: Number of needed API and FDF sites.

Drug	Number of API sites	Number of FDF sites
Ciprofloxacin	22	536
Atorvastatin calcium	44	865
Omeprazole	87	768
Modafinil	29	70
Metformin HCl	77	752
Metoprolol	41	338

continuous improvement can be beneficial for brand and generics if overall filing and reporting processes can be improved [64, 65].

2.6 Pharma's business model

Pharma's current business model, mortar/pestle model, is over 60 years old, which has evolved and worked fine. It is serving the needs of patients with chronic and acute diseases (diabetes, heart disease, high blood pressure, high cholesterol, obesity, etc.) well. Significant portion (~6.2 billion = 80% of 7.8) of the global population [66] does not have mutually subsidized healthcare system. They pay for the drugs from their own pocket and at times have to make a choice between food and drugs [67]. Patients even within the mutually subsidized system due to different apprehensions do not take their medication on a regular basis [68].

Serving medicine needs of ~ 6.2 billion population at affordable prices is an unprecedented opportunity that has not been offered to any industry on the planet [63]. To serve the needs of this population, current manufacturing technologies and practices that result in low asset utilization and low sigma quality will have to be replaced by efficient technologies and methods which could include continuous or modular API manufacturing and continuous formulations [69–83]. Rigorous financial analysis will have to be done to justify every investment, operational strategy, supply chains, and distribution. Principles of chemistry and chemical engineering, which have been applied to other chemical manufacturing, should be precisely applied. Profits from the large customer base will far exceed the current levels.

Continuous processes do not have constraints of equipment availability. Quality can reach six sigma level on properly designed processes. In addition, production can be ramped up and down to meet the market demand. Since continuous process technology is better than batch technology, the overall factory costs are generally lower. Thus, the profitability is higher. Strategies and methods of how many of the APIs can be manufactured using continuous processes are discussed in later chapters.

In this effort, pharma companies and the regulators will have to regroup and review the current regulations, filing, and approval processes and simplify them. Effort would be needed. Interests of many vested groups will have to be dealt with. Even if the regulators change/simplify ANDA and NDA approval and cGMP processes and pathways, industry still has to prove to itself that its innovations and better manufacturing practices are sustainable. Regulators will have to significantly curtail use of 483 citations [84] from their vocabulary. In addition, pharma companies have to pay attention to their impact on global warming [85]. It has been totally ignored.

Drug distribution's current model might also have to be revamped. As said earlier, compared to sending a man to the Moon and bringing him back and landing Cassini on Saturn, producing repeatable quality drugs with technologies that are better than the current technologies and methodologies, reducing drug shortages, improving drug availability and profitability of the companies would be much easier if principles of engineering, science, and economics are applied.

Practices reviewed in later chapters can revamp the supply chain and regulatory landscape.

Abbreviations

FDF	Finished dosage form
API	Active pharmaceutical ingredient
NDA	New drug application
ANDA	Abbreviated new drug application
FDA	Food and Drug Administration
CFR	Code of Federal Register
IND	Investigational new drug
CMC	Chemistry, manufacturing, and controls
ICH	International Conference on Harmonization
PQS	Pharmaceutical quality system
MAH	Marketing authorization holder
PBM	Pharmacy benefit manager

References

[1]   How Do Drugs and Biologics Differ?, https://archive.bio.org/articles/how-do-drugs-and-biologics-differ, Accessed April 17, 2020.

[2]   2019 Guide to the Business of Chemistry, https://www.americanchemistry.com/GBC2019.pdf, accessed April 10, 2020.

[3]   Sullivan, T., A Tough Road: Cost to Develop One New Drug Is $2.6 Billion; Approval Rate for Drugs Entering Clinical Development is Less than 12%, https://www.policymed.com/2014/12/a-tough-road-cost-to-develop-one-new-drug-is-26-billion-approval-rate-for-drugs-entering-clinical-de.html, March 21, 2019.

[4]   Current Good Manufacturing Practices, https://www.fda.gov/drugs/pharmaceutical-quality-resources/current-good-manufacturing-practice-cgmp-regulations, March 30, 2018, Accessed April 3, 2020.

[5]   New Drug Application, https://www.fda.gov/drugs/types-applications/new-drug-application-nda June 10, 2019, Accessed April 1, 2020.

[6]   The Drug Development Process, https://www.fda.gov/patients/learn-about-drug-and-device-approvals/drug-development-process January 4, 2018, Accessed March 20, 2020.

[7]   In Vitro vs. In Vivo Preclinical Drug Testing, https://www.td2inc.com/news/in-vitro-vs.-in-vivo-preclinical-drug-testing, October 1, 2018, Accessed June 29, 2020.

[8]   21 CFR Part 58.1: Good Laboratory Practice for Nonclinical Laboratory Studies, http://www.accessdata.fda.gov/scripts/cdrh/cfdocs/cfcfr/CFRSearch.cfm?CFRPart=58, Accessed April 2, 2020.

[9]   S. 1538 – An act to amend the Federal Food, Drug, and Cosmetic Act, https://www.congress.gov/bill/98th-congress/senate-bill/1538, Accessed April 20, 2020.

[10] Karki, L., Review of FDA Law Related to Pharmaceuticals: The Hatch-Waxman Act, Regulatory Amendments and Implications for Drug Patent Enforcement. https://www.neifeld.com/pubs/2005_reviewfdalaw.pdf, Accessed March 20, 2020.

[11] Guidance for Industry 180-Day Exclusivity When Multiple ANDAs Are Submitted on the Same Day, https://www.fda.gov/media/71304/download, July 2003, Accessed April 24, 2020.

[12] Abbreviated New Drug Application (ANDA), https://www.fda.gov/drugs/types-applications/abbreviated-new-drug-application-anda, Accessed May 10, 2020.

[13] Daphne, E. S. M., Bioequivalence and Interchangeability of Generic Drugs, https://www.merckmanuals.com/home/drugs/brand-name-and-generic-drugs/bioequivalence-and-interchangeability-of-generic-drugs, Accessed May 11, 2020.

[14] Raines, K. W., A Primer on Generic Drugs and Bioequivalence: an overview of the generic drug approval process, https://www.fda.gov/media/89135/download, Accessed May 10, 2020.

[15] Hornecker, J. R. Generic Drugs: History, Approval Process, and Current Challenges, U. S. Pharmacist, 2009, 34, 6, Generic Drug Review suppl., pgs. 26–30. Accessed January 10, 2020.

[16] Malhotra, G.: Can the Review and Approval Process for ANDA at USFDA be Reduced from Ten Months to Three Months? https://pharmachemicalscoatings.blogspot.com/2017/03/can-review-and-approval-process-for.html March 25, 2017, Accessed October 20, 2021.

[17] Malhotra, G.: GDFUA II ANDA (Abbreviated New Drug Application) Review Target of 8–10 Months should be a Cause of Concern, https://pharmachemicalscoatings.blogspot.com/2020/03/gdfua-ii-anda-abbreviated-new-drug.html March 24, 2020 Accessed October 20, 2021.

[18] ICH guideline Q12 on technical and regulatory considerations for pharmaceutical product lifecycle management https://www.ema.europa.eu/en/documents/scientific-guideline/ich-guideline-q12-technical-regulatory-considerations-pharmaceutical-product-lifecycle-management_en.pdf 4 March 2020, Accessed May 21, 2021.

[19] ICH Q13: Continuous Manufacturing of Drug Substances and Drug Product, https://database.ich.org/sites/default/files/Q13_EWG_Concept_Paper.pdf 14 November 2018 Accessed October 22 2021.

[20] Quality Considerations for Continuous Manufacturing Guidance for Industry https://www.fda.gov/media/121314/download February 2019 Accessed March 22, 2020.).

[21] Continuous Production, https://en.wikipedia.org/wiki/Continuous_production, Accessed June 1, 2019.

[22] The Pharmaceutical Manufacturing Process – Steps, Tools, and Considerations https://www.thomasnet.com/articles/chemicals/the-pharmaceutical-manufacturing-process-steps-tools-and-considerations/ Accessed July 5, 2020.

[23] Douglas, L., Erwin, P. E. Industrial Chemical Process Design, Second. McGraw-Hill Education, New York, 2014, https://www.accessengineeringlibrary.com/content/book/9780071819800. Accessed June 29, 2020.

[24] Encyclopedia of Chemical Technology, https://onlinelibrary.wiley.com/doi/book/10.1002/0471238961 Accessed June 29, 2020,.

[25] Mody, D., Strong David, S.: An Overview of Chemical Process Design Engineering, https://doi.org/10.24908/pceea.v0i0.3824, Accessed June 29, 2020.

[26] McGraw Hill Chemical engineering series, https://www.librarything.com/tag/McGraw+Hill+Chemical+engineering+series Accessed June 16, 2020.

[27] Batch production, https://en.wikipedia.org/wiki/Batch_production, Accessed June 1, 2019.

[28] Batch Production Vs. Mass Production: https://simplicable.com/new/batch-production-vs-mass-production, February 1, 2017, Accessed July 3, 2020.

[29] Buthmann, A., Cost of Quality: Not Only Failure Costs, https://www.isixsigma.com/implementation/financial-analysis/cost-quality-not-only-failure-costs/, Accessed April 7, 2020.

[30] Breyfogle III, F. W., Cupello, J. M., Meadows, B. Managing Six Sigma, Wiley-Interscience, John Wiley & Sons, Oct 11 2000, ISBN-10: 0471396737 pgs. 47–50..

[31] Wheeler, J. M. Getting Started: Six-Sigma Control of Chemical Operations, Chemical Engineering Progress, June 2002, pgs. 76–81, Accessed June 20, 2020.

[32] Shanley, A., Will the Pharmaceutical Industry Ever Get to Six Sigma? https://www.pharmtech.com/will-pharmaceutical-industry-ever-get-six-sigma, July 12, 2017, Accessed April 8, 2020.

[33] Malhotra, G.: The Good, the Bad, the Ugly (1) Complexities of Pharmaceutical Manufacturing, Profitability through Simplicity, April 8, 2018,.

[34] Hussain, A. S.: Pharmaceutical 6-Sigma Quality by Design, The 28th Annual Midwest Biopharmaceutical Statistical Workshop, Ball State University, Muncie, IN, May 23–25, 2005, Page 20, Accessed May 19, 2019, Accessed May 19, 2020.

[35] Malhotra, G.: Opportunities for Generic Pharma to Clear the Quality Stigma, https://pharma chemicalscoatings.blogspot.com/2019/05/opportunities-for-generic-pharma-to.html, May 23, 2019.

[36] Malhotra, G.: Square Plug In A Round Hole: Does This Scenario Exist in Pharmaceuticals? https://pharmachemicalscoatings.blogspot.com/2010/08/square-peg-in-round-hole-does-this.html, August 10, 2010, Accessed May 19, 2020.

[37] Malhotra, G.: Why Fitting a Square Plug in a Round hole is Profitable for Pharma and Most Likely Will Stay? https://pharmachemicalscoatings.blogspot.com/2014/08/why-fitting-square-plug-in-round-hole.html, August 1, 2014, Accessed April 5, 2020.

[38] OEE Benefits for Pharmaceutical, BioTech and Medical Device Manufacturers in Life Sciences, https://www.pharmaceutical-technology.com/contractors/process_automation/zenith% 20technologies/pressreleases/pressoee-benefits-life-sciences/ August 27, 2015 Accessed May 19, 2020.

[39] Benchmarking Shows Need to Improve Uptime, Capacity Utilization, https://www.pharmama nufacturing.com/articles/2007/144/ Sep 20, 2007, Accessed May 19, 2020.

[40] Malhotra, G.: Continuous Pharmaceutical Processes and Their Demands, Contract Pharma, https://www.contractpharma.com/issues/2016-04-01/view_features/continuous-pharmaceutical-processes-and-their-demands/?userloggedin=true, April 5, 2016, Accessed May 15, 2020.

[41] Tyson, T.: Solving the Industry's Capacity Utilization Problem, https://www.pharmasalmanac. com/articles/solving-the-industrys-capacity-utilization-problem, October 28, 2019.

[42] Malhotra, G.: Opportunities for Generic Pharma to Clear the Quality Stigma, https://pharma chemicalscoatings.blogspot.com/2019/05/opportunities-for-generic-pharma-to.html, May 23, 2019, Accessed April 25, 2020.

[43] Couper, J. R. et al., Chemical Process Equipment, Selection and Design, Third. Elsevier, **ISBN**: 9780123969590, 2010.

[44] Dunbar, K. http://jwilson.coe.uga.edu/EMAT6680/Dunbar/Assignment1/sine_curves_KD. html Accessed July 8, 2019.

[45] Malhotra, G.: Pharmaceutical Quality: Concepts, Misconceptions, Realities and Remedies https://pharmachemicalscoatings.blogspot.com/2019/11/pharmaceutical-quality-concepts. html, November 4, 2019, Accessed June 6, 2020.

[46] Continuous Formulation Manufacturing Unit, https://www.continuous-production.com/contin uous-manufacturing, Accessed April 10, 2020.

[47] Byrn, S., Futran, M., Thomas, H. et al, Achieving Continuous Manufacturing for Final Dosage Formation: Challenges and How to Meet Them, Journal of Pharmaceutical Sciences, 104, 792–802, 2015.

[48] Drugs@FDA: FDA-Approved Drugs, https://www.accessdata.fda.gov/scripts/cder/daf/index. cfm?event=browseByLetter.page&productLetter=M&ai=0, Accessed June 3, 2020.

[49] Common Drug Combinations, https://www.goodrx.com/blog/10-most-common-drug-combinations/ Accessed July 7, 2020.

[50] McMillan, G. K., Vagas, P. E., Hunter, H. P. Process/Industrial Instruments and Controls Handbook, Sixth Edition, McGraw-Hill Education, New York, 2019, https://www.accessengi neeringlibrary.com/content/book/9781260117974.

[51] Stephanopoulos, G. Chemical Process Control: An Introduction to Theory and Practice, Prentice-Hall, Englewood Cliffs, New Jersey, 1984, ISBN 0-13-128629-3.

[52] Dimethyl Fumarate, https://en.wikipedia.org/wiki/Dimethyl_fumarate, Accessed May 28, 2020.

[53] https://m.alibaba.com/products/dimethyl_fumarate.html Accessed May 28, 2020.

[54] https://www.pharmacompass.com/active-pharmaceutical-ingredients/dimethyl-fumarate May 28, 2020.

[55] Dimethyl Fumarate Selling price https://www.reuters.com/article/us-health-ms/u-s-prices-for-multiple-sclerosis-drugs-are-on-the-rise-idUSKCN1VH2I5 May 28, 2020.

[56] Multiple sclerosis Global Population https://www.healthline.com/health/multiple-sclerosis/facts-statistics-infographic#1 June 4, 2020.

[57] Mylan Wins District Court Decision Against Biogen's Tecfidera® Patent, https://www.prnews wire.com/news-releases/mylan-wins-district-court-decision-against-biogens-tecfidera-patent-301079843.html, June 18, 2020, Accessed July 7, 2020.

[58] https://www.pharmacompass.com/active-pharmaceutical-ingredients/dimethyl-fumarate Accessed July 7, 2020.

[59] Malhotra, G.: Comparison of Drugs Prices: US vs. India; Their Manufacturing Costs & Opportunities to Improve Affordability, https://pharmachemicalscoatings.blogspot.com/2018/01/comparison-of-drugs-prices-us-vs-india_89.html, January 18, 2018, Accessed May 20, 2020.

[60] Malhotra, G.: Impact of Regulations, Manufacturing and Pharmaceutical Supply Chain (PBMs) on Drug Shortages and Affordability Part 2, https://pharmachemicalscoatings.blogspot.com/2019/04/impact-of-regulations-manufacturing-and.html, April 3, 2019, Accessed May 28, 2020.

[61] Malhotra, G.: Systematic Demystification of Drug Price Mystique and the Needed Creative Destruction, https://pharmachemicalscoatings.blogspot.com/2019/10/systematic-demystification-of-drug.html, October 2, 2019, Accessed July 7, 2020.

[62] Malhotra, G.: Chemical Process Simplification: Improving Productivity and Sustainability John Wiley & Sons, February 2011,.

[63] Malhotra, G.: Alphabet Shuffle – Moving from QbA to QbD, Pharmaceutical Processing World, https://www.pharmaceuticalprocessingworld.com/alphabet-shuffle-moving-from-qba-to-qbd/ February 23, 2009, Accessed June 10, 2020.

[64] Malhotra, G.: Process of Continuous Improvement and Pharmaceuticals, https://pharmache micalscoatings.blogspot.com/2009/06/process-of-continuous-improvement-and.html, June 1, 2009, Accessed June 3, 2020.

[65] Malhotra, G.: Calling for Change, https://pharmachemicalscoatings.blogspot.com/2015/11/calling-for-change.html, November 17, 2015, Accessed June 3, 2020.

[66] An alternate look at pharmaceutical world revenues and drug affordability, https://www.man ufacturingchemist.com/news/article_page/An_alternate_look_at_pharmaceutical_world_rev enues_and_drug_affordability/135165, October 23, 2017, Accessed June 8, 2020.

[67] Malhotra, G.: Drug Prices: Food vs. Medicine – A Difficult Choice for Some https://pharmache micalscoatings.blogspot.com/2011/06/drug-prices-food-vs-medicine-difficult.html, June 16, 2011, Accessed June 11, 2020.

[68] Patel, N.: Why Don't Patients Take Their Medications? https://www.usciences.edu/blog/note worthy/posts/why-dont-patients-take-their-medications.html, October, 24, 2018, Accessed June 11, 2020.

[69] Malhotra, G.: Batch or A Continuous Process: A Choice; Pharmaceutical Processing, March 2005, Page 16.

[70] Malhotra, G.: Hesitation in the Drive to A Continuous Pharmaceutical Manufacturing Process: Real or Imaginary? Pharmaceutical Processing, July 2009 pgs.- 12–15.

[71] Malhotra, G.: A Radical Approach to Fine/Specialty API Manufacturing, Profitability through Simplicity, https://pharmachemicalscoatings.blogspot.com/2010/01/radical-approach-to-finespecialty-api.html, January 20, 2010, Accessed June 29, 2020.

[72] Malhotra, G.: Considerations to Simplify Organic Molecule (API) Manufacturing Processes: My perspective, Profitability through Simplicity, https://pharmachemicalscoatings.blogspot.com/2019/04/considerations-to-simplify-organic_77.html, April 20, 2019, Accessed June 29, 2020.

[73] Malhotra, G.: The Path Towards Continuous Processing, Pharmaceutical Processing, August 2010, pgs. 16–20.

[74] Malhotra, G.: Is Continuous Processing in Pharma's Future? Profitability through Simplicity, July 24, 2012.

[75] Malhotra, G.: Continuous Process in Pharmaceutical Manufacturing: Considerations, Nuances and Challenges, Contract Pharma, June 2, 2015.

[76] Malhotra, G.: My Perspective for Pharmaceutical Manufacturing Technologies/Processes and Continuous Improvements, CPhI 2015, http://epcotint.com/Pharmaevolution/GirishMalhotra September2015CPhIreport.pdf.

[77] Malhotra, G.: Continuous Pharmaceutical Processes and Their Demands, Contract Pharma, pgs. 37–40, April 5, 2016.

[78] Malhotra, G.: Manufacturing Processes Require Financial Justification, Contract Pharma, September 2016.

[79] Malhotra, G.: Strategies for Improving Batch or Creating Continuous Active Pharmaceutical Ingredient (API) Manufacturing Processes, Profitability through Simplicity, March 20, 2017.

[80] Malhotra, G.: Reality and Un-Reality: Continuous Processing in Pharmaceutical Manufacturing, Contract Pharma, April 3, 2017.

[81] Malhotra, G. Batch, Continuous or "Fake/false" Continuous Processes, Profitability through Simplicity, July 20, 2017, American Pharmaceutical Review, 20, 6, September/October 2017. pgs 86–91. Contract Pharma, Nov./Dec. 2017 pgs. 56–58, Chemistry Today, November/December 2017, 35(6) pgs. 62–65..

[82] Schaber, S. D., et al.: Economic Analysis of Integrated Continuous and Batch Pharmaceutical Manufacturing: A Case Study https://pubs.acs.org/doi/pdf/10.1021/ie2006752, Accessed April 8, 2020.

[83] Srai, J. S., et al.: Evaluating the potential for the continuous processing of pharmaceutical products, https://core.ac.uk/download/pdf/131079381.pdf, July 27, 2011, Accessed April 8, 2020.

[84] Malhotra, G.: Are US FDA 483 Citations a "Medal of Honor" or "Rite of Passage" to Disgrace for the Pharma companies? https://pharmachemicalscoatings.blogspot.com/2019/10/are-usfda-483-citations-medal-of-honor.html October 16, 2019, Accessed April 29, 2020.

[85] Malhotra, G: Climate Change and Greening of Pharmaceutical Manufacturing, https://pharma chemicalscoatings.blogspot.com/2022/01/climate-change-and-greening-of_24.html January 24, 2022 Accessed January 27, 2022.

Chapter 3
Physical and chemical properties

Every raw material used in the manufacture of chemicals, pharmaceuticals, and their formulations come in three forms. Each of these forms has their own characteristics and nuances. It is best to review characteristics of each chemical's physical forms and how their mutual behavior can be exploited and used in reactive and blending processes:

1. Liquid
2. Solid
3. Gas

Some are compatible with each other and others are not. Compatibility can be the degree of miscibility (solubility) that can vary from none to completely soluble. Safe handling of chemicals is necessary. Flammability is a characteristic of the chemicals. This has to be extremely carefully addressed as most organic chemicals have varying degrees of flammability. It is determined by the flash point of a material. Flash point is the minimum temperature at which a liquid forms a vapor above its surface in sufficient concentration that it can be ignited. Flammable liquids have a flash point of less than 100 °F. Liquids with lower flash points ignite easier [1]. Each chemical has certain toxicity and is discussed later in this chapter.

Process developers and designers have to know everything about the chemicals used and produced in every reaction. Some of the needed physical properties are listed in Tab. 3.1 [2]. Additional properties for specific need might also be needed. If they are not readily available, they will have to be generated.

Properties [2] listed in Tab. 3.1 are of high value, and when used result in process design that will produce quality products. It is not necessary that each of the property mentioned may be needed or used in the design. Properties of raw materials used in production of active pharmaceutical ingredients (APIs), produced intermediates and

Tab. 3.1: Chemical and physical properties of chemicals [2].

Physical properties	Chemical properties
– Mass	– Heat of formation
– Density	– Heat of reaction
– Melting/freezing point	– pH
– Boiling point	– Surface tension
– Viscosity	– Flammability
– Solubility	– Toxicity
– Azeotrope	– Hygroscopic

https://doi.org/10.1515/9783110702842-003

formulations that may not be commonly available. Their physical property data might have to be generated theoretically and checked experimentally.

Before internet, many trade publications and companies published physical and chemical property data. In 1960s and 1970s, chemical engineering [3] and hydrocarbon processing [4] magazines published 24 and 44 articles, respectively, about physical and thermodynamic properties of hydrocarbons. Other publications [5] detailing physical properties are also available. Companies do provide physical properties and safe handling procedures [6]. Most of this information is available to potential customers only.

With the advent of internet, physical and chemical property data can be accessed from these sources [7–10]. Sometimes data needed for specific needs has to be generated by the user in the laboratory.

Most companies provide material safety data sheets (MSDS) that meet the regulatory requirements but do not have detailed properties which are needed and used for meaningful and safe process design. If additional information is needed, it might be available if one is a customer. Figure 3.1 shows parts of a typical safety data sheet (SDS) [11]. Each supplier has to provide SDS.

There are companies that will provide excellent physical property, safety, and handling data, for example, data for sodium hydroxide [6].

They can also be generated mathematically as has been illustrated in different journals of American Chemical Society [12–14] and *Chemical Engineering Thermodynamics* book [15].

The need for all of physical and chemical properties might seem redundant or unnecessary at the initial stages of chemistry development and process considerations. However, as explained later and in different chapters, such information is necessary for the process development and design and lets the process designers be creative and imaginative in product development. It also lets the process engineers, operating personnel, and raw material buyers have total understanding and command of the processes. This can be beneficial in troubleshooting process that needs attention. If data such as solubility and viscosity are not available, it has to be internally generated.

3.1 Molecular weight

Molecular weight is not included in the property table as it is expected that each developer (chemist/chemical engineer) has to know the molecular formula and the molecular weight of each chemical that is used and/or produced in every reaction or the blending process. They tell us everything about each chemical that will be produced on our route to the final desired product and how they will be handled. Chemical formula is a fingerprint of each chemical. They have their distinct numerical number, and CAS number [16] can be used to know almost everything about the chemical.

Molecular weights are also used to mass balance the chemical reaction equations. These in turn are used to calculate each reaction step yield. Reaction parameters

p-Aminophenol CAS No 123-30-8	MATERIAL SAFETY DATA SHEET SDS/MSDS

SECTION 1: Identification of the substance/mixture and of the company/undertaking

1.1 **Product identifiers**
Product name   :   **p-Aminophenol**

CAS-No.   :   123-30-8

1.2 **Relevant identified uses of the substance or mixture and uses advised against**

Identified uses   :   Laboratory chemicals, Industrial & for professional use only.

1.3 **Details of the supplier of the safety data sheet**

Company   :   Central Drug House (P) Ltd
                                7/28 Vardaan House
                                New Delhi-10002
                                INDIA

Telephone   :   +91 11 49404040
Email   :   care@cdhfinechemical.com

1.4 **Emergency telephone number**
Emergency Phone #   :   +91 11 49404040 (9:00am - 6:00 pm) [Office hours]

SECTION 2: Hazards identification

2.1 **Classification of the substance or mixture**

Classification according to Regulation (EC) No 1272/2008
Acute toxicity, Oral (Category 4), H302
Acute toxicity, Inhalation (Category 4), H332
Germ cell mutagenicity (Category 2), H341
Acute aquatic toxicity (Category 1), H400
Chronic aquatic toxicity (Category 1), H410

For the full text of the H-Statements mentioned in this Section, see Section 16.

2.2 **Label elements**

Labelling according Regulation (EC) No 1272/2008
Pictogram

Signal word   Warning

Hazard statement(s)
H302 + H332   Harmful if swallowed or if inhaled
H341   Suspected of causing genetic defects.

SECTION 3: Composition/information on ingredients

3.1 **Substances**
Synonyms   :   4-Hydroxyaniline

Formula   : C_6H_7NO
Molecular weight   :   109.13 g/mol
CAS-No.   :   123-30-8
EC-No.   :   204-616-2
Index-No.   :   612-128-00-X

Hazardous ingredients according to Regulation (EC) No 1272/2008

Component		Classification	Concentration
4-Aminophenol			
CAS-No.	123-30-8	Acute Tox. 4; Muta. 2; Aquatic	<= 100 %
EC-No.	204-616-2	Acute 1; Aquatic Chronic 1;	
Index-No.	612-128-00-X	H302, H332, H341, H400, H410	
		M-Factor - Aquatic Acute: 10	

Fig. 3.1: Parts of a typical MSDS.

SECTION 9: Physical and chemical properties

9.1   **Information on basic physical and chemical properties**

a) Appearance   Form: powder
                                   Colour: beige

b) Odour   No data available

c) Odour Threshold   No data available

d) pH   No data available

e) Melting point/freezing   Melting point/range: 185 - 189 °C - lit.
    point

f) Initial boiling point and   No data available
    boiling range

g) Flash point   195 °C - closed cup

h) Evaporation rate   No data available

i) Flammability (solid, gas) No data available

j) Upper/lower   No data available
    flammability or
    explosive limits

k) Vapour pressure   No data available

l) Vapour density   No data available

m) Relative density   No data available

n) Water solubility   slightly soluble

o) Partition coefficient: n-   log Pow: 0.04
    octanol/water

p) Auto-ignition   No data available
    temperature

q) Decomposition   284 °C -
    temperature

r) Viscosity   No data available

s) Explosive properties   No data available

t) Oxidizing properties   No data available

Fig. 3.1 (continued)

can be optimized to improve the process yield. Better yield means lower cost and higher profits.

3.2 Physical state

Each chemical at room temperature has either of the three natural physical forms. Each form has its value and they are reviewed:

1. Liquid
2. Solid
3. Gas

If we understand and capitalize on mutual behavior of chemicals, we will have an improved design. Learning and understanding physical and chemical properties of chemicals that are used and produced (intermediates and final) in the process influence process unit processes and unit operations [17–20].

3.3 Liquid

Of the three forms, liquids are process developer's best friend. Chemicals that flow facilitate processing. Liquids also assist in the creation of other dispensable forms such as tablets, ointments, drops, and injections. Flow of liquids can be precisely controlled, a must for creating an excellent economic process. If need be, they can be solidified and gasified. However, this transformation is rarely needed. Reaction intermediates or final product creates processing opportunities. Generally, intermediates will not be purified. However, the final liquid product will be purified using single or combination of different unit operations [18]. Each has to be dealt accordingly.

3.4 Solid

Chemicals that are solid at room temperature can be handled in their natural state but compared to liquids, they present handling challenges. Solids are easier to handle in solutions. Solubility of the solid in different solvents at different temperatures is extremely useful information. Most of the time, solubility information is not available. It would have to be generated in the laboratory. Solids can be charged to the reactors and would necessitate proper solid handling measures. These depend on the process. Economics and solid toxicity dictate the method of handling and respective investment. Other alternate to handle solid is the molten form, but as discussed later, melting point, product demand, supply, logistics, equipment design, and economics are major considerations for this route.

Solids that are formed during the reaction could be filtered, if they are a product and purified. If further processing to a product is needed, proper unit operations and processes have to be used to convert to the desired product. Physical properties such as melting point and solubility come into play. Since most of the reactions are at elevated temperatures and solvents are present, viscosity of the mixture seldom interferes the reaction progress.

3.5 Gas

Gases are best handled in their compressed liquid form. They can be handled as solution but, discussed later, they have their own challenges.

3.6 Material handling

Melting, boiling, and freezing points determine the physical state of each chemical. At room temperature, each chemical comes in either of the three states: solid, liquid, or gas. This sets how each material will be handled for batch and continuous processes. Irrespective of the state of the chemical and quantity used, it is absolutely necessary that each chemical, intermediate and by-product be handled with respect. Toxicity of many chemicals is documented [21–24].

It is still possible that long-term toxicity impact of many chemicals or their mixtures on aquatic, human, and bird life may not be known for an extended time. Toxicity of chemicals can have significant ecotoxicological and environmental impact [25, 26]. It is necessary that an effort to improve the process yields be part of the continuous process improvement regimen. In addition, adequate care be taken to properly design and care for waste disposal and treatment. It is most important that the products and by-products of API, solvents, and their formulated mixtures be respected [27].

The method of handling materials in small quantities in the laboratory is significantly different from the ways for larger quantities for a batch or a continuous process. Raw materials that are liquid at room temperature can be pumped and metered precisely in scale-up and for a commercial process. They are the easiest to handle.

Raw materials that are solid at room temperature in the laboratory are handled in the laboratory hood or a similar enclosure following the suggested safety guidelines. They can be slurried or dissolved in a liquid and added to the reaction vessel. Since solids in the laboratory are dissolved or directly added to the reactor, developers generally do not consider how they would be added to the reactors on a commercial scale. Their addition methods change for scale-up experiments to commercial operations. Scale-up methods, if commercial operation emulates, can be extremely beneficial as they can be used as a proving ground.

Figure 3.2 reviews considerations for handling gas, liquid, and solid raw materials on a commercial scale. Each raw material state besides presenting challenges also have value if handled properly.

Generally fed as liquid. Flow rate can be controlled.
Ammonia can be fed as NH_4OH solution. However, unless needed, water takes up reactor volume and generally lowers process productivity.

GAS

Easiest and safest to handle. Flow can be totally controlled.

LIQUID

Generally supplied in bags or bulk. Bulk transfer and dust control system needed.

For batch processes individual bags are used. Dust control system needed. Dust control equipment has to be product specific to avoid contamination and hazard.

Solid as Solid

SOLID

For continuous process bulk solid handling and dust control system are needed. Supplier and company have to have compatible bulk transfer systems. Such process would be a single product feed system. For controlled feed solids are metered in the process.

Solid as molten liquid

If product is campaigned during the year, solid handling and dust control systems would be product specific to avoid contamination. Asset utilization would be less than optimum.

Solid can feed as a liquid. It is used for continuous processes. Liquid can be used for a batch process but it has to be need specific. It can be negotiated between supplier and buyer.

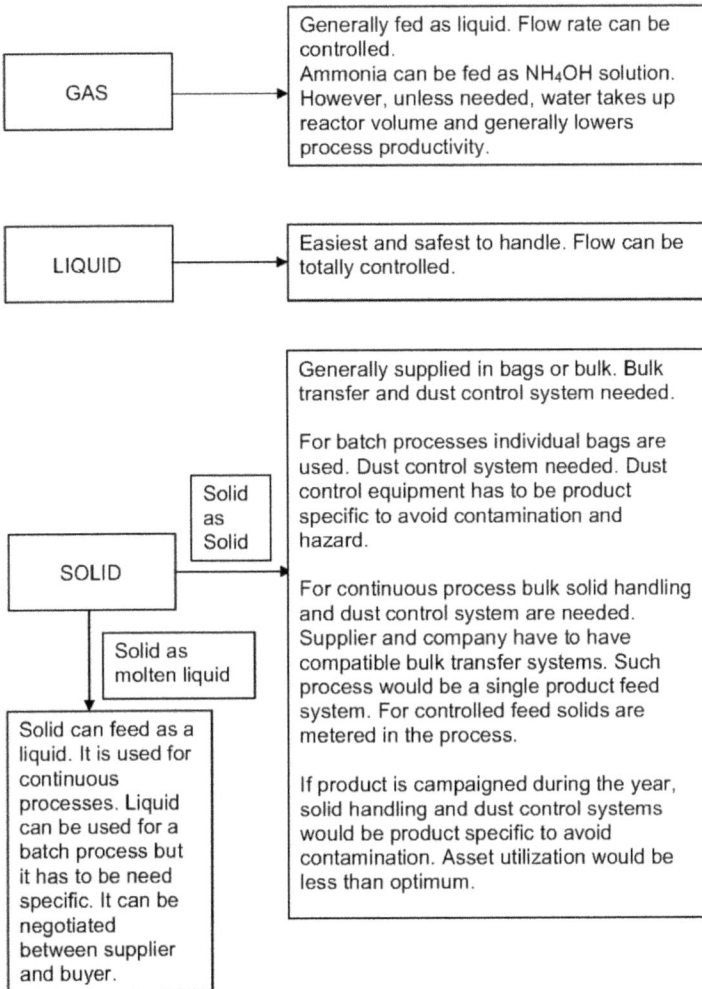

Fig. 3.2: Gas, liquid, and solid raw material handling options.

3.6.1 Gas as a raw material

In every commercial reaction, if a gas is used as a raw material, it will be generally fed as a liquid. Its flow rate can be precisely controlled and managed. Gas, fed as a liquid, is also an asset as it absorbs the heat of reaction/exotherm that is generated during reaction and essentially speeds the reaction rate. Reduced reaction time can reduce the size of the equipment, that is, capital investment. In laboratory, ammonia can be used as ammonium hydroxide but on a commercial scale, it is best used as 100% liquid. Reaction productivity loss is significant if it is used as ammonium hydroxide solution. In special cases, use of ammonium hydroxide solution might be necessary but such cases would be an exception.

3.6.2 Solids as a raw material

Since different solid raw materials are going to be used in process reactions and formulations, they will have to have their separate feed and dust control systems. If the same feed and dust control equipment is to be used, it will have to be cleaned in between uses to avoid cross-contamination. This practice will extend batch cycle times, lower asset utilization, and will also lower return on investment. During process design, judicious manufacturing planning has to be done to maximize asset utilization. Batch and continuous processes have different process requirements and have to be addressed.

Generally, raw materials for laboratory development are acquired from specialty chemical supply houses. This practice has the benefit of immediate acquisition but has significant drawbacks. Of them, biggest is that this raw material will not be used in a commercial operation as it will be expensive. It is best to acquire the material from commercial supplier. Its benefits are competitive pricing, and the actual material will be used for process development. Since many of the raw materials are used to produce other nondrug products, one should consider these sources for better quality, pricing, and customer service.

Solid raw materials can be solubilized or slurried in a solvent/liquid for a batch or a continuous process and fed to the reaction system. Figure 3.3 is a simple illustration. Solid would be metered in a circulating liquid/solvent stream and added to the reactor. Feed systems that meter solids accurately are available. Advantage of such a system can be solubilizing the solid in a reaction solvent or reactant so that the reaction time can be reduced for a batch process. Such a system would become an integral part of the continuous process. Heat exchanger (plate and frame or shell and tube) could be used. Of advances in heat exchange technology, electrically heated [28] can be of extreme value as they are simpler compared to conventional heat transfer fluid technologies. Conventional heat exchangers use Dowtherm [29] and Therminol [30]. Electrically heated heat exchangers can significantly reduce the investment, and their modular portability can be of value.

If a single-feed system similar to Fig. 3.3 is going to be used for many different API products, cross-contamination, safety, waste disposal, and economics will have to be considered. Such systems will work for continuous API manufacturing but there are not too many products on the pharma landscape that will require large single product slurry and solution making processes.

For formulations (batch and semicontinuous), commercially available pneumatic and screw feeders that can control the solid flow and feed rates would be useful. Cleaning between products can result in significant downtime and will be a major

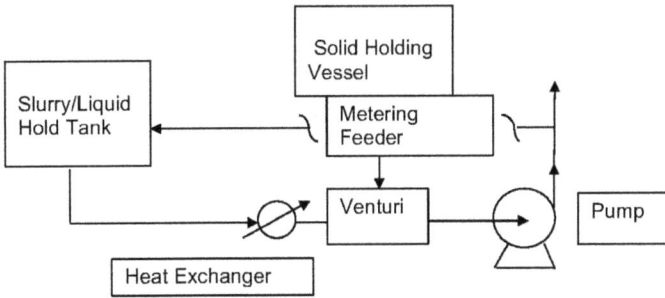

Fig. 3.3: Solid slurry or dissolution flow diagram.

investment consideration. Investing in dedicated product feed, dust handling, and disposal systems committed for single solid feed (single product batch processes) would be an expensive investment since such systems are not going to be used 100% of the time. However, these issues do not matter if the process is a single product process.

For API processes, process designers can come up with ways by which solid feed for batch processes can be simplified and improved. They will improve batch cycle time, improve product quality, improve asset utilization, and lower costs. However, many different process design and business considerations will have to fall in place, including company's business plan (short and long term), manufacturing strategies, process design, and supplier logistics.

3.6.2.1 Solid feed for batch processes

Schematic suggested in Fig. 3.3 can be of value if API batch campaigns of the same product are of significant length (large production volume). If the solid can be brought to the manufacturing site using trucks that can pneumatically off-load the product to manufacturer's solid holding vessel, there would be advantages. Supplier's truck could also be used by the manufacturer as the holding vessel during product's campaign.

Another option would be for the supplier bringing the API solid raw material/s as a melt in heated trucks or tank cars that would hook up with the buyer's process. Such systems have been and are used for continuous processes for more than 60 years.

However, under pharma's current business model, use of solid feed as melt may not be feasible. This is due to the fact that many companies are producing the same products [31]. The value of economies of scale is lost as the processes are fitted in the existing equipment [32, 33], and the opportunities to use better and more efficient technologies are also lost.

Another reason for not using solid as a melt is due to the fact that there are not too many products that have the volume requirement to be used as molten raw materials. Suppliers are also not prepared for supplying solid raw material as a melt. Consolidation, which is highly unlikely, might create limited number of plants for certain API but lack of assurance of continued raw material supply might interfere commercialization of such plants. There are few APIs (Tab. 3.2) that have the volume where raw material/s can be used as a melt. Since these are produced using a batch process, a different business model would be necessary for molten raw materials.

Tab. 3.2: Key raw materials and their melting points.

Product	Key raw material/s	Melting point (°C)
Paracetamol/acetaminophen	*para*-Aminophenol	~187
Metformin hydrochloride	Dicyandiamide	209.5
	Dimethylamine *hydrochloride*	160
Omeprazole and isomers	Pyrmethyl alcohol	~60

3.6.3 Liquid as a raw material and as a solvent

Liquids as a raw material are the easiest to handle even if they are corrosive, dangerous, and hazardous. Proper handling systems have been developed and are commercially available. Necessary material technologies exist for safely handling, storing, and transporting liquids. However, still adequate safety measures have to be incorporated in the process design.

3.6.4 Density

Density of raw materials involved in chemical reactions and formulations is of great value. Since density changes with temperature, it is important to know its value at different temperatures. Since liquids have different densities, it can be effectively used to separate immiscible liquids. This difference can be very effectively used in chemical reactive processes.

It is important to have density data for each chemicals involved, used, and produced in pharmaceutical manufacturing. Various sources are available. Vendors of different chemicals can be an excellent resource [6]. National Institute of Science and Technology [34] is also an excellent source. If the density data is not available, chemistry specific data will have to be generated internally.

3.6.5 Viscosity

Viscosity can be mathematically estimated [35] but would have to be verified in the laboratory. Chemists and chemical engineers have choices on how to use such liquids.

They can dissolve the viscous liquid in the solvent being used in the process or heat the liquid to an acceptable temperature where the liquid can be easily pumped.

3.6.6 Solubility

Solubility of chemicals is an important property in many ways. Mutual solubility or lack of it is useful in reactive chemical processes. It is extremely important to have solubility information for reactions as this property can be exploited to simplify and create excellent processes. Water or organic chemicals are used as solvents. We have to recognize that solvents are a process facilitator. Methods to predict solubility of organic compounds in organic solvents [36] have been proposed. However, there is no substitute for actual data.

If reactants and reaction products dissolve in the solvent used in the reaction, the reaction rate can improve through improved mass transfer. Process productivity can be improved if the reactants can be used at their highest solution concentration. Solubility of chemicals is used in purification, separation of products, and crystallization.

Basis of crystallization unit operation is solubility, a very important unit operation in pharmaceutical manufacturing. Lack of or lower solubility, through easier phase separation, can be exploited to create excellent processes. Overall, it is extremely beneficial that mutual solubility or lack of it is determined for the solvents and reactants at the initial process development stage as this knowledge can be used in process design. Combined use of lack of solubility and difference in densities has been illustrated later in this chapter.

3.6.7 Specific heat

Specific heat of chemicals [37, 38] is an important property. This information is required for the developer and process designer to control the process heat and design the necessary process equipment. This property is used in the design of heat exchangers, crystallizers, evaporators, and distillation equipment needed for the process, basically to control the heat needs of every process. The value of specific heat impacts the size of the equipment, that is, the investment.

3.6.8 Azeotrope behavior

Azeotropic behavior [39] of chemicals might not look to be of value but it is important. Minimum boiling point azeotropes have significant value in chemical processing.

Simply, they are a mixture of two or more liquids that have the lowest boiling point of the mixture than of any pure liquid. Azeotropic behavior of chemicals can be used very creatively in manipulating reaction processes. This is especially true when the two liquids of the azeotrope are immiscible or have low solubility. Imagine a reaction process where the liquid mix has two components: organic and aqueous and they form an azeotrope. Combination of this property and density difference as explained later, can be very effectively used to improve the reaction exotherm, reaction rate, and yield.

3.6.9 Heat of formation/reaction

The heat of reaction is an important property for every chemical reaction. For every reaction, it is necessary that we know it. Similarly, it is important to know the heat capacity of the reactants and reaction products. They are of value in the process equipment design. The difference between the heat of formation of products and heat of reactants suggests the amount of heat that has to be removed. This can be calculated and determined [15, 40–42]:

$$\Delta H \text{ of reaction} = \sum \Delta H \text{ of (products)} - \sum \Delta H \text{ of (reactants)} \qquad (3.1)$$

Most reactive processes are either endothermic or exothermic. How heat of reaction is controlled can significantly impact the rate of reaction and size of the equipment. Heat of formation is also influenced and controlled by the method and sequence of raw material addition. It is critical to control the heat of reaction. If the heat of reaction is not controlled adequately, the reaction temperature can rapidly rise and can result in explosive and/or hazardous situations. If the heat of formation can be effectively controlled, it can reduce the reaction residence time, that is, the size of the equipment (i.e., investment).

3.6.10 Flammability

Flammability [1, 43] of chemicals is important in every chemical process for process safety. One has to make sure that chemicals, individually and collectively, do not have the conditions where they will ignite and lead to explosion.

3.7 Exploitation of chemical and physical properties

Physical properties are of great value to the process developer and designer as s/he can use them to simplify and create excellent processes. Combining and exploiting different physical properties to create a simple process can be a challenge [44–46]. In the end, how the process is executed in a commercial scenario does matter [47–52] as it results in a process that has highest consistent quality and profitability. It is exhilarating and has its winning moments along with moments of "agony of defeat" [53]. However, there are no failures but learning experiences that allow us to use our knowledge for other applications.

Process developers should exploit and incorporate anything and everything they can imagine. Imagination and creativity allow us to exploit unique properties of chemicals and how they behave with each other, and the result would be an excellent process. This is very similar to creation of an excellent musical composition. Some of the examples of how physical properties can be manipulated are reviewed.

3.7.1 Exploitation of solvents

Solvents have multiple values. Most are generally known but are not used collectively and selectively. These values come from their density, boiling or freezing point, and mutual solubilities and their behavior when multiple chemicals are present. They do not get exploited to the fullest for lack of recognition, lack of developer's imagination, and possibilities.

3.7.2 Solubility and solubility differences

Solvents, including water, solubilize chemicals to different degrees. Mutual solubility of chemicals has a large impact on the process design and economics. If the data is not available for the chemicals involved, it has to be generated in the laboratory.

For example, in USP 7078524 [54], the reaction produces two isomers. One isomer is processed further to produce the desired product. Second isomer does not produce the desired product. These isomers have different solubilities. Their difference is used by the inventors to recycle the second isomer and improve the yield of the desired isomer. Having the knowledge of solubility was used to simplify processes and improve yield.

3.7.3 Density differences (phase separation)

In a reactive process, density differences are a process facilitator, especially when liquids are immiscible as is the case in organic and aqueous liquids, and the desirable and undesirable products need to be separated. Properly designed equipment, called decanter, separates two liquids without any elaborate process control and in-strumentation. Figure 3.4 is a schematic of a decanter design [19]. This is also called a gravity decanter. Although this decanter is generally used for continuous pro-cesses but also work for batch processes. Careful process design is necessary.

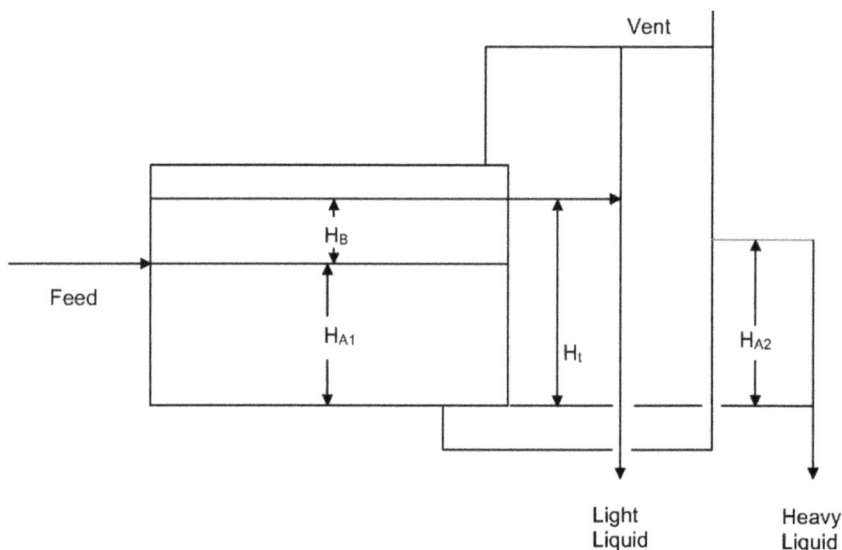

Fig. 3.4: Decanter design.

A hydrostatic balance between heavy liquid (density ρ_A) and light liquid (density ρ_B) is expressed as follows:

$$H_B \cdot \rho_B + H_{A1} \cdot \rho_A = H_{A2} \cdot \rho_A. \tag{3.2}$$

With $H_t = H_{A1} + H_B$, the height of interface can be calculated:

$$H_{A1} = [H_{A2} - H_t(\rho_B/\rho_A)]/(1 - \rho_B/\rho_A) \tag{3.3}$$

The overflow leg is made movable so that the adjustment for separation can be done, if necessary. The movable leg gives manufacturing personnel an additional advantage to use this process equipment in many different processes. These can fit extremely well in modular plants.

Density and solubility difference between liquids can be used to optimize pro-cesses. An illustration is given as follows. This concept can be adopted for any reac-tive process where such a need exists.

A chemical process described in Fig. 3.5 shows the solubility and density differ-ences used in an ingenious way to improve the process yield. If the described process was not used, the reaction yield would have been about ~50%. By using the scheme as described, the reaction yield improved to about 90%. This is a continuous process.

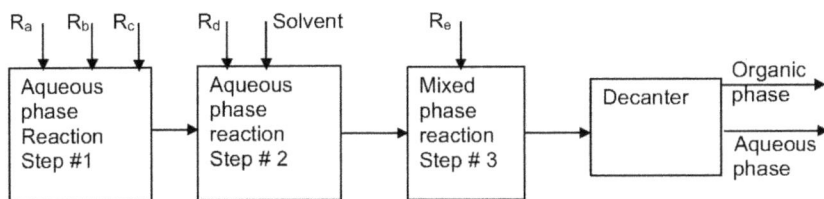

Fig. 3.5: Schematic reaction scheme.

Step #1 is a total aqueous phase reaction. In step #2, a solvent is added with a reactant and a catalyst. Reaction produces two products, a product and a precursor that needs to be converted to the product. Solvent extracts the product and allows reactant R_e in step #3, to convert the precursor from step #2 to the product which is then dissolved in the solvent. Organic and aqueous phases are separated. Organic phase which contains the desired product is processed further to produce the finished product. A combination of dissolution and extraction improves the process yield. Aqueous phase is treated to recover and recycle the catalyst. Similar strategies can be exploited and used for batch processes also. However, complete understanding of process chemistry and physical properties is necessary.

3.7.4 Exotherm/heat of reaction

It is well known that the rate of reaction doubles when the reaction temperature is increased by 10 °C [55]. This phenomenon can be used to improve the reaction rate, that is, improve/reduce the conversion time. This is of extreme value for batch as well as continuous processes. Many of the reactions are exothermic and need to be controlled to prevent a runaway reaction which can be hazardous if not managed.

In a batch reaction, it is a common practice that the components are generally added directly to the reactors. However, if the addition results in a localized exotherm, it can have less than the desirable impact on product quality. Localized heating can impart unwanted color to the product or could produce an unwanted product. There are ways to control the addition and the exotherm. One of the ways is to add the exotherm imparting chemical to the suction of the pump and recirculate the reactor contents followed by an inline heat exchanger or add the pumped liquid appropriately so that the heat of exotherm due to dilution is dispersed and used to speed the reaction [20].

3.7.5 Melting point

Melting point is one of the least exploited properties. There is a reason and good rationale for it. Majority of the time at the initial process development stage do not have sufficient information about the chemicals that will be used. As has been suggested earlier, it is necessary to gather and review every physical and chemical properties of each chemical and intermediate produced. There is good reason and rationale for it. Information collected here will be useful for the laboratory process development to scale up, commercialization, and raw material handling and storage.

Generally, in the first reaction step, there will be two solids or two liquids or their combination that will be reacted to produce an intermediate or a product. Most likely in a solid–liquid reaction, the solid concentration would be ~20% to <40%. This range is influenced by solubility, solid's melt point, and reaction temperature. This is a common practice due to limitations of the laboratory equipment. If the reaction product's melting point is not too far from the reaction temperature, raising the reaction temperature closer to the reaction product's melting point can improve the reaction rate, as discussed earlier [54]. This can lower production costs. The value of exploitation of melting point is further discussed in later chapters.

3.8 Toxicity

With time, toxicity and carcinogenicity of fine/specialty chemicals and its subset's APIs have been recognized. Improved handling and undue exposure of these chemicals have become part of the daily handling. Data about carcinogenicity of many chemicals is available [21–23]. How chemicals impact soil, human, and aquatic and bird life are not known and not readily available, especially for the API. Knowing the ecotoxicology of the chemicals involved and generated is important as it assists in designing an eco-friendly process and chemistry from the start of the process development.

As discussed in this chapter, combination (mutual behavior of chemicals or sociochemicology) of melting/boiling point, solubility, manipulation of heat of reaction/exotherms, azeotropic behavior, and use of no solvent or one additional solvent besides water can lead to improved productivity (reduced solvent use), better yield (lower waste), and lower cost (higher profits) products. Chemists and chemical engineers can also use US EPA's EPI suite [56] for estimation of physical/chemical and environmental fate estimation program for their process development.

All these are not a speculation or wishful thinking but achievable in the synthesis of fine/specialty chemicals that have APIs as their subsets. Chemists and chemical engineers use these in different unit processes [17] and unit operations [18] to produce a consistent quality product.

Abbreviations

MSDS	Material safety data sheet
SDS	Safety data sheet
EC No.	European Community No.
CAS	Chemical Abstract Service
API	Active pharmaceutical ingredient
EPA	Environmental Protection Agency
EPI	Estimation Programs Interface

References

[1] Flammability https://blink.ucsd.edu/safety/research-lab/chemical/liquids/index.html Accessed July 19, 2020.

[2] Examples of Chemical and Physical Properties, https://sciencenotes.org/chemical-and-physical-properties Accessed April 22, 2020.

[3] Chemical Engineering, Physical & Thermodynamic Properties, June 1974 pgs. 74–78..

[4] Gallant, R. W. Physical Properties of Hydrocarbons, Hydrocarbon Processing, July 65, 44, 7, pgs., 95–103. Accessed April 15, 2020.

[5] Yaws, C. L. The Yaws Handbook of Physical Properties for Hydrocarbons and Chemicals, Second Edition, Gulf Professional Publishing, January, 2015, ISBN: ISBN: 9780128008348.

[6] Oxychem Caustic Soda Handbook https://www.oxy.com/OurBusinesses/Chemicals/Products/Documents/CausticSoda/caustic.pdf Accessed June 25, 2020.

[7] ChemSpider, http://www.chemspider.com, Accessed April 29, 2020.

[8] ChemBlink, https://www.chemblink.com/cas_list.htm, Accessed April 29, 2020.

[9] A Smart Chem-Search Engine, https://www.chemsrc.com/en/ April 29, 2020.

[10] https://www.chemicalbook.com/ProductIndex_EN.aspx.

[11] Material Safety Data Sheet, p-Amino phenol, https://www.cdhfinechemical.com/images/product/msds/37_3549149_p-Aminophenol-CASNO-123-30-8-MSDS.pdf, Accessed July 13, 2020.

[12] Verma, K. K., Doraiswamy, Estimation of Heats of Formation of Organic Compounds, I & E C Fundamentals, 4, 4 November 1965 pgs. 389–396, Accessed April 10, 2020.

[13] Rossini, F. D., Heats of Formation of Simple Organic Molecules Industrial and Engineering Chemistry, December 1937, pgs. 1424–1430, Accessed May 4, 2020.

[14] Zavitsas, A. A., Rogers, D. F., Matsunaga, N., J. Heats of Formation of Organic Compounds by a Simple Calculation, Organic Chemistry 201075, 6502–6515, Accessed January 10, 2020.

[15] Dodge, B. F. Chemical Engineering Thermodynamics, McGraw-Hill Book Company, 1944.

[16] CAS number, CAS Registry, https://www.cas.org/support/documentation/chemical-substances, Accessed July 5, 2020.

[17] Unit Processes https://encyclopedia2.thefreedictionary.com/Unit+processes, Accessed June 22, 2020.

[18] Unit Operations, https://encyclopedia2.thefreedictionary.com/unit+operations, Accessed June 22, 2020.

[19] McCabe, W. L., Smith, J. C. Unit Operations of Chemical Engineering, Second, McGraw-Hill Book Company, 1967, Accessed April 30,2020.

[20] Malhotra, G.: Chemical Process Simplification: Improving Productivity and Sustainability John Wiley & Sons, February 2011.

[21] Toxic Chemicals https://www.nrdc.org/issues/toxic-chemicals, Accessed August 10, 2020,.

[22] Hoff, D., et al. Predicting the Toxicities of Chemicals to Aquatic Animal Species, https://www.epa.gov/sites/production/files/2015-08/documents/predicting_the_toxicities_of_chemicals_to_aquatic_animal_species.pdf, October 27, 2010, Accessed August 10, 2020.

[23] UNDERSTANDING TOXIC SUBSTANCES, https://www.purdue.edu/research/docs/pdf/Introduction%20to%20Chemical%20Hazards%20in%20the%20Workplace.pdf, Accessed August 10, 2020.

[24] Toxicity and Assessment of Chemical Mixtures, https://ec.europa.eu/health/scientific_committees/environmental_risks/docs/scher_o_155.pdf November 2011, Accessed August 10, 2020.

[25] Larsson, D. G., Joakim et al. Effluent from Drug Manufactures Contains Extremely High Levels of Pharmaceuticals, Journal of Hazardous Materials, 148, 3, 30 September, 2007, 751–755..

[26] Malhotra, G.: Pharmaceuticals, Their Manufacturing Methods, Ecotoxicology, and Human Life Relationship, Pharmaceutical Processing, November 2007, pgs.24–26, Accessed August 10, 2009.

[27] American Chemical Society. "Pharma's Potential Impact on Water Quality". ScienceDaily. www.sciencedaily.com/releases/2020/03/200325110904.htm, March 25, 2020, Accessed August 10, 2020.

[28] Process Technology, https://processtechnology.com/index.html, Accessed April 25, 2020.

[29] Dowtherm, https://www.chempoint.com/products/dow/dowtherm-synthetic-thermal-fluids/dowtherm-synthetic-thermal-fluids, Accessed July 25, 2020.

[30] Therminol, https://www.therminol.com/heat-transfer-fluids Accessed July 25, 2020.

[31] Malhotra, G.: Pharmaceutical Quality: Concepts, Misconceptions, Realities and Remedies, https://pharmachemicalscoatings.blogspot.com/2019/11/pharmaceutical-quality-concepts.html November 4, 2019, Accessed July 25, 2020.

[32] Malhotra, G.: Square Plug In A Round Hole: Does This Scenario Exist in Pharmaceuticals? https://pharmachemicalscoatings.blogspot.com/2010/08/square-peg-in-round-hole-does-this.html, August 10, 2010 Accessed May 19, 2020.

[33] Malhotra, G.: Why Fitting a Square Plug in a Round hole is Profitable for Pharma and Most Likely Will Stay? https://pharmachemicalscoatings.blogspot.com/2014/08/why-fitting-square-plug-in-round-hole.html, August 1, 2014, Accessed April 5, 2020.

[34] Journal of Physical and Chemical Reference Data, https://www.nist.gov/srd/journal-physical-and-chemical-reference-data, Accessed August 12, 2020.

[35] Ivanciuc, O. et al. Estimation of the Liquid Viscosity of Organic Compounds, Journal of Chemical Information and Computer Sciences, 1999, 39, 3, 515–524. Accessed September 17, 2020.

[36] Svärd, M., Rasmuson, Ä. C. (Solid+ Liquid) Solubility of Organic Compounds in Organic Solvents – Correlation and Extrapolation, Journal of Chemistry Thermodynamics, 2014, 76, 124–133. Accessed August 28, 2020.

[37] Garvin, J.: Determine Liquid Specific Heat For Organic Compounds, Chemical Engineering Progress, May 2002 pgs. 48–50, Accessed May 10, 2020.

[38] Shaw, R. Heat Capacities of Liquids, Journal of Chemical Engineering and Data, 14, 4, October, 1969, pgs. 461–465. Accessed May 10, 2020.

[39] Horsley, L. H. AZEOTROPIC DATA, Advances in Chemistry, American Chemical Society, 1973, 10.1021/ba-1952-0006, Accessed May 17, 2020.

[40] Thermochemistry, https://chem.libretexts.org/Bookshelves/Physical_and_Theoretical_Chemistry_Textbook_Maps/Map%3A_Physical_Chemistry_for_the_Biosciences_(Chang)/03%3A_The_First_Law_of_Thermodynamics/3.6%3A_Thermochemistry, Accessed August 27, 2020.

[41] Rossini, F. D. Heat of Formation of Simple Molecules, Industrial and Engineering Chemistry, 1937 pg. 1424–1430, Accessed May 4, 2020.
[42] Zavitsas, A. A., Rogers, D. W., Matsunaga Heats of Formation of Organic Compounds by a Simple Calculation, Journal of Organic Chemistry, 2010, 75, pg., 6502–6515. Accessed May 4, 2020.
[43] Chemical Laboratory Safety and Security: A Guide to Developing Standard Operating Procedures, National Academies Press (US); 2016 Jul 7, Accessed September 7, 2020.
[44] Johnson, D. W., Colver, C. P., Mixture Properties by Computer- Density, Hydrocarbon Processing, December 1968, 47, 12 pgs. 79–84 Accessed January 29, 2020.
[45] Johnson, D. W., Colver, C. P., Mixture Properties by Computer- Density, Hydrocarbon Processing, January 1969, 48, 1, pgs. 127–133 Accessed January 29, 2020.
[46] Johnson, D. W., Colver, C. P., Mixture Properties by Computer- Viscosity, Thermal Conductivity and Diffusivity, Hydrocarbon Processing, March 1969, 48, 3, pgs. 113–122 Accessed January 29, 2020.
[47] Malhotra, G.: Chemical Process Simplification, John Wiley & Sons, February 2011, Accessed August 27, 2020s.
[48] Malhotra, G.: Focus on Physical Properties to Improve Processes. Chemical Engineering, April 2002, pgs. 63–66, Accessed August 10, 2020.
[49] Maxwell, J. B. Data Book on Hydrocarbons, D. Van Nostrand Company, Inc. Seventh Printing, November 1962.
[50] Liley, P. E., et al. Physical and Chemical Data, https://terpconnect.umd.edu/~nsw/chbe301/chap02.pdf, McGraw Hill Companies Ins. Accessed May 10, 2020.
[51] Perry Robert, H. et al. Chemical Engineers' Handbook, 4th, McGraw-Hill Book Company.
[52] Solubilities of Organic Compounds in Organic Solvents https://www.nature.com/articles/npre.2010.4243.2.pdf?origin=ppub, December 12, 2009 Accessed August 24, 2020.
[53] Wide World of Sports Intro 1978, https://www.youtube.com/watch?v=P2AZH4FeGsc, April 29, 2010, Accessed August 27, 2020.
[54] USP 7,078, 524: Process for the synthesis of ganciclovir Accessed February 12, 2007.
[55] The Arrhenius Law, https://chem.libretexts.org/Bookshelves/Physical_and_Theoretical_Chemistry_Textbook_Maps/Supplemental_Modules_(Physical_and_Theoretical_Chemistry)/Kinetics/Modeling_Reaction_Kinetics/Temperature_Dependence_of_Reaction_Rates/The_Arrhenius_Law/The_Arrhenius_Law_-_Arrhenius_Plots, August 24, 2020, Accessed August 29, 2020.
[56] EPI Suite™-Estimation Program Interface: https://www.epa.gov/tsca-screening-tools/epi-suitetm-estimation-program-interface#what Accessed October 26, 2020.

Chapter 4
Pharma chemistry/chemical development

Fundamentally, the difference between active pharmaceutical ingredients (API) and fine/specialty chemicals lies in how they are used. Each is produced using the same or similar synthesis chemistries (unit processes [1]), same chemicals, and same process equipment (unit operations [2]) to create the desired molecule. The same or nearly similar synthesis methods create disease-curing chemicals (API), flavors and fragrances, additives for rubber and plastics, food colorants, preservatives, chemicals for the electronic industry, dyes for textiles, printing inks, cosmetics, and coatings. The difference between APIs and other chemicals besides use is their quality and manufacturing standards. APIs and their formulations have to abide by the standards established by the respective drug regulatory bodies.

Chemical process development has a streamlined methodology, which, if followed, leads to the development of simple and economic processes for the API and their formulations. APIs are dispensed in multiple finished dosage forms [3]. APIs dosage is generally in milligrams. They are converted to an oral dosage for the required consumption. Each tablet is a blend of inert organic and inorganic excipients [binders, disintegrant, fillers, lubricants, dyes, sweeteners, preservatives, antioxidants, dispersants, and coating materials] [4–7]. Other inert excipients can be calcium carbonate, titanium di oxide, and binders (cellulose derivatives: hydroxy propyl, hydroxyl propyl methyl) [8, 9], to name a few.

Like API, every inert excipient has its own established manufacturing process and is produced under its own repeatable quality standards [7, 10].

Manufacturing processes of organic inert excipients are similar to the synthesis of fine chemicals, which are the same as or similar to API synthesis methods. Unit processes and unit operations used to manufacture API and organic excipients are also the same or similar [1, 2, 11, 12].

APIs are mixed with inert excipients and converted to respective tablets – one of the finished dosage forms, FDF [3]. Currently, FDA has good manufacturing practice regulations that need to be followed and practiced in the production [13] of every drug.

4.1 Examples of organic excipients

Saccharin [14] and BHT (butylated hydroxy toluene) [15] are selected just as examples. They are fine/specialty chemicals. If either of them had disease-curing value, they would be called API. The following are illustrations of their synthesis.

https://doi.org/10.1515/9783110702842-004

4.1.1 Saccharin

Saccharin, an artificial sweetener, is used as an excipient in some tablets. One of the ways it can be produced is from methyl anthranilate. Methyl anthranilate is also an artificial grape flavor that is added to drug tablets, chewing gum, candy, and soft drinks as a flavoring agent. It has another unique application, too. It is used as a bird repellent on golf courses. Both of these products are produced from phthalic anhydride, a commodity chemical. Figure 4.1 illustrates the process.

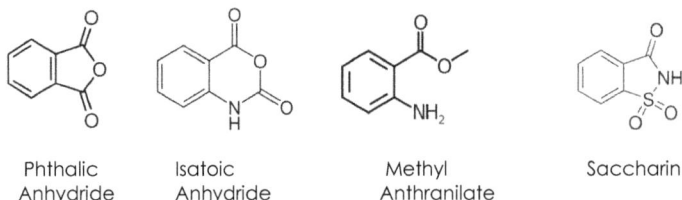

| Phthalic Anhydride | Isatoic Anhydride | Methyl Anthranilate | Saccharin |

Fig. 4.1: Phthalic anhydride and its derivatives.

Tab. 4.1: Simple process description of phthalic anhydride to its derivatives.

Process description		
Phthalic anhydride to isatoic anhydride	Isatoic anhydride to methyl anthranilate	Methyl anthranilate to saccharin [14]
USP 3324119 [16] is one of the ways. There are other process patents. Phthalic anhydride is reacted with ammonia and bleach to produce the product. Isatoic anhydride is a solid. For its continuous production, drum filtration and spray drying are used.	USP 3123631[17] Conversion of isatoic anhydride to methyl anthranilate is a simple process of reaction with methanol in alkali media and purification by distillation. Depending on product demand, it can be a batch or continuous process. Methyl anthranilate is an artificial grape flavor and a bird repellant [18].	Saccharin is produced by continuous diazotization, sulfation, chlorination, amidation, and precipitation, filtration, crystallization, and drying using vacuum or spray drier.

Each of the products above is generally produced using continuous processes, operating for about 7,500 h/year.

Table 4.1 is a brief description of the production of saccharin from phthalic anhydride.

4.1.2 Butylated hydroxy toluene

BHT, an antioxidant, is used in formulated tablets. It is a fine/specialty chemical that is produced by butylation of o-cresol (Fig. 4.2), which can be produced from phenol and its catalytic methylation or sulfonation of toluene, two widely available commodity chemicals. BHT is also an additive for various foods, fuels, oils, and plastics.

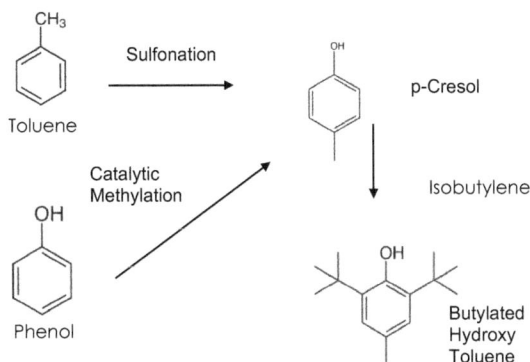

Fig. 4.2: Alternate routes [19, 20] for *p*-cresol and conversion to butylated hydroxy toluene [15].

Microcrystalline silica, calcium carbonate, and titanium dioxide [21] and other inorganics are used as inert excipients [9] in tablets. The characteristics and the manufacturing processes of the inorganic excipients are not part of the discussion in this book.

4.2 Chemical synthesis product/processes/development

Branded and generic drugs are similar, but there are differences in their process development methods. Once a new molecule is identified for its disease-curing value and efficacy, process development starts in earnest, in order to have a viable, safe and economic manufacturing process that will produce a quality product.

The thrust of companies with brands is to get a product of defined acceptable quality and performance approved by the regulators as quickly as possible. Generally, process economics is not a consideration but speed to market is. It needs to be recognized that the unit processes and unit operations used to produce the disease-curing molecule are exactly the same as used in fine/specialty chemical manufacturing.

Most of the work is done by a team of chemist/s and chemical engineer/s. The mindset of the process developer and the designer involved in the development of a pharmaceutical API and its FDF from the "get go" has to be that their product/process will be commercialized. It has to produce a quality product that meets established regulatory standards and customer needs. In addition, the process has to be safe and meet the prevailing safety and environmental standards.

Methods and plans for the development of a process for a fine/specialty chemical and its cousin, API), are taught in most chemistry and chemical engineering curriculums. There are publications discussing these techniques [22, 23]. Development teams at different companies have their own favored procedures. Practitioners can modify and create their own methods to develop an optimum process.

Steps outlined in Tab. 4.2 are generally followed, but other sequences that achieve the same end result are acceptable.

Tab. 4.2: Steps for new chemical entity evaluation.

- Chemistry of the product
- Mutual behavior of chemicals
- All potential synthesis routes
- Commercial availability of raw materials
- Laboratory experiments
- Process scheme
- Process economic and product cost

Chemists generally use published articles, their knowledge, and experience, along with the global patent base to consider relevant process chemistry. Synthetic routes for many chemicals are elaborated in the United States [24], World Intellectual Property Organization [25], and published literature are a treasure trove. For creative chemists and chemical engineers, they are an opportunity to exploit previously developed methods and processes to develop their own processes. Value of literature and patent search is very important, as it teaches many alternate processes. They also show competitors how to use the teachings to create better processes.

Since each brand product eventually becomes generic, generic companies that are interested in producing the brand products start their process development work as soon as they find out about the new drug molecule. Thus, the processes discussed here are applicable to every API developer and producer.

The conceptual phase of process development involves identification of the chemistry and potential synthesis routes. Most likely, one or more synthesis routes could be identified. Based on experience, knowledge, raw material cost information, and preliminary process route/s, a cost analysis should be performed to select the most economic route. In parallel, safety and toxicity evaluation, ease of processing, material handling, and environmental considerations should begin. Some might consider this too early, but it is best to know all the issues (toxicity, material handling, storage, etc.) as soon as possible. Knowledge and familiarity with raw materials and intermediates are useful in creating a safe and environmentally economic process. Regulatory filing processes are facilitated. Replacements or redesign later cost more money.

The process of developing a synthesis route starts with the route selection process, their feasibility, and economics. Generally, chemists start the process development, but it is helpful and best if the chemical engineers, who are well versed in costing processes, scale-up, and commercialization get involved from the initial process development stage. This consideration is helpful, as process simplification starts from the inception of the process development and scale-up. The team's knowledge is very helpful in resolving any processing issue.

Irrespective of methods used, the ultimate goal of the process developer and the designer involved in the development of a pharmaceutical API and its FDF, from the "onset," has to be that their process/product will be commercialized. It has to produce a quality product that the customer needs and has to be optimum. In addition, the process has to be safe and meet the prevailing safety and environmental standards.

Paper synthesis route/s should be reviewed and costed. It is important that for costs, commercial raw material prices be used in every cost estimate. Costs based on prices from the chemical supply house are useless, as they will not give realistic estimates. Costing of the product using the respective synthesis methods and product volume illustrate the impact of yield, stoichiometry, and manufacturing-related costs on the factory cost of the product. This is further discussed later in this chapter.

It is equally important to get the cost accountants (if the chemist and the chemical engineer are not familiar or well-versed in costing products) and the purchasing team involved in the process, at the earliest. This could be considered premature, but in this process, besides team building, there is value in considering different processing and operating options. Using the best source of the lowest priced quality raw materials is important. A chemist who is familiar with the process and a chemical engineer who is familiar with the equipment and manufacturing will be able to review different options that will result in the best and the lowest cost process for commercialization.

Pharma companies do not invest in new process equipment for every product, and they use the existing equipment to produce new products [26, 27]. While the process selection and its costing is being done, engineers need to review how the process for the new product will fit in, meet the cGMP requirements, and abide by all of the necessary safety practices. For this, it is necessary that the chemists and chemical engineers have a thorough understanding of chemical and physical properties, and their mutual behavior, and be capable of exploiting them to create and design an optimum commercial process.

Solvent selection plays an important part in the process development and simplification. Process development and their commercialization is a combination of our teachings at the universities, personal experiences at chemical and pharmaceutical manufacturing facilities, creativity, and imagination.

Knowledge and understanding of process chemistry, stoichiometry, process yield, reaction products, and raw materials would lead to capitalizing on mutual behavior of chemicals. This is essential. They are the tools and means by which processes can be simplified. Mutual behavior of raw materials and reaction products and by-products (sociochemicology) [28] has to be understood, manipulated, and exploited to create excellent processes.

A review of various patents for the APIs indicates that besides water, additional different solvents are used in the synthesis of a single molecule. This may be necessary, and an increased number of solvents in the laboratory may be acceptable in demonstrating the viability of a process and product, but such processes, when translated for commercial production of a product, add complexity to the whole process, in multiple ways. This can be through inventory management, recovery and reuse of solvents, and product yield loss through solubility of process intermediates and the final product.

Process yields directly impact the product costs. Efforts are made to maximize the yield in each process step, so that the cumulative yield for the process is the highest. Table 4.3 illustrates the relationship of different process steps and cumulative yield of the whole process. Lower yields per step result in higher product costs, not only due to reduced amount of product produced compared to theoretical yield, but also due to disposal/recovery of the generated waste.

Tab. 4.3: Yield and number of reaction steps correlation.

	Overall yield				
	Steps				
Yield per step	2	3	4	5	6
95%	90%	86%	81%	77%	74%
90%	81%	73%	66%	59%	53%
80%	64%	51%	41%	33%	26%
70%	49%	34%	24%	17%	12%

In pharmaceutical API and FDF product and process development, and their commercialization, stringent requirements of toxicity have to be recognized and addressed, as the products are consumed to cure diseases. Ecotoxicology is equally important, as the impact of effluents from the pharmaceutical plants on humans that includes other mammals, aquatic, and bird life is increasingly recognized. Efforts have to be made to minimize the impact [29–31]. Table 4.3 presents opportunities to improve yields. Higher yields result in higher profits, as the monies are not needed for remediation.

For every product's development, chemists' and engineers' understanding of the process and product demand are crucial. At brand stage, generally, selling prices can be high, thereby keeping the real demand volume (API Kilos needed) low, till the time the drug becomes generic or affordable. This generally results in the product being manufactured in equipment that is not specifically designed for the product. Such processes [26, 27] are not optimum.

Table 4.4 illustrates the relationship among the API demand, population, and number of tablets needed. Mathematically, at 100% conversion, 1 kg can produce 1 million 1 mg tablets. Basically, it suggests that small quantities of API, depending on the dose, can serve a large population. It needs to be recognized that the number of tablets used per day can vary and accordingly change FDF demand for the total API and drugs.

Tab. 4.4: API demand, population, and tablets needed per year.

Case	Dose (mg)	Tablets per day	Days per year	Population	Kilogram API need per year	Tablets needed/year
One kg produces 1 million of 1 mg tablet						
A	10	1	365	1,000,000	3,650	3,650,000,000
B	50	1	365	1,000,000	18,250	18,250,000,000
C	100	1	365	10,000,000	365,000	3,650,000,000,000
D	500	1	365	10,000,000	1,825,000	18,250,000,000,000

Globally, drug selling price is independent of the API sale price. API cost is part of the drug price, and any process improvement or cost reduction at API level may or may not influence the final drug dose sale price unless the API sale price is increased. Table 4.5 illustrates the monetary contribution of different priced APIs at different dosages. Sale price of API has financial value to the API producer, but to the finished drug dose seller, its contribution could be in single digits.

Tab. 4.5: API content cost contribution to different dose drugs.

API selling price ($/kg)	25	45	65	100
Dose (mg)	API content cost ($/tablet)			
10	0.00025	0.00045	0.00065	0.001
50	0.00125	0.00225	0.00325	0.005
100	0.0025	0.0045	0.0065	0.01
500	0.0125	0.0225	0.0325	0.05

On review of Tab. 4.5, one can speculate that the profitability of a 10-mg tablet selling at ten cents per tablet would not affect the drug seller's profitability in any way, even if the API content cost doubles. It is important to recognize and understand that the selling prices of the tablets are independent of API manufacturing costs or their selling price.

The API manufacturer/seller has to carefully evaluate its own effort for any cost reduction. Each API producer has to justify its own return on investment for such investment. Due to limited patent life, brand API producers do not want to invest in any cost reduction efforts. For such an effort, they may have to get regulatory approval, which costs additional money, too. Similarly, every generic API producer will evaluate cost reduction advantage of its patent-protected process and compliance costs. Table 4.6 illustrates the value of such effort.

Tab. 4.6: Months for return and revenue.

API volume (kg/ year/site)	API factory selling price ($/kg)	API revenue ($)	Savings 5%	Savings ($)	Return months
1,000	$25.00	$25,000.00	$2.25	$2,250.00	133
2,000	$50.00	$100,000.00	$4.50	$9,000.00	33
5,000	$75.00	$375,000.00	$6.75	$33,750.00	9
10,000	$100.00	$1,000,000.00	$9.00	$90,000.00	3
Arbitrary R&D effort cost, $25,000					

All of the above that has been discussed applies to branded and generic products. It is well known that almost all branded products move to generic products. Safety and consistent product quality from the very beginning are the price drivers.

The following are brief reviews of some selected products. Review of these patents suggests that inventors of the processes have progressively improved and simplified processes and made efforts to reduce manufacturing cost, using improved chemistries and processes, and reduce use and production of toxic chemicals in the chemistries.

The purpose of this is to emphasize that significant chemistry and processing information is available that can be used to create better chemistries/processes at lower costs.

4.2.1 Mycophenolate mofetil

Mycophenolate mofetil is sold under the brand name Cellcept. Once it became generic, many companies filed patents for it. USP 7019133 B2 [32] and US 2010/029560 A1 [33] are two alternate methods of manufacture of mycophenolate mofetil.

It is important that the chemist and chemical engineer in the process development be familiar with chemical and physical properties of each chemical (Tab. 4.7) involved and produced. These are not only needed for safe process design but are also used in manipulating and improving productivity, and can be used in lowering of cost of each product.

Tab. 4.7: Physical and chemical properties of reactants used in USP 7019133 B2 [32] and US 2010/0298560 A1 [33].

Chemical	CAS #	Formula	Mol. wt	Melting point (°C)	Boiling point (°C)
Methyl mycophenolate	31,858-66-9	$C_{18}H_{22}O_6$	334	105	540
2-Morpholinoethanol	622-40-2	$C_6H_{13}NO_2$	131	−1	224
Dibutyltin oxide	818-08-6	$C_8H_{18}OSn$	248	105	161.0
Toluene	108-88-3	C_7H_8	92		110.6
Ethyl acetate	141-78-6	$C_4H_8O_2$	88	−84	73.9
Hydrochloric acid	7647-01-0	HCl	36.5	−35	85
Caustic soda	1310-73-2	NaOH	40		145
Mycophenolic acid	24,280-93-1	$C_{17}H_{20}O_6$	320	141	612
Thionyl chloride	7719-09-7	$SOCl_2$	119	−105	79
Triethylamine	121-44-8	$C_6H_{15}N$	101	−115	91
MMF	128794-94-5	$C_{23}H_{31}NO_7$	433	95	637

The corresponding chemical and physical properties of intermediates can be gathered from the literature and Chemical Abstract Services once the compounds are defined.

In USP 7019133 B2 [32], mycophenolate mofetil is manufactured by transesterification of methyl mycophenolate in the presence of a catalyst. In US 2010/0298560 A1 [33], mycophenolic acid is reacted with an aliphatic or aromatic amine to produce an amine salt. Amine salt is halogenated to an acid halide, which is esterified to produce the desired product.

Syntheses in the above patents are the inventors' perspective and their attempt to simplify manufacture, using the mutual behavior of each chemical used and produced. Tables 4.8 and 4.9 illustrate manufacturing costs of two different chemistries [32, 33] of mycophenolate mofetil. S&E (service and expense) costs relate to conversion costs, which include hourly and salaried labor, utilities, depreciation, and other fixed and variable charges that companies allocate to each product. Most companies use their own ways to calculate factory cost of each product.

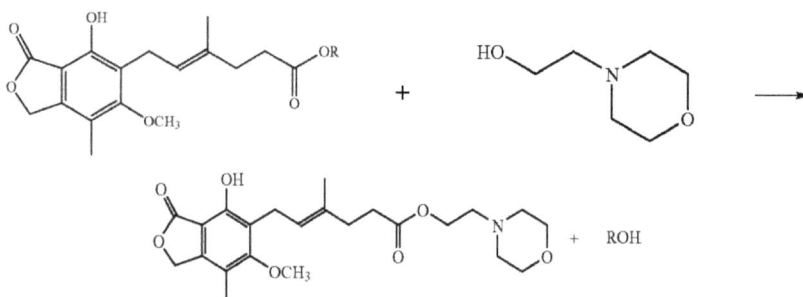

Fig. 4.3: Mycophenolate mofetil USP 7109133 B2 [32].

Tables 4.8 and 4.9 are an illustration of a product cost using USP 7109133 [32] and US 2010/0298560 A1 [33]. These costs are based on synthesis examples from respective patents and are not optimized.

Process for the manufacture of mycophenolate mofetil described in US 2010/ 0298560 A1 [33] can be further simplified and improved, and the global API need can be produced in a single plant. It is currently produced by 41 producers and sold at $232/kg [34]. Factory cost of mycophenolate mofetil based on USP 7019133 B2 [32] (Tab. 4.2) is about $165.09/kg and per US 2010/0298560 A1 [33] is about $133.31/kg. Each cost was calculated by getting the prices of each raw material from their vendors. Both processes are not optimized. In addition, cost of solvents used in the process is not included.

In the above cost estimates, S&E (service and expense) charge of 30% of the raw material cost is used. S&E is, essentially, conversion cost to cover hourly, salaried labor, utility cost, maintenance charges, depreciation, etc. This number could be considered a low number, by some. If the process is complex, S&E can be increased, based on each company's norms.

Costing of products from the paper chemistry stage to the laboratory chemistry stage is necessary. Such an exercise makes the chemistry developers aware of how the complexity of process can impact the factory cost. It gives them an opportunity to improve processing. Brand and generic API developers can create excellent processes, as they progress. Costing would include material recovery and disposal cost of toxic and nontoxic materials. These are also reviewed in Chapter 5.

Other APIs and fine/specialty chemicals are also reviewed. The purpose is to emphasize that significant processing and chemistry information, along with physical properties, can be used to develop excellent processes at lower costs.

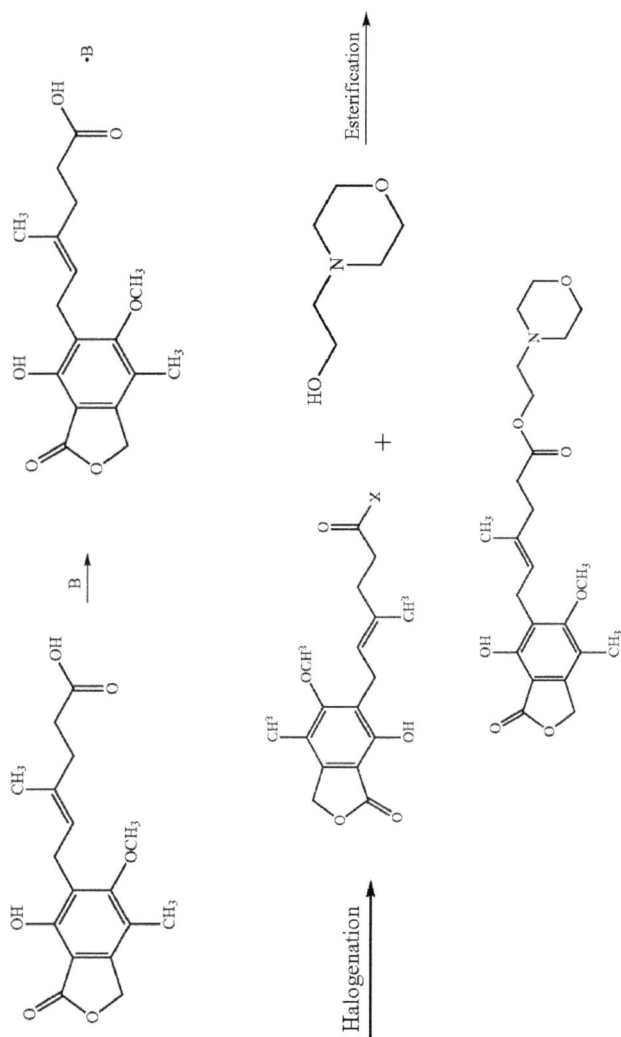

Fig. 4.4: Mycophenolate mofetil US 2010/0298560 A1 [32] [33].

Tab. 4.8: Mycophenolate mofetil factory costs USP 7109133 B2 [32].

USP 7019133 B2 [32]	CAS #	Mol. wt	Kilogram	Cost ($/kg)	$
Methyl mycophenolate	31,858-66-9	334	86.21	132.00	11,379.31
2-Morpholinoethanol	622-40-2	131	35.34	11.20	395.86
Dibutyltin oxide	818-08-6	248	31.90	33.00	1052.59
Toluene	108-88-3	92	10	2.20	22.00
Ethyl acetate	141-78-6	88	15	2.75	41.25
Hydrochloric acid	7647-01-0	36.5	8.43	0.53	4.45
Caustic soda	1310-73-2	40	9.24	0.33	3.05
MMF	128794-94-5	433	100		12,898.51
	Yield	89.5%			128.98
	S&E				36.11
	Factory cost			$/kilo	165.09

Tab. 4.9: Factory cost of mycophenolate mofetil US 2010/0298560 A1 [33].

US 2010/0298560 A1 [33]	CAS #	Mol. wt	Kilogram	Cost ($/kg)	$, total
Mycophenolic acid	24280-93-1	320	80.65	120.01	9678.23
Thionyl chloride	7719-09-7	119	35.48	2.44	93.68
Triethylamine	121-44-8	101	35.48	4.4	156.13
2-Morpholinoethanol	622-40-2	131	40.32	11.2	451.61
Ethyl acetate	141-78-6	88	10	2.75	27.50
Hydrochloric acid	7647-01-0	36.5	8.43	0.53	4.45
Caustic soda	1310-73-2	40	9.24	0.33	3.05
MMF	128794-94-5	433	100		10,414.64
	Yield	91.6%	$/kg		104.15
	S&E				29.16
	Factory cost			$/kg	133.31

4.2.2 Nevirapine

The following patents discuss the manufacture of Nevirapine. These patents give a telltale story of how process developers recognize complexities of preceding synthesis and toxicity of the chemicals used and produced.

1. USP 5366972 (1994) Boehringer Ingelheim [35]
2. USP 5569760 (1996) Boehringer Ingelheim [36]
3. USP 6680383 (2004) and EP 1 519 936 B1 (2008) Boehringer Ingelheim [37]
4. WO 2007/010352 A1 (2007) Emcure Pharmaceuticals [38]
5. WO 2012/168949 A2 (2012) Laurus Labs [39]
6. USP 8212025 [40]

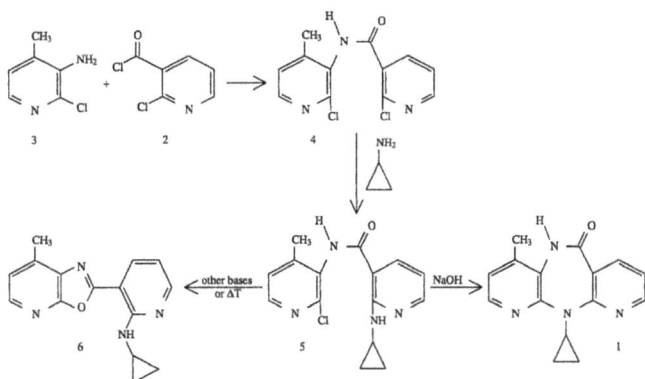

Fig. 4.5: Nevirapine synthesis in USP 5366972 [35].

Figure 4.5 is illustration of USP 5366972 [35] for the preparation of nevirapine.

The process described in USP 5366972 [35] is complex. It would have low productivity if commercialized as developed in the laboratory. The overall yield of the four steps in Example 1 is about 40%. The shortcomings of this patent are acknowledged in subsequent patents. If this process is commercialized as is, most likely, the processing costs to recover solvents and disposal of waste would be as much as, or more than, the actual product cost from this step. This would also be due to low productivity of the process. Example 1 of USP 5366972 [35] is as follows.

Example 1

5,11-Dihydro-6H-dipyrido[3,2-b:2′,3′-e][1,4]diazepin-6-one (Nevirapine)

There are four steps. The overall yield of the following is about 40%.

Step a) 2-Chloro-*N*-(2-chloro-3-pyridinyl)-3-pyridinecarboxamide
In a three-necked round-bottomed flask, fitted with an efficient reflux condenser, mechanical stirrer and dropping funnel were placed 215 g (1.672 mol) of 3-amino-2-chloropyridine dissolved in a mixture of 400 mL dioxane, 500 mL cyclohexane, and 130 mL pyridine. The solution of 299.2 g (1.7 mol) of freshly prepared 2-chloro-3-pyridinecarboxylic acid chloride in 200 mL dioxane was added at such a rate as to keep the vigorous reaction under control. Thereafter, the reaction mixture was allowed to cool to room temperature, and the resulting crystalline precipitate was filtered off and washed successively with cyclohexane and ether.

The dark brown product was dissolved in 5 L of a 3% aqueous solution of sodium hydroxide. The resulting solution was treated with charcoal, suction filtered, and the filtrate was acidified by addition of 50% aqueous acetic acid. The resulting precipitate was collected by filtration and thoroughly washed with water. After being dried overnight in a stream of nitrogen at room temperature, the almost colorless product had an m.p. of 156–159 °C and was sufficiently pure for further reactions. The yield was 376.0 g (84% of theory).

Step b) *N*-(2-Chloro-3-pyridinyl)-2-[[(4-methoxyphenyl)methyl]amino]-3-pyridinecarb oxamide
13.4 g (0.05 mol) of the product obtained in step a) was dissolved in 20 mL of xylene, and the resulting solution was admixed with 13.8 g (0.1 mol) of *p*-methoxybenzylamine. Thereafter, the mixture was refluxed for 2 h. The reaction mixture was then evaporated in vacuo, and the residue was purified by column chromatography on silica gel (0.2–0.5 mm), using dichloromethane/ethyl acetate 10/1 (v/v) as an eluent. Concentration afforded colorless crystals, melting at 122–124 °C (after recrystallization from acetonitrile). The yield was 17.2 g (93% of theory).

Step c) 5,11-Dihydro-11-[(4-methoxyphenyl)methyl]-6H-dipyrido[3,2-b:2',3'-e][1,4]d iazepin-6-one
16.7 g (0.0453 mol) of the product obtained in step b) was dissolved in 150 mL of absolute dioxane, and the resulting solution was admixed with 6.7 g (0.14 mol) of a 50% dispersion of sodium hydride in mineral oil. Thereafter, the mixture – while protected against the external atmosphere by a low flow of nitrogen – was refluxed until no starting material could be detected by TLC. The surplus of sodium hydride was decomposed by cautious addition of 10 mL of a mixture of methanol and tetrahydrofuran (50/50 v/v). The reaction mixture was neutralized by addition of acetic acid and then was evaporated in vacuo. The residue was purified by column chromatography on silica gel (0.2–0.5 mm), using successively dichloromethane/ethyl acetate 10/1 (v/v) and dichloromethane/ethyl acetate 1/1 (v/v) as eluents. The crystalline product obtained by evaporation of suitable fractions was recrystallized from acetonitrile and 2-propanol. The product had an m.p. of 213–215 °C and was identified as 5,11-dihydro -11-[(4-methoxyphenyl)-methyl]-6H-dipyrido[3,2-b:2',3'-e][1,4]diazepin-6-one. The yield was 10.3 g (68% of theory).

Step d) 5,11-Dihydro-6H-dipyrido[3,2-b:2',3'-e][1,4]diazepin-6-one
10.0 g (0.3 mol) of the product obtained in step c) was dissolved in 50 mL of trifluoroacetic acid, whereby the mixture became slightly warm. Thereafter, the reaction mixture was stirred at 60 °C for 1 h. No starting material could be detected by TLC at that time. The mixture was then evaporated in vacuo. The residue thus obtained was thoroughly stirred with 0.5% aqueous ammonia and then, was filtered by suction. The raw product was recrystallized from 150 ml of dimethyl sulfoxide to provide colorless crystals of m.p. >340 °C. The yield was 4.8 g (75% of theory). The product was identified as 5,11-dihydro-6H-dipyrido[3,2-b:2',3'-e][1,4]diazepin-6-one.

Patentee in its USP 5569760 [36] acknowledges complexities of USP 5366972 [35].

If the process of '760 [36] is commercialized as is, most likely, the processing costs to recover solvents and disposal of waste would be as much as, or more than, the actual product from this step. This would also be due to low productivity of the process.

Table 4.10 is an illustration of Step 1 of Example 1 of USP 5366972 [35]. Similar information must be generated for each reaction step. This process step uses multiple solvents and the solid concentration is less than 30%. Such an exercise might look unnecessary and complicated, but it is needed, as it would lead to the development of an economic process that will produce a quality product from the get go. Table 4.10 does not illustrate stoichiometry. It is discussed later in this chapter.

Tab. 4.10: Step 1 of Example 1 of USP 5366972 [35].

	CAS	Formula	Mol. wt	Melting point (°C)	Boiling point (°C)
3-Amino-2-chloropyridine	94770-75-9	$C_5H_6Cl_2N_2$	165	76–78	
Cyclohexane	110-82-7	C_6H_{12}	84	6.5	80.7
Pyridine	110-86-1	C_5H_5N	79	−42	115.3
2-Chloro-3-pyridinecarboxylic acid chloride	121495-79-2	$C_6H_3Cl_2NO$	176		
Dioxane	5703-46-8	$C_4H_8O_2$	88		60.3

WO 2007/010352 A1 [38] very well reviews scheme 1 and the shortcomings of the USP 5366972 [35] for the preparation of nevirapine.
–   Reduction of 2-chloro-4-methyl-3-nitro pyridine with stannic chloride, employing acetic acid as a solvent to give 3-amino-2-chloro-4-methyl pyridine. The reaction is characterized by long reaction times of at least about 12 h and utilizes a hazardous reagent like stannic chloride, which is difficult to handle on an industrial scale due to its fuming and corrosive nature. Further, the process is not environment friendly, since there is considerable load on the effluent treatment plant for removal of sludge material formed during work up. The preparation of 3-amino-2-chloro-4-methyl pyridine is the first step of the reaction. This has to be prepared in large quantity, and this proportionately creates a large volume of the sludge, which becomes difficult to remove on a commercial scale.
–   Acylation of 3-amino-2-chloro-4-methyl pyridine with 2-chloronicotinoyl chloride in the presence of a base like pyridine, and in a mixture of solvents like cyclohexane and dioxane to yield 2-chloro-N-(2-chloro-4-methyl-3-pyridinyl)-3-pyridine carboxamide. The reaction employs a solvent like dioxane, which is highly toxic for inhalation, and hence, cannot be utilized on an industrial scale. Further, dioxane needs elaborate safety precautions during storage and handling in view of its explosive nature. Further, the yield of the compound of formula obtained is only about 10%, which is very low for commercial utilization. Also, dioxane solvent present in the effluent, if released into soil, is difficult to remove, as dioxane is not biodegradable.

- Reaction of 2-chloro-N-(2-chloro-4-methyl-3-pyridinyl)-3 -pyridine carboxamide with cyclopropyl amine to give 2-N'-cyclopropyl amino-N-(2-chloro-4-methyl- 3-pyridinyi)-3-pyridine carboxamide.
- Cyclization of 2-N'-cyclopropylamino-N-(2-chloro-4-methyl-3-pyridinyl)-3-pyridine carboxamide of formula by employing a strong base like sodium hydride in dimethyl formamide as solvent, to give nevirapine. Further, this method mentions that isolation of nevirapine involves quenching with water, which is quite dangerous, especially when a reagent like sodium hydride is used, as sodium hydride reacts violently and explosively with water and is highly flammable.

WO 2007/010352 [38] also reviews USP '760 [36]. It suggests that USP '760 [36] is a time-consuming, complex process requiring multiple reactors. It discloses an improved method for preparing 2-N'-cyclopropylamino-N-(2-chloro-4-methyl-3-pyridinyl)-3-pyridine carboxamide, wherein the reaction of 2-chloro-N-(2-chloro-4-methyl-3-pyridinyl)-3-pyridine carboxamide with cyclopropyl amine is carried out in the presence of an oxide or hydroxide of an element of the second group of the periodic table, with diglyme as solvent. Further, 2-N'-cyclopropylamino-N-(2-chloro-4-methyl-3-pyridinyl)-3-pyridine carboxamide, thus formed is not isolated but is converted after removal of the inorganic base to nevirapine.

According to the inventors of WO 2007/010352 [38], USP '760 [36] suffers from the following drawbacks:

- 2-N'-Cyclopropylamino-N-(2-chloro-4-methyl-3-pyridinyl)-3-pyridine carboxamide formed by reaction with cyclopropyl amine is not isolated; therefore, it is very likely that an impurity of formula (VIII) disclosed in US 5569760 [36] and likely to be formed in this reaction will be carried forward to the final stage and would be isolated along with nevirapine, thereby affecting its purity profile.
- Uses a strong base like sodium hydride, which is difficult to handle on an industrial scale and also requires stringent conditions such as careful quenching with water.

Figure 4.6 describes Scheme-II of US 6680383 [37] for the preparation of nevirapine. This process utilizes a costly base such as sodium hexamethyl disilazane (HMDS) for the production of nevirapine.

The method disclosed in Scheme-II USP '383 [37] is slightly different from that disclosed in Scheme-I of USP 5366972 [35], but this route also utilizes a fuming and costly base such as sodium hexamethyl disilazane in the final step, for preparation of nevirapine.

In addition, USP 6680383 [37] utilizes a very toxic and hazardous raw material like 2-halo-3-cyano pyridine, compound IX (Fig. 4.6), which is difficult to handle on a large scale due to its high toxicity, since compound IX contains a cyano functional group.

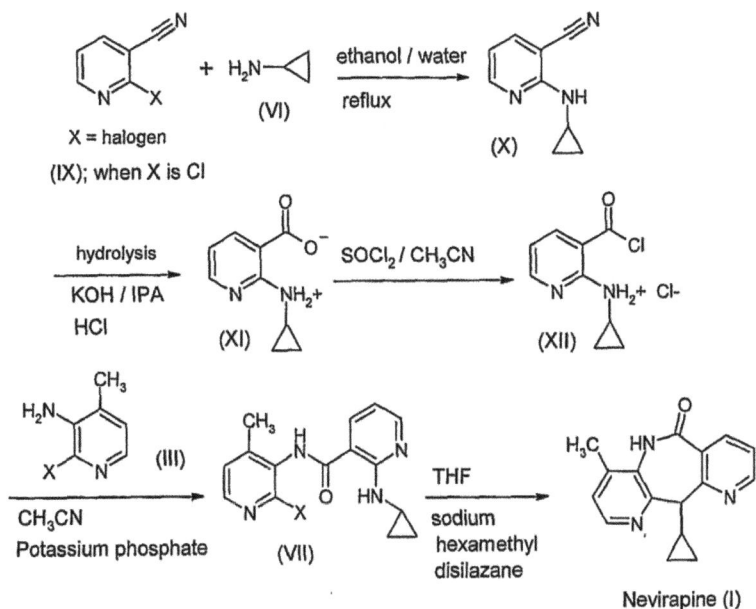

Fig. 4.6: Scheme-ll: Method for the preparation of nevirapine USP 6680383 [37].

In view of the above shortcomings, there is a need for a method for preparing nevirapine, which would not only make the process safe and environment friendly, but would also be simple and cost-effective for commercial utilization.

Patent WO 2007/010352 [38] describes its process in Fig. 4.7. It claims the following:

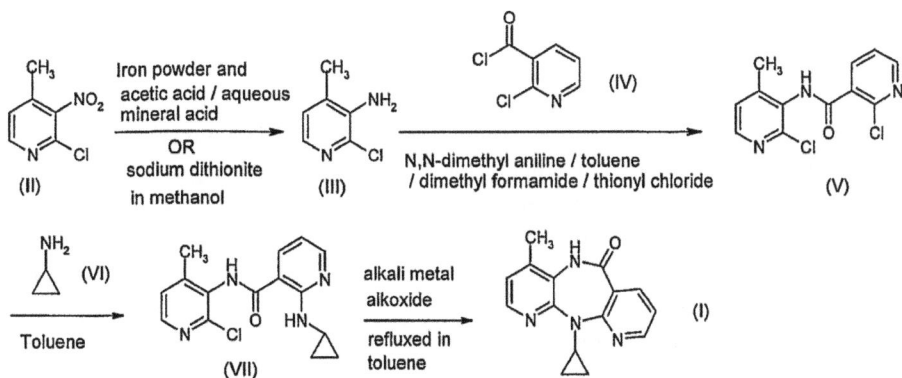

Fig. 4.7: Method of preparation of nevirapine per patent WO 2007/010352 [38].

1. An aspect of the invention relates to a cost-effective method for the preparation of 3-amino-2-chloro-4-methyl pyridine of formula (III), which comprises reduction of 2-chloro-4-methyl-3-nitro pyridine of formula (II) with iron powder, in the presence of an acid, either organic or inorganic.

2. Another aspect of the invention relates to an alternate method for preparing 3-amino-2-chloro-4-methyl pyridine, which comprises reduction of 2-chloro-4-methyl-3-nitro pyridine of formula with sodium dithionite, utilizing an organic solvent.

3. Another aspect of the invention relates to the use of safer environment-friendly reagents like a carbonate of an alkali metal as base and an alkyl acetate or an aromatic hydrocarbon as solvent, for preparation of 2-chloro-N-(2-chloro-4-methyl-3-pyridinyl)-3-pyridine carboxamide, avoiding the use of hazardous solvents like dioxane and pyridine.

4. Another aspect of the invention relates to preparation of nevirapine, comprising cyclization of 2-N'-cyclopropylamino-N-(2-chloro-4-methyl-3-pyridinyl)-3-pyridine carboxamide of formula in the presence of an alkali metal alkoxide and in an inert organic solvent, followed by neutralization with an acid and isolation of nevirapine by crystallization from an organic solvent.

5. Utilization of a single solvent for the preparation of nevirapine, thereby providing a cost-effective process.

Patent WO 2007/010352 [38] claims to overcome the following shortcomings with respect to prior art:

– Avoids use of stannic chloride for reduction of the nitro group in the first step of preparation of 3-amino-2-chloro-4-methyl pyridine.

– Aqueous medium is used during the process of the invention, and use of organic solvent is avoided during the preparation of 2-chloro-4-methyl-3-amino pyridine, making the process cost-effective and reducing load on effluent treatment.

– Circumvents utilization of pyridine and dioxane in the second step of preparation 2-chloro-N-(2-chloro-4-methyl-3-pyridinyl)-3-puridine carboxamide.

– Avoids use of strong bases like sodium hexamethyl disilazane and sodium hydride used in prior art, for preparation of nevirapine, thereby making the process safe, cost-effective, and environment-friendly.

– Avoids load on the effluent treatment plant.

– Does not utilize dioxane as solvent, thereby making the process environment-friendly.

– Provides a process, which utilizes toluene as solvent in Step-II, Step-III, and Step-IV. The advantages accrued, therefore, would be low inventory of solvents, easy recovery, and recycling of toluene. These factor, make the process of the instant invention more cost-effective.

– Provides an API with acceptable limits of impurity profile.

WO 2012/168949 A2 [39] claims to use less of expensive cyclopropylamine, if the re-action is carried out with cheaper bases. The same claim is made in USP 8281025 [40]. Reaction time can also be optimized and it yields higher purity product.

USP 8212025 [40] is a further improvement on earlier processes through reduction of excessive amounts of cyclopropylamine, if the reaction is carried out in potassium fluoride or trisodium dodecahydrate.

It is interesting to note that Gates Foundation funded an effort [41] to lower nevirapine manufacturing cost, even after 40 years. This work resulted in patent WO 2016/118586 A1 [42].

Generally, any cost reduction work, as stated earlier, in any company that is producing the product would not be funded, due to two reasons: 1) too many companies producing the same API [43]; and 2) extremely low return on investment (ROI) per site, as shown in Tab. 4.6.

At the time of Gates funding, nevirapine demand was projected to be around 1,000+ metric ton per year [44]. However, due to better combination of other HIV drugs, the demand for nevirapine has significantly reduced. In addition, as there are many nevirapine manufacturing sites [45], the benefits of better process, most likely, will not be fully realized.

Patent WO 2016/118586 A1 [42] illustrates nevirapine synthesis Fig. 4.8.

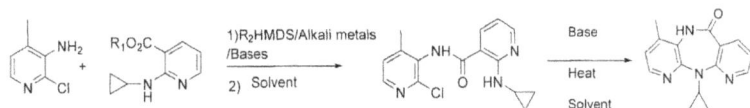

CAPIC + CYCLOPROPYLAMINONICOTINATE → CYCLOR → NEPIRAVINE

CAPIC = 2-chloro-3-amilo-4-picoline

CYCLOR = 2-(cyclopropylamino) nicotinamido-3'-amino-2'=chloro-4'-methylpyridine

Fig. 4.8: Nevirapine synthesis [42].

Compared to other patents chemistry used in WO 2016/118586 A1 [42] is significantly improved. However, even with chemistry improvements, productivity of the overall process can be improved further, if the solvent use could be reduced. This can happen in different ways and depends on the selected solvent and the equipment used. Again, since the chemistry could be practiced at many plants, every plant may not offer this advantage that is offered by WO 2016/118586 A1 [42].

As has been stated earlier, especially in the manufacture of APIs, stoichiometry and process optimization done on the laboratory scale unlike other fine/specialty chemicals is the best and, may be, the only opportunity for the brand company commercializing the product. Any drug that gets in the clinical trials can make optimization a challenge. Generic companies will improve the chemistry to maximize their profits. They only have to prove the bioequivalence and watch the product impurity profile.

All of the above examples suggest that the chemistries developed for every new API are continuously improved as they transition from brand (patent-protected) synthesis to a generic drug. In addition to chemistry, how their physical properties are exploited to simplify processing is also important. This combination directly impacts the product's manufactured cost. Metformin hydrochloride is used as an illustration in later chapters.

4.2.3 Chlorosulfonylation reaction

The following is a comparison of two chlorosulfonylation reactions, As suggested earlier, it is important that developers review the literature and issued patents, as they develop economic processes. Stoichiometry of two chlorosulfonylation reactions is compared in Tab. 4.11.

USP 7109203 [46] suggests use of acetic acid as a solvent. Unless acetic acid is used further in any downstream reaction, it will be lost as waste. Use of a re-usable solvent could be beneficial. This patent uses more-than-needed sulfur dioxide. Unreacted sulfur dioxide will have to be neutralized, and its salt would be disposed as waste. USP 4464537 [14], on the other hand, suggests more economical stoichiometry with a reusable solvent. Thus, it is necessary for a laboratory development process to consider the scenario "does the process developed in the laboratory has minimum waste and is economic?"

Review of each of Tab. 4.11 chemistries illustrates that stoichiometry and solvent use in the processes are significantly higher than needed. In USP 4464537 [14], aqueous phase diazotization (step #1) can be easily controlled at 30–35 °C. A chloro-hydrocarbon is used as a solvent in the second reaction step. This process can be a batch or a continuous process. The reaction illustrated in Tab. 4.11 (synthetic route of methyl 2-(chlorosulfonyl) benzoate) [14] has been practiced commercially for the continuous manufacture of saccharin, since the early 1970s.

The same could be done for USP 7109203 [46], and a higher reaction temperature could accelerate the process. This is definitely feasible with today's process control technologies and equipment.

It is important to understand and recognize how sociochemicological [28] behavior of chemicals can be exploited to simplify manufacturing processes.

The following information, from the 1940s, is to illustrate that chemistry and manufacturing technologies to produce drugs useful in treating various diseases had a strong foundation. Process developers were cognizant of yield, quality, toxic and hazardous nature of chemicals, materials of construction, and different unit operations.

Tab. 4.11: Example comparing two similar chlorosulfonylation chemistries.

	USP 7109203 [46]	USP 4464537 [14]
	4-Bromo-2-chloroaniline	Methyl Anthranilate

	Mole ratio	
Key: amine	1.0	1.0
HCl	14.0	3.22
NaNO$_2$	0.97	1.0
SO$_2$	25.9	1.3
Solvent	Acetic acid	1,2-Dichloroethane
CuCl$_2$	0.5	0.028
Key: chemical	4-Bromo-2-chloroaniline	Methyl anthranilate
Properties		
Mol. wt	206.5	153
Melting point (°C)	~70	24
Boiling point (°C)		256
Reaction product, MP (°C)	75–80	59–64

4.2.4 8-Hydroxyquinoline

Figure 4.9 represents 8-hydroxyquinoline and Figure 4.10 is its synthesis procedure practiced by I. G. Eiberfeld Wuppertal. German process description and its English translation are presented.

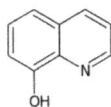

Fig. 4.9: 8-Hydroxyquinoline.

OFFICE OF MILITARY GOVERNMENT FOR GERMANY (US)

FIAT FINAL REPORT NO. 940   30 September 1946

THE MANUFACTURE OF NICOTINAMIDE AND

RELATED INTERMEDIATES IN THE FRENCH

BRITISH AND AMERICAN ZONES

OF

OCCUPATION IN GERMANY

BY

V.L. KING

AND

R.P. PARKER

TECHNICAL INDUSTRIAL INTELLIGENCE DIVISION

U.S. DEPARTMENT OF COMMERCE

THIS REPORT IS ISSUED WITH THE WARNING THAT IF THE SUBJECT MATTER SHOULD BE
PROTECTED BY U.S. PATENTS OR PATENT APPLICATION, THIS PUBLICATION CANNOT BE
HELD TO GIVE ANY PROTECTION AGAINST ACTION FOR INFRINGEMENT.

FIELD INFORMATION AGENCY, TECHNICAL

1

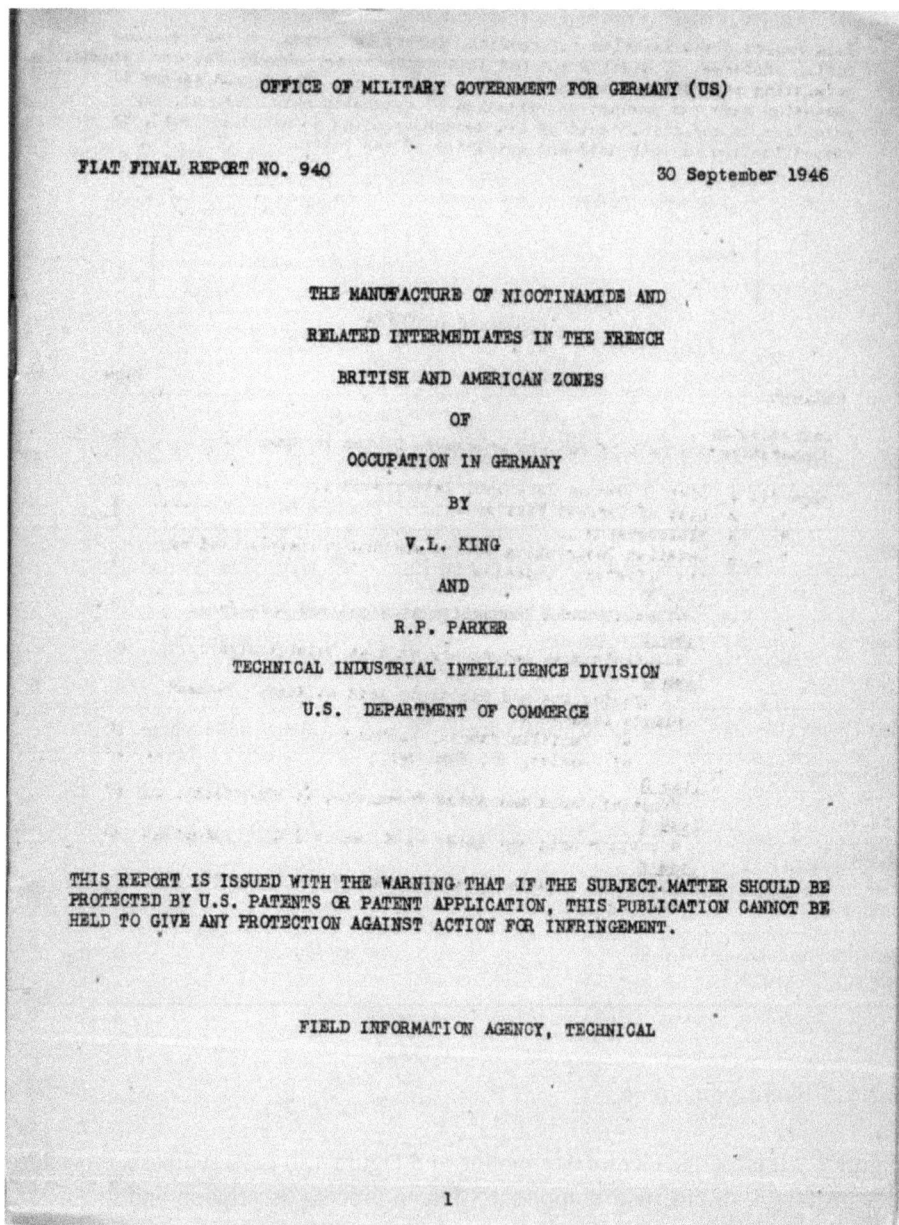

Fig. 4.10: Hydroxyquinoline synthesis in German.

NICOTINIC ACID AMIDE

Item D-1   8-HYDROXY QUINOLINE (OXYCHIN)

Procedure from I.G. Elberfeld Wuppertal

O X Y C H I N.

Betriebsvorschrift

Stand vom Oktober 1941.
Produktionsmoeglichkeit: 1000 kg monatlich.

                8-Oxychinolin
                C$_9$H$_7$ON

                Mol. Gew.: 145
                F.P. 72 - 74°
                loeslich in Dipropylaether Benzol, Aceton,
                                                      Methanol.

Reaktionsverlauf:

Nach der Skraup'schen Chinolinsynthese wird aus o-Amidophenol Glycerin,conc.
Schwefelsaeure und o-Nitrophenol das 8-Oxychinolin hergestellt. Die Schwefelsaeure wird mit Natronlauge und Soda neutralisiert, das Oxychinolin abgesaugt
und durch Destillation gereinigt.

Noetige Rohprodukte:

                100 kg 8-Oxychinolin benoetigen:

                 57,6 kg o-Nitrophenol
                229,5 kg Glycerin
                224   kg Schwefelsaeure 94 - 96 %ig
                 90   kg o-Amidophenol
                900   kg Eis
                600   kg Natronlauge 30%ig
                 52   kg Soda

18

Fig. 4.10 (continued)

Item D-1   Skraup-Apparatur Geb. 18b

zur Herstellung von Oxychin.

500 1 email. Ruehrkessel mit Dampfmantel, Reaktionskessel
100 1 eisernes Messgefaess fuer conc. Schwefelsaeure mit eingebautem
      kleinerem Messgefaess,
  3 qm Kuehler, Aluminium u. V₂A, u. Reservekuehler ders.Art.
1000 1 email. Ruehrkessel, Faellkessel
 600 1 eisernes Messgefaess fuer Natronlauge
 300 1 eisernes Messgefaess fuer Wasser
  2 qm Nutsche ausgemauert mit Steinen q 70 u. Asplit A,
2000 1 eiserne Vorlage ausgemauert
  4 Igetro-Trockenschraenke mit email. Trockenplatten 60 x 80 cm,

Umloese-Apparatur:Geb. 20a

1000 1 email.Ruehrkessel mit Dampfmantel
   1 qm Aluminium-Kuehler
     Scheibler Filter
1000 1 email. Ruehrkessel im Wasserbad
     Aluminium-Kuehler, 2 qm Kuehlflaeche,
     Aluminium-Nutsche, 1 m Durchmesser.
1000 1. email. Vorlage.

Herstellung:
       1.) Kondensation.

Apparatur: (Nichinapparatur).

Gesamteinsatz:
        36,3 kg o-Nitrophenol
        143,4 kg Glycerin
        155,5 kg Schwefelsaeure 94 – 96%ig
        56,25 kg o-Amidophenol

19

Fig. 4.10 (continued)

Item D-1

Arbeitsweise:
36,3 kg o-Nitrophenol und 143,4 kg Glyzerin werden vorgelegt und auf 118-120° angeheizt. Es werden dann 12,5 l = 23,- kg Schwefelsaeure (96%ig) auf einmal zugegeben, wobei die Temperatur auf 130-133° steigt. Nun werden 60 mal je etwa 0,94 kg o-Amidophenol und 1,2 l Schwefelsaeure (96%ig) abwechselnd zugegeben. (Dauer 3 Stunden). Die Temperatur steigt hierbei auf 144-147° und wird durch langsamere oder schnellere Zugabe der Portionen auf dieser Temperatur gehalten. Hoehere Temperaturen ergeben geringere Ausbeuten. Wenn alles eingetragen ist, wird die Partie noch 4 Stunden auf Siedetemperatur gehalten. Infolge Abspaltung von Wasser sinkt die Temperatur wieder auf 130-133°.

2.) Aufarbeitung des Kondensationsproduktes.

Gesamteinsatz:
600 kg Eis
etwa 300-400 kg Natronlauge 30%ig
26 - 27 kg Soda
600 kg Eis werden vorgelegt und soviel Wasser zugegeben, dass der Ruehrer angesetzt werden kann (etwa 100 l). Unter Ruehren wird die auf 40 - 50° abgekuehlte Partie auf das Eis gedrueckt und mit etwa 300-400 kg 30%iger Natronlauge schwach sauer gestellt. Der letzte Rest der Saeure wird mit Soda gebunden. Die Reaktion muss zuletzt soda alkalisch sein (m tron alkalisch schlecht saugbar). Die Temperatur darf 40° nicht ueberschreiten, da das ausfallende Oxychin (E.P. des reinen Oxychins: 73°) andernfalls klumpig ausfaellt und zuviel Wasser und Soda einschliesst. Die Partie wird abgesaugt, mit Wasser nachgewaschen und gut trocken gesaugt.

3.) Destillation.

Apparatur:
1 1000 l eiserner Destillierkessel mit direkter Gasfeuerung und absteigendem eisernem Kuehler.
2 100 l eiserne Vorlagen mit Schauglas
1 100 l eiserne Sicherheitsvorlagen mit Wasserbad (Sammelgefaess fuer uebersublimiertes Oxychin)
Die Destillation kann vorerst nicht durchgefuehrt werden; das Destillat aus dem Eisenkessel ist nicht rein genug, und der kleine Cu-Destillierkessel in Geb.82 ist vollkommen verschlissen.   gez.Dr.Gundlach

Einsatz:
2 Partien Rohoxychin.
Arbeitsweise:
Zunaechst wird ohne Vakuum bis zu einer Innentemperatur von 130° das Wasser abdestilliert. Nun wird Vakuum angeschlossen und das Oxychin bei 25 mm zwischen 165 bis 180° ueberdestilliert. Dauer 7-8 Stunden. Die Innentemperatur liegt zum Schluss ueber 360°.
Ausbeute: 90 bis 100 kg
pro Partie also: 45 - 50 kg = 50 - 56% d.Th.,
bezogen auf Amido- u. Nitrophenol = 75 - 83% d.Th., nur auf Amidophenol bezogen.
100 kg o-Amidophenol geben 111 kg 8-Oxychinolin.

20

Fig. 4.10 (continued)

Synthesis of 8-hydroxyquinoline, Figure 4.9, is a simple reaction between *O*-aminophenol and *O*-nitrophenol. Synthesis developed and commercialized by I. G. Elberfeld Wuppertal is detailed in FIAT Final Report 940, September 30, 1946 [47]. Figure 4.10 is a copy of the synthesis in German. Its English translation, best effort, is included. Work illustrated here tells us the extent of chemical technology development in the era. It is produced by reacting *O*-aminophenol with *O*-nitrophenol using glycerin as a solvent, in the presence of 94–96% sulfuric acid with glycerin as a solvent.

The process briefly described in this publication is a reaction of ortho nitrophenol with ortho aminophenol, in the presence of concentrated sulfuric acid, with glycerin acting as a solvent.

Translation of 8-hydroxyquinoline process

Oxychin 8-hydroxyquinoline (C_9H_7ON)

Procedure from I. G. Elberfeld Wuppertal October 1941
Production possibility 1,000 kg per month
Molecular weight 145 MP 72–74 °C soluble in dipropyl ether, benzene, acetone, methanol
According to Skraup's quinoline synthesis, o-amido phenol becomes glycerin conc. sulfuric acid and o-nitrophenol produced 8-oxyquinoline. The sulfuric acid is neutralized with sodium hydroxide solution and soda, the oxyquinoline is suctioned off and purified by distillation.

Necessary raw materials
100 kg 8-hydroxyquinoline needs:
57.6 kg o-nitrophenol
229.5 kg glycerin
224 kg sulfuric acid 94–96%
90 kg o-amido phenol
900 kg ICE
600 kg caustic soda
52 kg sodium carbonate
Skraup apparatus (Geb. 18 B)
For production of 8-hydroxyquinoline
500 L reaction vessel with steam jacket reaction vessel
100 L iron measuring vessel with built-in smaller measuring vessel
3 sq. m. aluminum and V2A, and reserve cooler of same type
1,000 L reaction kettle fall boiler?
600 L iron measuring vessel for caustic soda
300 L iron measuring vessel for water
2 m2 Nutsche lined with stones
2,000 L bricked iron template
4 L getro dryers
Enclosed apparatus
1,000 L reaction vessel with steam jacket
1,000 L reaction vessel with water bath
Aluminum cooler 2 sq. m. cooling surface
Aluminum nutsche of 1 meter diameter
1,000 L template

Manufacturing:

1) Condensation:

36.3 kg o-nitrophenol
143.4 kg glycerin
155.5 kg sulfuric acid 94–96%
56.25 kg o-amido phenol
36.3 kg o-nitrophenol and 143.4 kg glycerin
These were placed in a reaction kettle and heated to 118–120 °C. Then 12.5 L = 23 kg sulfuric acid was added, the temperature rising to 130–133 °C. Now o-amido phenol 0.94 kg and sulfuric acid 1.2 L are added alternately 60 times over three hours. The temperature rose to 144–147 °C, and was achieved by adding longer or faster portions maintained at this temperature. Higher temperatures result in lower yields. When everything had been entered, the batch was kept at the boiling temperature for another 4 h. As a result of the elimination of water, the temperature drops back to 130–133 °C.

2) Processing of the condensation products

600 kg ice
About 300–400 kg of caustic soda
26–27 kg sodium carbonate
600 kg of ice are presented and enough water is added that can be left at rest (approx. 100 L). While stirring, the portion that has cooled to 40–50 °C is pressed onto the ice and made slightly acidic with about 300–400 kg of 30% sodium hydroxide solution. The last remainder of the acid is bound with soda. The reaction must finally be soda alkaline (mtron alkaline poorly absorbable). The temperature must not exceed 40 °C, otherwise the precipitated oxyquin (e. p. of pure oxychine: 73 °C) falls out in lumps and includes too much water and soda. The area is vacuumed, washed with water and vacuumed dry.

3) Distillation

Apparatus:
1 1,000 L iron distillation kettle with direct gas firing and descending iron cooler
2 100 L iron templates with sight glass
1 100 L iron safety templates with water bath (collecting vessel for over sublimated oxyquin)
The distillation cannot be carried out for the time being; the distillate from the iron kettle is not pure enough, and the small copper distillate in geb. 82 is completely worn out.

Commitment:
2 lots of raw hydroxyquine
Working method:
First, the water is distilled off without vacuum up to an internal temperature of 130 °C. Now, the vacuum is connected and the oxyquin is distilled at 25 mm between 165 and 180 °C for 7–8 h. The interior temperature is finally over 360 °C.
Yield 90–100 kg
45–50 kg per batch = 50–56% d. theoretical, based on amido u. nitrophenol = 75–83% of theory th. only related to amido phenol. About 100 kg of o-amidophenol give 111 kg of 8-oxyquinoline

USP 4044011 [48] suggests the use of acrolein [49], a highly toxic material for the synthesis of 8-hydroxyquinoline as a solvent. Due to acrolein's toxicity, most process developers, most likely, will use alternate solvents. Use of acrolein, a toxic chemical, as a solvent, needs to be avoided. Alternate manufacturing processes can be developed by using some of the processing equipment reviewed in Chapter 8 and exploiting physical properties of the raw materials.

Preparation of phthalimide from phthalic anhydride is a simple reaction [50–53]. However, a patent review suggests that it can be synthesized in different ways leading to different end results. The simplest method would be feeding molten phthalic anhydride and bubbling very slight excess of liquid ammonia in a small reactor within a reactor, to give sufficient residence time for reaction completion Figure 4.11. Liquid ammonia also reacts as the heat sink for this exothermic reaction, and melt temperature accelerates the reaction rate.

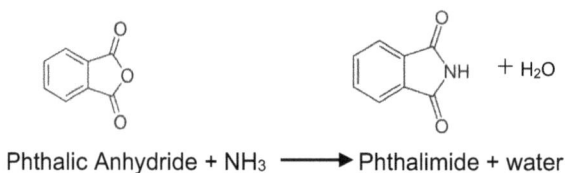

Phthalic Anhydride + NH₃ ⟶ Phthalimide + water

Fig. 4.11: Single step conversion of Phthalic anhydride to Phthalimide.

4.3 Green process development

Based on the review of brand and generic company API patents earlier in this chapter, a conclusion can be drawn that additional effort is needed to create environment-friendly processes. Effort has been applied to many chemistries/processes that have been developed and commercialized. They are classified as "green chemistry and/or green manufacturing processes." Such processes are simple and have significant financial benefit [54, 55]. However, an effort to improve yield and solvent use can be made as soon as the excessive use of solvent and less-than-expected yield are recognized in the laboratory.

The process of using chemical reactants and solvents that are environment-friendly begins in the laboratory from the very beginning of the development process. This is a must for branded drugs as the opportunity to improve their process after the molecule has entered drug trials seldom comes and can be a challenge, as the changes might alter the molecule's efficacy and performance. Process changes might require regulatory review and that can be an expensive and cumbersome course.

Green processes might not be considered critical for the brand drug chemistries, as the objective is to get the developed molecule to clinical trials and to the market, as soon as possible. If an effort is not put in for environment-friendly processes from the inception, the impact of lack of "green processes" on our environment can be significant. There will be consequences if such processes are practiced [29–31]. Being "green" is not an ill-conceived mission but a noble cause. Such efforts have been practiced by some, since toxicity and environmental impact of chemicals have been recognized. They were nor labeled "green" but were called conservation and yield improvements. Creation of a "green process" will be the legacy chemists and chemical engineers leave behind, even if some are skeptical about the impact of effluents on water, air, soil, animals and aquatic life. Road maps are discussed in the following chapters.

Compared to brand drug producers, generic manufacturers have a much bigger opportunity to create green processes. They have an advantage as well as a disadvantage. The advantage comes in the way of time in which they have to create an environment-friendly process compared to the brand producers. Their manufacturing cost will be lower than the brand molecule producer costs. The disadvantage comes from the fact that there can be many producers of the same molecule, and they cannot capitalize on economies of scale to take full advantage of their effort [43]. Still, many generic companies strive to create better processes than brand API producers. We see this scenario with the spade of patents that are filed with better and economic chemistries. Generics can capitalize on the opportunity of economies of scale. This is further discussed in Chapter 8 Road Map.

Being green requires dedicated thinking and application of fundamentals of science and engineering, which are taught in most chemistry and chemical engineering curricula. It also requires creativity and imagination to manipulate/modify unit processes [1] and unit operations [2] that are needed to create or simplify chemical processes.

As was stated earlier, developers have to think that every chemistry and the corresponding process will be a commercial success. If a process is developed with such a mindset, most of the processes will be environment-friendly. A road map for green API process development is outlined. The fundamentals of the map to green (environment-friendly) chemistry will stay the same in most situations, though they can certainly be modified to suit different process development needs.

4.3.1 Unit processes and unit operations

In the development of a "green manufacturing process" every unit process [1] and unit operation [2] of a chemical process is crucial. We have to be cognizant of the development work in the laboratory. Every reaction step in the development of a chemical product has to be simple, so that each unit process [1] and unit operation [2] is the simplest and results in the highest yield. It is all right to check the progress of laboratory chemistry and process development by monitoring the conversion of each reaction step and testing product quality.

However, this does not mean that one has to follow the same practice to monitor commercial processes. If frequent process analysis has to be relied on for each step in a commercial operation to produce a quality product, then it shows that developers did not completely understand the process and do not have command over the process. There are opportunities to improve the process. Referenced articles [56–62] discuss some of the dos and don'ts of process development and improvement.

For establishing simple and efficient processes, I learnt that a basic rule of thumb is that majority of the non-optimized multiple reaction processes will have a yield of about 60–68%. This yield range might seem odd, but I came across this, over and over. We were able to improve yields of such processes. Table 4.3 shows a relationship of number of steps and their yields. Lower yields suggest that processes are overly complex and need revised chemistries, conditions, and methods. They present an opportunity to improve processes and their yields.

Lower yield means that the key raw material and associated reactants are producing undesirable products that have to be properly treated and disposed. This also suggests that the process stoichiometry needs to be reviewed and corrected. Since chemicals have different levels of toxicity, their proper disposal becomes important, as they can harm the environment.

Reaction steps of a synthesis need to be reviewed individually and collectively to develop a green process. Items listed in Tab. 4.12 can be used for such a review. These are not necessarily in any order and are to be used as a guideline, not as a rule. These have been discussed in Chapter 3 but are reviewed again, as they facilitate process development.

Tab. 4.12: Considerations for chemical process development.

1.	Do we have the physical properties of each reactant and intermediate? Collection of this information can be difficult. However, these are necessary for the design of every unit process and operation.
2.	Do we understand the total process feasibility? To do so, each unit process step has to be reviewed individually and collectively.
3.	Do we know the solubility of our chemicals? This applies to raw materials used and chemicals produced in every reaction. It is an important processing criterion and can influence separation and cycle times, significantly.
4.	Is the stoichiometry optimized? This can be difficult initially, but with experience, becomes second nature. Product cost and profitability are reflection of stoichiometry.
5.	Is the heat and mass balance known? Again, this is necessary, as they will be used to design respective unit operations. We can use the heat of reaction (exothermic and/or endothermic) in the reaction process and the subsequent unit operation.
6.	Are the reaction kinetics understood and applied to simplify the process? Our effort should be to achieve zero-order reaction. It will reduce reaction time and improve conversion yield.
7.	Can a single solvent in addition to water be used for the whole process? This economizes solvent recovery and the related investment. Relative physical properties of the solvent and water can facilitate the separation process.

Tab. 4.12 (continued)

8.	Can we eliminate the isolation of intermediates?
9.	Are the raw materials to be used easy to handle? Liquids are easy to handle, and if solid raw materials can be solubilized, batch or continuous processing will improve. Ref: Chapter 3.
10.	Are the safety requirements met, and is the process safe?
11.	Does the process meet all environmental standards?
12.	If the process developers were operating the process, what process modifications and/or additions would be included to make it the simplest? This may seem to be an unnecessary exercise, but it is important. Simplicity of processing is critical and allows us to reduce processing time and have an environment-friendly process. Having hands-on experiences in scaling up processes that are developed in our laboratories can be the most educative experience in process simplification and commercialization of a chemical synthesis.

4.3.2 Mass balance

Mass balance is the landscape of the process and gives the developers and engineers the fundamental information that is needed for process commercialization. Table 4.13 is a typical example. This information is necessary for every process and is needed to simplify, create, and design an economic process. It is also needed for product cost and waste disposal information.

Tab. 4.13: Typical mass balance.

Process diagram: Streams 1, 2, 3 → Mix → stream 4 → Reaction (with streams 5, 6 entering from top and streams 7, 8 exiting) .

Stream	1	2	3	4	5	6	7	8
Rate, Kg./Hr.	5000	2496	16325	23821	5360	1184	29945	420
Temp. ºC	50		25	47			85	
Material 1	5000			226				
Material 2		1248				592	7	
Material 3		1248	16325	18135		592	18754	
Material 4					5360			
Material 5				5460			574	
Material 6								
Material 7							10190	
Material 8								420
	5000	2496	16325	23821				

4.3.3 Solvent selection

As discussed in Chapter 3, solvents are a necessary part of every reaction. They are facilitators of every process. Efforts have to be made to minimize the solvent use in every reaction. This can happen if the solubility of every solid raw material and reaction product/s in the used solvents is known. Completely dissolved/solubilized chemicals facilitate thorough mixing and highest reaction conversion.

Water is the most obvious choice but, generally, in the development of API processes more than one solvent in addition to water are used. Patents reviewed [35–39, 46] are a typical illustration. If processes based on these patents were to be commercialized, they will, most likely, be the "ungreen" processes.

In order to have a "green process" in addition to water, every effort has to be made to use no more than one additional organic solvent. It may be considered a Herculean task, but is doable. To have a green process, it is not only necessary but should be demanded that the developers know, use, and capitalize on mutual behavior of solvents to minimize their number and variety. As discussed in Chapter 3, by having a complete understanding of chemical's mutual behavior, chemists and chemical engineers can simplify the manufacture of APIs. In Chapter 8, alternate processes/methods to lower the solvent use are reviewed.

Solvent selection also depends on how the used solvent will be recovered/recycled. There are three ways, and they are:
1. Filtration of solid products
2. Azeotropic distillation
3. Density-based separation

As stated earlier, solvents are reaction facilitators, and thus, their selection impacts the API manufacturing process.

4.3.4 Stoichiometry

Every reaction has a theoretical and actual mass balance. Actual mass balance tells us the real status of every manufacturing process. It is necessary to review the two to see how the process chemistry can be simplified. Certain excesses of raw materials are necessary to maximize the reaction step yield. However, too much excess of materials means that there will be unwanted byproducts that would have to be removed from the process to ensure that the desired product meets its quality expectations.

Sometimes rationale for excessive use of participating reactants is not known or understood. However, too much excess can add undue processing complexity. Comparison of USP '203 [46] and USP '537 [14] suggests excessive use of hydrochloric acid and sulfur dioxide. Excessive hydrochloric acid and sulfur dioxide used in USP '203 [46] would have to be neutralized and would be part of effluent. USP '203 [46] uses acetic acid as a solvent and would have to be neutralized and end up as part of the process waste. USP '537 [14] uses 1,2-dichloroethane as a solvent. Its solubility [63] in water is low, and it can be separated from the reaction mass easily, due to its density difference.

It is interesting to compare the theoretical and actual use of reactants in the first two steps of USP '203 [46]. Theoretical moles of hydrochloric acid, $NaNO_2$, and sulfur dioxide used in USP '203 are 2.0, 1.0, and 1.0. However, in the patent, the respective moles used are 14.0, 0.97, and 25.9, suggesting opportunities to simplify the process. In addition, comparison of two similar reactions, 'USP '537 [14] and '203 [46] suggests that 'USP '537 [14] is definitely an eco-friendlier process than USP '203 [46]. Such analysis and review of every single processing step individually and collectively is necessary. They suggest how the processes can be simplified.

Creating a "green chemistry/process" is not just about having knowledge, understanding, and command over mutual behavior of chemicals and their interaction to produce APIs. All these have to be applied to have green chemistry or green process. Effort, creativity, and imagination are needed to apply them in combination for a green process. An observation that cannot be missed is that most of the chemistry patents developed by generics for the brand drugs have progressively moved to using less toxic and environment-friendly solvents. Such efforts have to continue. In addition, the chemistry has to be such that it can be easily translated in to a viable economic manufacturing process. Translation of the chemistry to a "green process" requires creativity and imagination besides having application knowledge of chemical engineering principles.

Abbreviations

FDF	Finished dosage form
BHT	Butylated hydroxy toluene
S&E	Service and expense
MMF	Mycophenolate mofetil
TLC	Thin-layer chromatography
HMDS	Hexamethyl disilazane

References

[1] Unit Processes https://encyclopedia2.thefreedictionary.com/Unit+processes, June 22, 2020].
[2] Unit Operations, https://encyclopedia2.thefreedictionary.com/unit+operations, Accessed June 22, 2020.

[3]   Dosage Forms, https://www.fda.gov/industry/structured-product-labeling-resources/dos age-forms Accessed July 14, 2020.

[4]   Inactive Ingredients https://www.drugs.com/inactive/ Accessed June 17, 2020.

[5]   Overview of pharmaceutical excipients used in tablets and capsules, https://www.drugtopics. com/hospitalhealth-system-pharmacy/overview-pharmaceutical-excipients-used-tablets-and-capsulesOctober 24, 2008, June 17, 2020, March 30, 2018, Accessed June 15, 2020.

[6]   Common Chemicals Used in Medicine – AZ Chemistry https://azchemistry.com/chemicals-used-in-medicine Accessed June 17, 2020.

[7]   Pharmaceutical Excipients-Some Definition, https://www.pharmaexcipients.com/pharmaceu tical-excipients-some-definition/, Accessed June 22, 2020.

[8]   Marques-Marinho, F. D., Vianna-Sosres, Cellulose and Its Derivatives Use in the Pharmaceutical Compounding Practice, https://www.intechopen.com/books/cellulose-medical-pharmaceutical-and-electronic-applications/cellulose-and-its-derivatives-use-in-the-pharmaceutical-compounding-practice, August 29, 2013, Accessed June 21, 2020.

[9]   Chaerunisaa, A. Y., Sriwidodo, S., Abdassah, M. (July 19th 2019). Microcrystalline Cellulose as Pharmaceutical Excipient, IntechOpen, DOI: 10.5772/intechopen.88092. Available from: https://www.intechopen.com/books/pharmaceutical-formulation-design-recent-practices/mi crocrystalline-cellulose-as-pharmaceutical-excipient, Accessed June 22, 2020.

[10]   Pharmaceutical Inert Excipients Good Manufacturing Practice Guide, https://ipec-federation. org/wp-content/uploads/2019/05/20170323%20IPEC-PQG%20GMP%20Guide_Final.pdf, 2017 Accessed June 22, 2020.

[11]   McCabe, W. L., Smith, J. C. Unit Operations of Chemical Engineering, Second, McGraw-Hill Book Company, 1967.

[12]   Malhotra, G.: Chemical Process Simplification: Improving Productivity and Sustainability John Wiley & Sons, February 2011.

[13]   Current Good Manufacturing Practice (cGMP) Regulations https://www.fda.gov/drugs/phar maceutical-quality-resources/current-good-manufacturing-practice-cgmp-regulations, March 30, 2018, Accessed June 15, 2020.

[14]   USP 4464537, Tonne P. et. al. Preparation of Saccharin, BASF Corporation, Accessed September 10, 2020.

[15]   US Patent 2841623, Norton, D. G. et. al., Shell Development Company, Accessed September 10, 2020.

[16]   USP 3324119, Hill D. R. et. el.: Production of Isatoic Anhydride and certain Halo Derivatives, Maumee Chemical Company, Accessed September 10, 2020.

[17]   USP 3123631, Staiger, R. P. et.al.: Process for processing esters of Anthranilic Acid, Maumee Chemical Company, Accessed September 10, 2020.

[18]   Haz-Map, Information on Hazardous Chemicals and Occupational Diseases https://haz-map. com/Agents/6068, Accessed Sept 25, 2020.

[19]   Englund, S. W. et. al. Synthesis of Cresol, Industrial and Engineering Chemistry, 45, 1, pgs, 189–193.

[20]   Sad, M. E. et.al.: Selective synthesis of p-cresol by methylation of phenol, Applied Catalysis A: General, 342, 1–2, 30 June 2008, Pgs. 40–48.

[21]   The crucial role of titanium dioxide in modern pharmaceuticals https://tdma.info/the-crucial-role-of-titanium-dioxide-in-modern-pharmaceuticals/, Accessed June 23, 2020.

[22]   Mc Donnell, D. C. Development and Scale up of API Manufacture, http://eleceng.dit.ie/gavin/ DT275/cppt9009/Dev%20and%20Scale%20up%201%20CMcD%2022-09-09.pdf], September 22, 2009 Accessed August 22, 2020.

[23] Tirronen, E. et.al Process Development in the Fine Chemical Industry, Chemical Engineering Journal, 2003, 91, 103–114. https://www.sciencedirect.com/science/article/abs/pii/S1385894702001432. accessed July10, 2020.

[24] United States Patent and Trademark Office http://patft.uspto.gov.

[25] World Intellectual Property Organization https://ipportal.wipo.int.

[26] Malhotra, G.: Square Plug In A Round Hole: Does This Scenario Exist in Pharmaceuticals? Profitability through Simplicity, https://pharmachemicalscoatings.blogspot.com/2010/08/square-peg-in-round-hole-does-this.html, August 17, 2010 Accessed September 15, 2020.

[27] Malhotra, G.: Why Fitting a Square Plug in a Round hole is Profitable for Pharma and Most Likely Will Stay?, Profitability through Simplicity, https://pharmachemicalscoatings.blogspot.com/2014/08/why-fitting-square-plug-in-round-hole.html, August 1, 2014 Accessed. September 15, 2020.

[28] Malhotra, G.: https://www.youtube.com/watch?v=48oZEA-az1U&list=PLiugGTl26EbnuDOChLtYs9ntR_VxUewgx&index=11 March 30, 2013, accessed September 28, 2020.

[29] Malhotra, G.: Pharmaceuticals, Their Manufacturing Methods, Ecotoxicology, and Human Life Relationship, Profitability through Simplicity, Originally published November 2007, Accessed August 10, 2020.

[30] Küster, A., Adler, N. Pharmaceuticals in the Environment: Scientific Evidence of Risks and Its Regulation, Philosophical Transactions of the Royal Society B, 369, 20130587. http://dx.doi.org/10.1098/rstb.2013.0587. accessed September 21, 2020.

[31] Larsson,, Et al, Effluent from Drug Manufactures Contains Extremely High Levels of Pharmaceuticals, Journal of Hazardous Materials, 148, 3, 30 September 2007, Pages 751–755, Accessed July 2007.

[32] USP 7019133 B2, Lee, K. et.al Process for making Mycophenolate Mofetil by Transesterification, Chunghwa Chemical Synthesis & Biotech Ltd.

[33] US 2010/029560 A1, Choi, B. T. et.al. Process for Preparing Mycophenolate Mofetil.

[34] Mycophenolate Mofetil www.Pharmacompass.com,https://www.pharmacompass.com/active-pharmaceutical-ingredients/mycophenolate-mofetil Accessed October 12, 2020.

[35] USP 5366972 Hargrave, K. D. et. al. 5,11-dihydro-6H-dipyrido(3,2-B:2′,3′-E)(1,4)diazepines and their use in the prevention or treatment of HIV infection, Boehringer Ingelheim.

[36] USP 5569760 Schneider, H., et al. Process for preparing nevirapine, Boehringer Ingelheim, Accessed September 2011.

[37] USP 6680383 B1 Boswell, et. al. Method of making Nevirapine, Boehringer Ingelheim Accessed 2011.

[38] WO 2007/010352 A1 Gharpure, et. al. An Improved Process for Industrial Production of Nevirapine, Emcure Pharmaceuticals Accessed 2020.

[39] WO 2012/168949 A2 Chava, et. al. A Process for Preparation of Nevirapine, Laurus Labs Accessed July 10, 2020.

[40] USP 8212025, Sanapureddy, et. al. Process for preparing Nevirapine, Aurobindo Pharma Ltd. Accessed July 10, 2020.

[41] VCU engineering professor receives $5 million grant, https://news.vcu.edu/article/VCU_engineering_professor_receives_5_million_grant_to_drive_down June 29, 2015 accessed September 24, 2020.

[42] WO 2016/118586 A1 Ahmad, et. al. Low cost, High Yield Synthesis of Nevirapine Accessed 2020.

[43] Malhotra, G.: Impact of Regulations, Manufacturing and Pharmaceutical Supply Chain (PBMs) on Drug Shortages and Affordability Part 2, Profitability through Simplicity, https://pharma

chemicalscoatings.blogspot.com/2019/04/impact-of-regulations-manufacturing-and.html
April 3, 2019 Accessed September 28, 2020.

[44] Longstreet, A. R. et al Investigating the Continuous Synthesis of a Nicotinonitrile Precursor to
Nevirapine, Beilstein Journal of Organic Chemistry, Beilstein Journal of Organic Chemistry,
2013, 9, 2570–2578. https://doi.org/10.3762/bjoc.9.292. Accessed September 28, 2020.

[45] Nevirapine, A. P. I. Average Price, https://www.pharmacompass.com/active-pharmaceutical-
ingredients/nevirapine/api-price-information/api-exports, Accessed September 24, 2020.

[46] USP 7109203, Sulfonamide Derivatives, Novartis, Accessed February 2009.

[47] FIAT Final Report No. 940 September 30, 1946 Accessed January 10, 2020.

[48] USP 4044011, Cognion, J., Process for the preparation of 8-hydroxyquinoline,Produits
Chimiques Ugine Kuhlmann, Accessed September 10, 2020.

[49] Agency for Toxic Substances and Disease Registry, https://www.atsdr.cdc.gov/index.html,
Accessed September 10, 2020.

[50] USP 2668326 Schlaudecker, G. S., Process for the Preparation of Phthalimide, Maumee
Development Company Accessed October, 19, 2020.

[51] USP 3819648 Boehme, W. R. Production of Phthalimide, Dawe's Laboratories Accessed
January 13, 2019.

[52] USP 4001272 Eckhard, et. al. Continuous manufacture of Phthalimide, BASF
Aktiengesellschaft Accessed July 21, 2020.

[53] Zagres, W. et al. USP 9701632 Process for the Production of Phthalimides, Lanxess
Deutchland GMBH Accessed July 21, 2020.

[54] Li, et al Green Chemistry for Chemical Synthesis, Proceedings of National Academy of
Sciences, September 9 2008, 105, 36, 13197–13202. https://doi.org/10.1073/
pnas.0804348105. Accessed October, 10, 2020.

[55] Basics of Green Chemistry, https://www.epa.gov/greenchemistry/basics-green-chemistry,
Accessed October, 10, 2020.

[56] Malhotra, G.: Less is More in API Process Development: Pharmaceutical Manufacturing,
https://www.pharmamanufacturing.com/articles/2005/294/ pgs. 50–51 July/August 2005
Accessed September 10, 2020.

[57] Malhotra, G.: QbD: Myth or Reality? Pharmaceutical Processing.

[58] https://www.pharmaceuticalprocessingworld.com/quality-by-design-qbd/ February 2007,
Pages 10–16 Accessed September 21, 2020.

[59] Malhotra, G. Big Pharma: Who's Your Role Model, Toyota or Edsel? Pharmaceutical
Manufacturing, https://www.pharmamanufacturing.com/articles/2007/106/ Pg. 40,
June 2007 Accessed September 21, 2020.

[60] Malhotra, G. Implementing QbD: A Step-by-Step Approach. https://www.pharmaceuticalpro
cessingworld.com/implementing-qbd-a-step-by-step-approach/ Pg. 16–18, February 2008
Accessed September 21, 2020.

[61] Malhotra, G. Pharmaceutical Manufacturing: Is It the Antithesis of Creative Destruction?
https://www.pharmamanufacturing.com/assets/wp_downloads/pdf/Creative_Destruction_
Malhotra.pdf July 2008, Accessed September 21, 2020.

[62] Economist Intelligence Unit. Quality manufacturing: A blockbuster opportunity for
pharmaceuticals (2005). http://graphics.eiu.com/files/ad_pdfs/eiu_oracle_pharma_wp.pdf
Accessed September 1, 2008.

[63] Solubility of Dichloroethane, https://en.wikipedia.org/wiki/1,2-Dichloroethane, Accessed
October 11, 2020.

Chapter 5
Manufacturing process development and case studies

It takes a whole village for an innovation to be developed, launched, and adopted.
— Edmund (Ned) Phelps, Nobel Laureate

In earlier chapters, the value of capitalizing on physical and chemical properties of raw materials and chemistries of different intermediates and manufactured products have been emphasized. In this chapter, discussion is about how different unit operations [1, 2] which through exploitation of physical and chemical properties of the reactants that are used and produced in different unit processes [1, 3] can create viable, economic, and green processes. Each chemistry must be optimized to be economic and green and must result in process that will produce the desired quality product. Reaction steps give us clues about how to simplify the manufacturing processes. These clues are also explored. APIs (active pharmaceutical ingredients) are fine/specialty chemicals that are used to cure diseases and dispensed in solid (tablet), liquid, or ointment form.

Theoretically, brand and generic API manufacturing processes should be optimized to have the highest yield, that is, be green. However, laboratory synthesis processes for the brand molecules, most of the time due to the need for speed to market, are not optimized. Yields as low as 10% are not out of the realm of reality [4]. This is due to brand API's monopoly for the length of the patent and their financial contribution to the drug's selling price (Tab. 5.1). Generic APIs, compared to brand APIs, are relatively more optimized and greener but not to the extent that is possible. Like brand API, generic API's financial influence on the generic drug selling price is minimal. Table 5.1 is an illustration of generic drug prices. Later, in this chapter, API and the formulated cost for Tecfidera (dimethyl fumarate) versus its list price are reviewed (Tab. 5.20). Similar cost estimates can be done for every API and their formulations.

Due to API's miniscule financial contribution on the drug selling price, illustrated in Tabs. 5.1, 4.5, and 2.4, it is not necessary to optimize their chemistries and manufacturing processes. This is also true for brand drugs with a priority to get to the market. However, many generic API producers generally commercialize their own developed process for the same API after the API patent has expired or lost its patentability. Since generic producers compete with other API producers, their processes must be optimized for the selected chemistry and have to be profitable. High yields and waste minimization are an excellent way to stay in business. Continuous process improvement is necessary and beneficial.

https://doi.org/10.1515/9783110702842-005

Tab. 5.1: Relationship of API prices to generic drug's selling prices [5].

Drug	Metformin HCl	Ciprofloxacin	Generic levothyroxine	Atorvastatin
API cost, $/kg [2]	4.00	25.00	4400.00	310.00
Inert excipients, $/kg (@40%API cost)	1.60	10.00	1760.00	124.00
Conversion cost, $/kg (@40%API cost)	1.60	10.00	1760.00	124.00
Profit (@ 40% above)	2.88	18.00	3168.00	223.20
Total, $/kg	10.08	63.00	11,088.00	781.20
Average dose	500 mg	500 mg	0.112 µg	20 mg
Formulator sale price per tablet, $	0.005	0.032	0.001	0.008
Patient purchase price ($/tablet)				
Walmart	0.07	1.04	0.11	0.30
Rite-Aid With insurance	0.07	0.2	0.17 ***	0.31
Rite-Aid Without insurance	0.7	4.77	0.82	3.97

Brand Levothyroxine with co-pay 100 µg price ≅$1.12 per 100 µg tablet.

Brand API producers, like generic producers, due to environmental and safety considerations may make an effort to optimize their processes, make them green, but that is not a priority. However, as stated earlier getting to the market is. Since the cost of API, as stated earlier, does not have much influence on the drug's selling price, most consider that process economics is not a factor in process development and optimization. In addition, it is well known that once the NDA (new drug application) application has been submitted to the regulators, it is expensive to make any changes to the process to economize or make the synthesis green [6]. Since companies have to prove that any change made to the process has not changed product's efficacy and performance, this becomes a deterrent for improvements.

5.1 Commercialization of APIs

Independent of who produces the brand or the generic drug API, to have an economic and green process (a new way of saying that the process has to have the highest yield and minimum waste) that will produce repeatable quality product, it takes a "VILLAGE" [7, 8] to master, exploit, and capitalize on every interaction and property of each chemical and translate them to viable and economic process.

"VILLAGE" includes a team of chemists and chemical engineers, financial analysts (accounting), supply chain professionals, quality control, maintenance, compliance/regulatory, and manufacturing at a company to think about and commercialize manufacturing processes. Chemical engineers and chemists are the Picasso, the Michelangelo, the Zaha Hadid, the Gaudi, or other artisans of their process development and designs. Their designs are emblematic of their learning, creativity, and imagination. They use their training including their capabilities to exploit mutual behavior of chemicals and creatively extend equipment performance to design excellent economic processes [9] that will produce quality products.

In addition, each development chemist from the onset has to consider that their developed process will be commercialized even when it is well known that most of the laboratory developed processes will need to be reconfigured for a commercial process. The process of continuous improvement gets instilled. For commercialization, necessary process engineering principles will have to be applied for scale up. Process centricity must be the focus [10, 11]. Some may not agree with this perspective and also might not agree that such an effort is necessary. However, such an effort is not only essential but critical as it expedites product commercialization and facilitates compliance with cGMP [12] and every other regulation. Members of the "VILLAGE" have to challenge the stoichiometry and unit operations from the start. This results in process optimization and simplification. Such an effort will also result in process that will produce good quality products repeatedly.

Review of drug molecule chemistries in Chapter 4 and elsewhere [4, 6, 13, 14] reinforces that most of the API syntheses as developed in the laboratory are complex and without simplification, and they will be cumbersome to scale up and commercialize. Many a times, the laboratory process will have to be modified to fit in the existing equipment [15, 16] that is not designed for the process. Use of the existing equipment can happen for any of the following reasons:

1. Most APIs have the production volume that can be produced at a single site to meet global patient needs (Tab. 4.4) but they are produced at many sites [17, 18]. Companies producing these products must modify their process and thus will lack the value of economies of scale. In addition, as stated earlier, API selling price does not have any impact on the drug selling price, preventing capitalizing on such values.
2. With the current industry overcapacity [19–22] and <50% asset utilization, no company wants to invest in dedicated equipment for a single API process that will operate for only a few moths per year. If they did, equipment would be idle as it would not be suitable for any other products. Alternates are available and discussed later.

Since the chemistries are fitted in the existing equipment [15, 16], they pose additional challenges for scale-up. These challenges can come from not having the most fitting or suitable unit operations (proper equipment, reactor type (size, materials of construction), heat exchange capacity, drying including spray drying, distillation, crystallization, etc.) needed for the production. Inventiveness, as explained later, can lead to consolidation and efficient manufacturing and can become the norm.

Involvement of chemists and chemical engineers, as suggested earlier, in the laboratory development can also facilitate continuous improvement processes for the API manufacturers as they know every nuance of the equipment used [6, 8, 23, 24]. Their expertise is of extreme value when there are manufacturing upsets or issues. The process of continuous improvement could also be used by the brand APIs, if regulators hold producers accountable for every improvement [25]. For this to happen, stricter regulatory measures would be necessary.

It is possible that the brand companies might not be interested in "continuous improvement process" due to the limited time left in the patent life after the drug has been approved for which they may have to spend additional money to assure regulators that the product efficacy has not changed.

Engineers and chemists involved in the process have to understand every reaction step, physical properties of the reactants and products produced, and translate them from the lab scale (grams per batch) to kilograms per batch or kilograms per hour scale. Many in the pharmaceutical industry might not agree. Development of small-molecule API starts and does not finish with the development chemistry at the lab bench. Only in rare cases this may not be true. In reality, the commercialization process just begins with the thought of synthesizing a new molecule. It continues till the final manufacturing process is commercialized.

Pathway and elements reviewed in Fig. 5.1 illustrate the laboratory to commercialization process. Each of the following elements is generally incorporated in product and process development, scale-up, commercialization, and simplification. Again, to reiterate these elements work in parallel to reduce production time, improve yield (reduce waste) and product quality, and lower product cost. Companies also have to be proactive to changes but on the pharmaceutical landscape the related costs can be a deterrent. Elements outlined in Fig. 5.1 can be modified to suit individual company needs.

5.2 Considerations for scale-up

There are considerations that are used in scaling up every chemical reaction product to the final product. Many books have been written on the subject [2, 3]. Engineers are taught to take their learning from the laboratory to commercial-scale production through scale-up exercises. Books do not teach how to exploit physical and chemical

Paper Process Review	→	Laboratory Chemistry Development	→	Pilot plant	→	Manufacturing

Compilation of physical & chemical properties of raw materials, Intermediates and final products Costing of alternate routes	Best route Reaction conditions definition for highest yield. Capitalize on mutual chemical behavior, chemical process and translation to mfg. Analytical tools and methods have to precisely defined, tested and used.	Translation of laboratory process to larger equipment that emulates manufacturing. Testing of process design concepts and control strategies.	Translation and testing of process operating conditions and control strategies. Tweaking of stoichiometry, operating conditions and parameters.

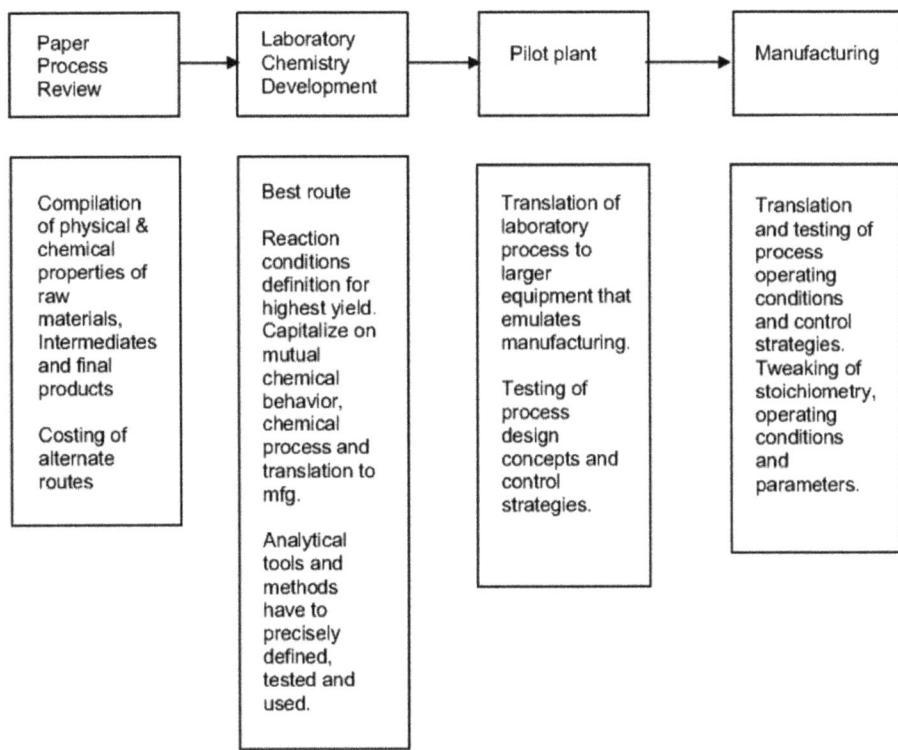

Fig. 5.1: Steps involved in API commercialization.

properties and mutual behavior of chemicals. This interaction is learned through experience and inference [7, 24]. Many other resources are available.

5.2.1 Role of process equipment and mutual interaction of chemicals in pharmaceuticals

Process equipment and chemicals used and produced in API manufacturing have their selective roles in the manufacture of API and each influences manufacturing. Similarly, chemicals used and produced in every reaction have mutual behavior that influences processing steps [2, 3, 8, 24]. It is necessary that the process developers understand and exploit mutual and collective behavior of chemicals and the processing equipment to create optimum cost-effective processes. Overemphasis is necessary.

5.2.2 Role of process equipment

Processing equipment evolution began over 200 years ago when liquids and solids had to be mixed and reacted to produce chemical products that had values. Most likely it all started with laboratory glassware. As explained in Chapter 1, many API producers and their formulators started as fine/specialty chemical companies. Some became pharmaceuticals and others remained chemical producers that facilitated lifestyles. Since the equipment and chemicals used in both industries are similar, commonalities of manufacturing have remained.

Evolution of metallurgy, glass technologies, and mechanical engineering led to better processing equipment for chemical production which has significantly improved manufacturing. Better understanding of physical properties, for example, boiling point, melting point, viscosities, solubilities along with agitation, heating, cooling, and other behaviors such as heat of reaction and reaction rates led to the creation of unit operations [1–3]. These have facilitated production of different chemicals. Even with these advances, transition from laboratory to a commercial manufacturing still requires understanding of mutual behavior of chemicals and how best to incorporate them to create economic and eco-friendly processes.

Once the chemical reaction mass balance equations are prepared, laboratory glassware is the first place where the chemistry and the reactions can be tested and experimented. This practice started over 150 years ago by fine/specialty chemical companies still is the way chemical syntheses are developed. Since most small-molecule APIs are fine/specialty chemicals, their synthesis chemistries and manufacturing methods are also similar. Due to this similarity, processing equipment that was developed and used in the fine/specialty chemical industry is also used in the pharmaceutical API manufacturing and their formulations.

Glassware used in the laboratory gives developers clues about the processing steps. They are an excellent place to explore possibilities but are the last place to create an excellent process. After proving feasibility, laboratory developed synthesis must be married with the process equipment to create an economic and eco-friendly process. Unit processes [1, 3] and unit operations [1, 2] used in the fine/specialty chemical industry are also used for the pharmaceutical ingredients also.

Since the unit operations/processes used in fine/specialty chemical manufacturing technology have been successfully worked in the manufacture of APIs, their use has continued. Pharma's focus has been to assure and retain product quality. Not much effort has been made by the pharmaceutical industry to develop or improve any of the established practices. In addition, once the product is commercialized, a need for continued process improvement leading to any cost reduction in API manufacturing and their formulations, as explained in earlier chapters, is not considered necessary.

Another reason for lack of manufacturing technology innovation in the pharmaceutical industry has been the excess of idle capacity that is available at various generic API manufacturing sites throughout the world [19, 20]. Not many companies

want to invest in capital equipment that does not have adequate return on investment. Thus, the processes are fitted [15, 16] in the existing equipment. Many a times,
such processes are not the most productive. As said earlier, API manufacturing costs
do not have much influence on the drug selling price. As a consequence, investment
in appropriate technology is difficult to justify for the brand and the generic API producers. However, there are opportunities that are discussed later in this chapter.

5.2.3 Stoichiometry, mutual interaction of chemicals, and reaction mechanism

Since the laboratory experiments are on milligram to kilogram scale, equipment
used in the laboratory, depending on chemistry, does not give sufficient information to exploit reaction mechanism/s, physical and chemical properties, and mutual
behavior of chemicals. Stoichiometry that makes the basis of a process also outlines
the reaction mechanism which gives the developers and designers clues on how the
commercial process should be executed. These do need to be tested in the laboratory. As suggested earlier, all of the interactions have to be understood, learned,
translated, and extrapolated to create excellent commercial processes that produce
quality products. Experience of chemical engineers and chemists is of high value,
as they begin the scale-up process. Application of principles of organic, inorganic,
physical chemistry and interactive behavior of chemicals is necessary.

Every chemical reaction has its own nuances which are not clearly obvious to
everyone who is not associated with the reaction chemistry and its development. At
times, even the experienced chemists and chemical engineers can miss subtle
clues. As discussed in Chapter 3, fundamental to chemistry is the changes occurring
when chemicals interact. Reaction mechanisms tell us how each reaction should be
conducted. They also suggests the sequence of raw material addition. Their recognition and exploitation is of value. They also tell us where the chemicals are added
and/or removed in a reaction process. This knowledge facilitates and simplifies
manufacturing processes. Such nuances get augmented by mutual solubility or the
lack of it. Even the reaction processing temperatures influence their behavior. All of
the information generated can be exploited in the scale-up process only.

In the laboratory experiments, raw materials are added to the glassware and
feasibility of reaction and approximate yields are determined. This is an excellent
start. However, the sequential raw material addition which can be done in a pilot
plant or a production plant cannot be tested in the laboratory glassware. Labora-
tory-developed stoichiometry has to be translated and scaled up to a real process. It
is necessary to compare the laboratory stoichiometry to an optimized stoichiometry
and costs to evaluate the real results are quantified when the process is designed,
scaled up, and commercialized. Following are examples of reactions where the reaction mechanism tells us how to take advantage of the reaction mechanism but

the size or methods emulate the laboratory glassware and it is not easy to replicate the reaction mechanism.

In most of the reactions, either of the following can happen. Concepts discussed here are applicable to the production of every API. One of the basic considerations must be how the liquid/solvent used in the process can be minimized. Some of the options are as follows:

1. Reactant/s acting as a solvent at room temperature
2. Reactant/s acting as a solvent at its melting point
3. Sequential addition of reactants to capitalize on reaction mechanism
4. Using reaction product/by-product as a solvent
5. Catalysts can be solid or liquid but need an appropriate liquid media for mixing

Melting points and mutual solubilities can be used to facilitate production. If their value is not recognized and not incorporated in the processes properly, the reaction yields can be significantly lower and reaction times significantly longer. Mastering and using the information give the process designers an opportunity to create simple and eco-friendly processes. Since many of the same APIs are produced by many companies, value of economies of scale is lost and many cannot fully [18, 26] exploit all of the nuances discussed.

Every reaction is carried out in liquid media. Solvents (water or organic) are the facilitator. Raw materials are either all liquid or solubilized or slurried solids. In scale-up from laboratory to larger scale, use of liquids, as a reactant or solvent, has a major impact on the reaction productivity and process. In most of the laboratory processes, solvent use generally ranges from 25% to 45%. This level of use in the laboratory is to prove feasibility of the process. Having proven reaction feasibility, it is scale-up chemist's or engineer's job to minimize or eliminate solvent use while maximizing productivity. Their use can be reduced and/or eliminated if the reactants can be used as a melt and the products are liquid also. This is discussed later. Laboratories do not have the right and/or the necessary equipment to explore and show value of physical properties of the chemicals involved. Engineers and chemists involved in the scale-up have to exploit these. How these get exploited depends on product volume, raw material melt temperature, its supply, and use methodology.

Water should be the preferred solvent [27]. If this is not possible, every attempt needs to be made to limit the use of organic solvents to one additional solvent besides water. Such an effort reduces effluent and minimizes environmental impact. If the reaction needs an organic solvent besides water, its separation methods (azeotropic distillation and water insolubility and separation as discussed in earlier chapters) should also be part of the selection process. Solvent selection due to pharmaceutical product quality can be a challenge.

Addition of reactants has a certain safe mixing sequence. Adverse consequences can result if the sequence is not followed. Simplest example is dilution of concentrated inorganic acid. Acid is added to water that is under agitation not the other way around.

Water added to concentrated acid due to the generated exotherm can lead to localized heat and eruption. If the localized exotherm is not dissipated or controlled, it can discolor the product. Runaway reactions can result in explosions. Each chemical reaction and addition sequence give us clues that we have to respect for safe process design.

Melting point of products can be of value if the reaction temperature is above the melting point of the products. An example is reviewed later in this chapter.

How the designers use and manipulate each of the above characteristics impacts the process design. These are extremely critical factors in the process selection and the process design. API production volume influences the process design.

Synthesis of a few API are reviewed to understand what they tell us and how the information discussed above can be used to simplify the scale-up and commercialization process. Some of these are high volume products but since many companies produce them, value of economies of scale is lost [18, 26]. Chemical engineers and chemists well versed in scale-up can further refine, design, and commercialize processes that are economic. Analysis done here can be extended to other products. The purpose of review is not to be critical of any method(s) but to remind ourselves that every chemical reaction has to be conducted safely.

5.3 Dimethyl fumarate

As reviewed in Chapter 1, dimethyl fumarate was initially used as a fungicide in Germany and now it is also used for psoriasis. Process chemistry and design for this product is a classical chemical engineering exercise that requires application of principles of chemistry, chemical engineering, and process control.

National Psoriasis Foundation estimates [28] that about 2–3% of the global population suffers from psoriasis, a skin disease. Using this one can guesstimate the global need for the API, illustrated in Tab. 5.2. Table 5.2 is a calculation of the dimethyl fumarate API volume needed per year to serve the needs of about 10% (pure speculation) of the global population suffering from psoriasis, provided it is affordable. With about 1.35 million kg/year market, the global demand can be produced at *one* or *maximum two* plants. However, based on traditions of the pharmaceutical industry, dimethyl fumarate is being produced in multiple plants [29].

Synthesis and stoichiometry outlined in patents and literature [30–34] using maleic anhydride or fumaric acid are reviewed. Each route is simple. Maleic anhydride (Fig. 5.2) and fumaric acid (Fig. 5.3) are converted to monomethyl fumarate which is subsequently converted to dimethyl fumarate. Either of the routes can be used in a continuous or a batch process. Fig. 5.4 is the stepwise reaction mechanism of fumaric acid to dimethyl fumarate conversion.

Table 5.3 illustrates physical properties of the major chemicals used and produced in either routes to the final product.

Tab. 5.2: Global dimethyl fumarate market size.

Dimethyl fumarate market size	
Global population	7,700,000,000
Psoriasis population percentage	2%
Psoriasis population	154,000,000
Potential user population	10%
Patient population	15,400,000
Use per day, mg	240
Total use and API need, kg/yr.	1,349,040

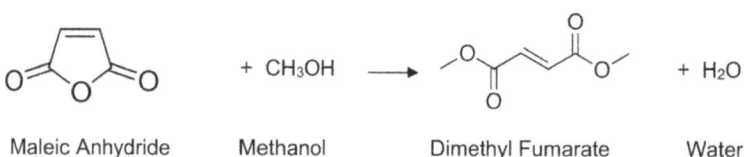

Maleic Anhydride   Methanol   Dimethyl Fumarate   Water

Fig. 5.2: Maleic anhydride to dimethyl fumarate.

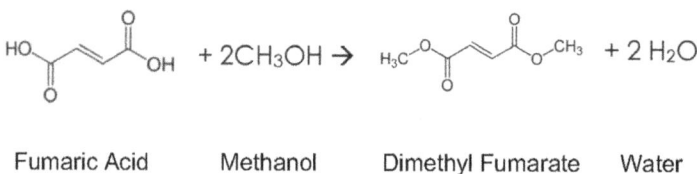

Fumaric Acid   Methanol   Dimethyl Fumarate   Water

Fig. 5.3: Fumaric acid to dimethyl fumarate.

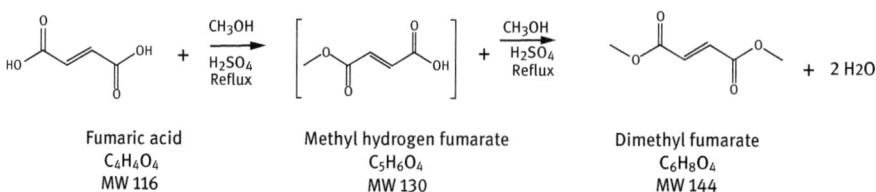

Fumaric acid   Methyl hydrogen fumarate   Dimethyl fumarate
$C_4H_4O_4$ $C_5H_6O_4$ $C_6H_8O_4$
MW 116   MW 130   MW 144

Fig. 5.4: Fumaric acid to dimethyl fumarate [30].

Two maleic anhydride-based chemistries are illustrated in Figs. 5.5 [31] and 5.6 [34] and show their reaction mechanisms.

Anyone knowledgeable and experienced can capitalize on physical properties to create a simple process. Fumaric acid route generates an extra mole of water than the maleic anhydride process. Additional water has to be properly disposed and would add cost to the process.

Tab. 5.3: Physical properties of chemicals for dimethyl fumarate.

Chemical	CAS #	Chemical formula	Mol. wt.	MP (°C)	BP (°C)
Fumaric acid	110-17-8	$C_4H_4O_4$	116	298	356
Maleic anhydride	108-31-6	$C_4H_2O_3$	98	55	202
Methanol	67-56-1	CH_4O	32	-98	48
Monomethyl furmarate	2756-87-8	$C_5H_6O_4$	130	144	~275
Thiourea	62-56-6	CH_4N_2S	76	171	Decomposes
H_2SO_4	7664-93-9	H_2SO_4	98	10	337
Thionyl chloride	7719-09-7	$SOCL_2$	119	-104.5	166
Dimethyl fumarate	624-49-7	$C_6H_8O_4$	144	104	193

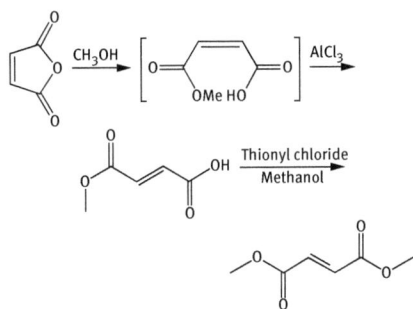

Fig. 5.5: Maleic anhydride to dimethyl fumarate [31].

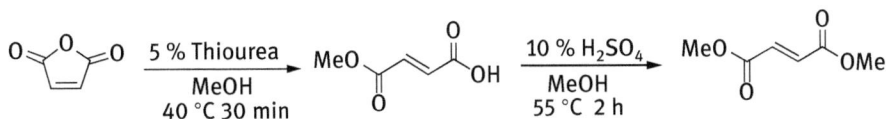

Fig. 5.6: Maleic anhydride route [34].

Methanol and sulfuric acid are common in either route. Price difference and availability between maleic anhydride and fumaric acid would be the cost driver. Since maleic anhydride is a widely used commodity chemical, one can speculate that its price will be lower significantly than fumaric acid, giving it a cost advantage.

In their respective laboratory processes, formic acid and maleic anhydride are slurried and converted to the product and purified to yield the final, desired material. Maleic anhydride due to much lower melting point offers the option of use as a melt but still requires specialized setup in the lab or the pilot plant to test the concept. Due to limitations of the laboratory equipment, this can be a challenge. It is possible that the suppliers would not have this information and it would have to be generated. Solubility of maleic anhydride in methanol would give this route a distinct process advantage.

Cost analysis of dimethyl fumarate based on maleic anhydride route [34] is illustrated in Tab. 5.4. Service and expense is the conversion cost that includes hourly and salaried labor, utilities, maintenance, and allocated charges. Cost of solvent per kilo is not included. Minimal and optimized use will not add significant cost. Selling price in Tab. 5.4 is the price if dimethyl fumarate was a fine/specialty chemical. Similar analysis can be done for the alternate routes considered for any API. Such analyses facilitate process selection choice. Actual selling price of dimethyl fumarate is much higher, Chapter Two [35] than shown in Tab. 5.4.

Tab. 5.4: Cost analysis of dimethyl fumarate.

	Mol. wt	Moles	Kilogram	Price ($/kg)	$/kg
Maleic anhydride	98	1	71.64	0.99	70.92
Methanol	32	2.2	51.46	0.66	33.97
Thio urea	76	0.1	5.56	1.98	11.00
Sulfuric acid	98	0.02	1.43	0.22	0.32
Dimethyl fumarate	144	Yield 95%	100		122.32
			Service and expense	40%	48.93
			Total		171.25
			$/kg		1.71
			Profit	30%	0.51
			Selling price		2.23

Chemistry and processing of Fig. 5.5 can be simplified by substituting sulfuric acid for thionyl chloride Fig. 5.6 [34] to produce dimethyl fumarate. This process eliminates handling of aluminum chloride ($AlCl_3$), a significantly more hazardous and challenging material to handle compared to sulfuric acid. A choice might have to be made between $AlCl_3$ and thiourea for the conversion of maleic anhydride to methyl hydrogen fumarate, Fig. 5.5 and Fig. 5.6.

Based on tradition, the process would be scaled up using the available reactors. For scale-up and commercialization, chemical engineers will recommend/choose the best process to meet the market need. Reaction kinetics will clearly define the process and the needed equipment. Process controls are available to control stoichiometry and operating conditions. Batch manufacturing process instructions would be prepared based on the selected chemistry. These would have to be tested on the selected equipment and may have to be modified for the desired yield and assure product quality.

Based on global API requirement (Tab. 5.2), all of the needed dimethyl fumarate can be produced at a single plant using a continuous process. Multiple sites would be needed for its formulation. Figure 5.7 is a schematic diagram of a potential commercial process.

Separate and concentrated slurries of maleic anhydride and methanol and thiourea and methanol [36] can be produced continuously. They will be metered in the stoichiometric ratios to reactor 1. Separately, methanol and sulfuric acid mixture is prepared and reacted with monomethyl fumarate from reactor 1 to produce dimethyl fumarate. Kinetics of each reaction step has to be determined in the laboratory and tested in the pilot plant to see how the reaction can be accelerated.

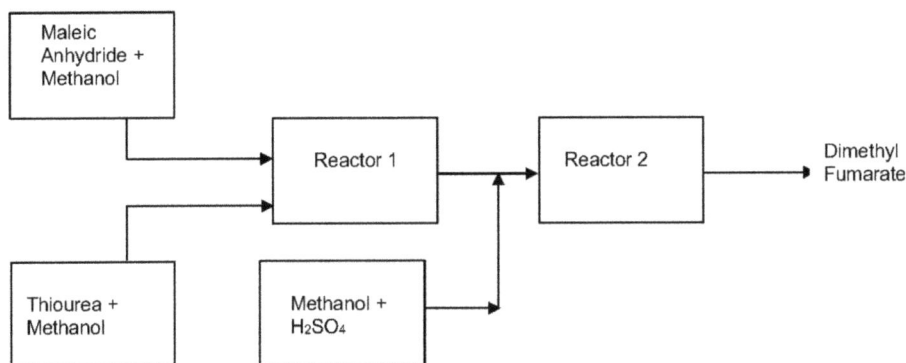

Fig. 5.7: Schematic flow diagram for continuous dimethyl fumarate.

Reactors 1 and 2 can be recirculating back mix pipeline reactors that have the needed residence time along with inline regular shell and tube or plate and frame or electrically heated heat exchangers to maintain the desired reaction temperature. Inline static mixers can be effectively used. The needed residence time will dictate the size. Such reaction schemes have been used since the mid-1960s for many reactions for many products.

The concept presented here is not new and can be refined, scaled up and commercialized. As stated earlier, the scale-up process begins in the laboratory and use of smaller scale process equipment due to significantly lower API volume needs can be used to produce most products. This concept will be further discussed in this and later chapters.

5.4 Hydrochlorothiazide

Hydrochlorothiazide (HCTZ) is another fine/specialty chemical that prevents body from absorbing too much salt, which can cause fluid retention. It is a high-volume diuretic drug that makes the top 20 drug list [37].

HCTZ synthesis is classical organic chemistry illustrated in Fig. 5.8. Physical properties of the chemicals used in the synthesis are illustrated in Tab. 5.5.

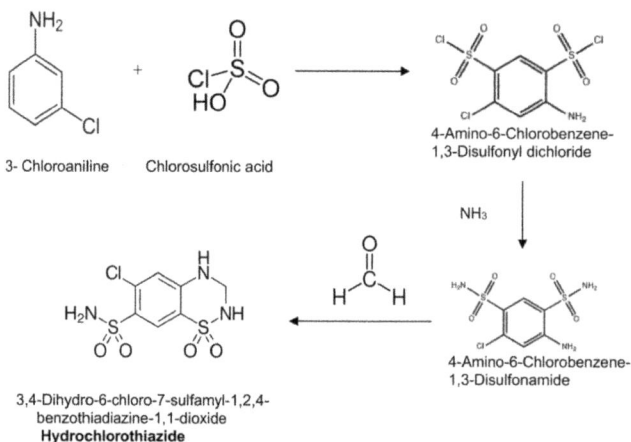

Fig. 5.8: Synthesis of hydrochlorothiazide [38].

Tab. 5.5: Physical properties of chemicals for hydrochlorothiazide.

Chemical	CAS	Chemical formula	Mol. wt.	MP (°C)	BP (°C)
3-Chloroaniline	108-42-9	C_6H_6ClN	127.5	−10.4	95–96
Chlorosulfonic acid	7790-94-5	$ClHO_3S$	116.5	−80	151–152
4-Amino-6-chlorobenzene-1,3 disulfonyl dichloride	671-89-6	$C_6H_4Cl_3NO_4S_2$	324.5	130–132	474
Ammonia	7664-41-7	NH_3	17	−78	−33
4-Amino-6-chlorobenzene-1,3-disulfonamide	121-30-2	$C_6H_8ClN_3O_4S_2$	285.5	257–261	
p-Formaldehyde	30525-89-4	$OH(CH_2O)_nH$ ($n = 8$–100)	90	120	
Hydrochlorothiazide	58-93-5	$C_7H_8ClN_3O_4S_2$	297.5		273

In Step 1 of the hydrochlorothiazide synthesis, theoretically 2 mol of chlorosulfonic acid reacted with 3-chloroaniline to produce 4-amino-6-chlorobenzene-1,3 disulfonyl dichloride. First, two reactants are liquids and result in the formation of a solid product (4-amino-6-chlorobenzene-1,3 disulfonyl dichloride). USP 2965675 [39] points to conducting the reaction at high temperature. This can be beneficial as it can minimize the solvent use. USP 3161675 [40] issued to Merck & Co. issued in 1964 and USP 3326908 [41] issued to Schering Corp. outline additional methods for the production of sulfonamide. A solvent might be necessary to assure a slurry for the reaction with liquid ammonia. In step 2, as the liquid ammonia evaporates, its feed rate can be metered to control the heat of exotherm and the reaction temperature. Step 3 may require a solvent [42, 43]. Chemists well versed in different syntheses can optimize the process. Testing of reactions at higher temperatures than suggested in the literature and patents needs to be explored as it can define better reaction conditions.

Since hydrochlorothiazide is a high-use product [37], it is of value to estimate its global need. Table 5.6 is an illustration. To meet the global needs of about 2% population, a single plant using a continuous process would suffice.

Tab. 5.6: Yearly HCTZ (API) need.

Global population	7,800,000,000
% using HCTZ	2.00
Dose, mg	12.5
Days/yr.	365
Total API, Kg./yr.	711,750.00

If the demand increases, the single plant can be optimized to meet the needs or an additional plant can be used. Currently, about 50 plants [44] are producing the API. None of them benefit from any economies of scale.

5.5 Benchmarking of chemistries

Benchmarking of similar chemistries is an excellent tool for the development of economic processes. It also implants various ideas and concepts that can be incorporated by chemists and chemical engineers in developing new chemistries. Two different chemistries that involve diazotization and SO_2 reaction steps for the production of sulfonyl chloride-type molecules are reviewed.

5.5.1 USP 7109203

USP 7109203's [45] chemistry is reviewed and compared to a similar reaction chemistry. Partial reaction of Example 1 of USP '203 is as follows and Fig. 5.8 describes the theoretical stoichiometry. Table 5.7 lists physical properties of chemicals used in step 1 of example 1.

Tab. 5.7: Physical properties of chemicals for step 1 of Example 1.

Chemical	CAS	Chemical formula	Mol. wt	MP (°C)	BP (°C)
4-Bromo-2-chloroaniline	38762-41-3	C_6H_5BrClN	206.5	70–72	241.8
Hydrochloric acid, 37%	7647-01-0	HCl	36.5	−35	57
Sodium nitrite, 40%	7632-00-0	$NaNO_2$	69	−7	120
Sulfur dioxide	7446-09-5	SO_2	64	−72	−10
Copper II chloride	7447-39-4	$CuCl_2$	134.5		

A stirred solution of 4-bromo-2-chloroaniline (3.99 g) in acetic acid (90 mL) at 15 °C is treated with concentrated hydrochloric acid (22 mL), followed by a solution of sodium nitrite (1.29 g) in water (4.5 mL) at 10 °C. After 30 min, the mixture is added to a stirred solution of sulfur dioxide (32 g) and copper II chloride (1.3 g) in acetic acid (128 mL) and water (6.4 mL), also at 10 °C. After stirring for a further 16 h, the mixture is diluted with ice-cold water (500 mL) and extracted with ethyl acetate (4× 100 mL). The combined extracts are washed successively with water (4× 100 mL) and then brine, dried over magnesium sulfate, filtered, and evaporated to give crude 2-chloro-4-bromobenzenesulfonyl chloride [CAS# 351003-52-6].

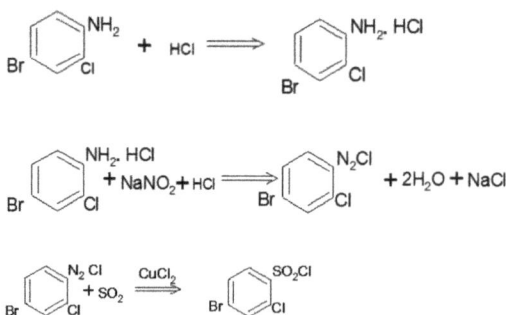

Fig. 5.9: Partial chemistry of Example 1 of USP 7109203.

Tab. 5.8: Stoichiometry of step 1 of USP 7109203.

Chemical	Chemical formula	Mol. wt	Grams	Moles	Mole ratio	Theoretical
4-Bromo-2-chloroaniline	C_6H_5BrClN	206.5	3.99	0.019	1.0	1.0
Hydrochloric acid, 37%	HCl	36.5	9.77	0.26	13.7	2.0
Sodium nitrite, 40%	$NaNO_2$	69	1.2	0.019	1.0	1.0
Sulfur dioxide	SO_2	64	32	0.5	83.3	1.0
Copper II chloride	$CuCl_2$	134.5	1.3	0.009	1.5	??

A review of the stoichiometry mentioned in USP '203 suggests that large stoichiometric excess of sodium nitrite, hydrochloric acid, sulfur di oxide, and copper(II) chloride are being used. In addition, acetic acid is being used as a solvent. It has no value in the reaction. If the process, as defined, is to be commercialized, a significant effort to redefine the chemistry would be needed. As is, there will be significant cost added due to acetic acid that adds no value to the reaction. It also adds cost to the process as it has to be disposed. Catalytic amount of copper(II) chloride is needed. Copper salt can be recovered and reused, provided the product warrants such recovery.

Stoichiometry of USP '203 can be compared with stoichiometry of USP '537 Tab.5.9.

5.5.2 USP 4464537 [46]

About 76 g (0.5 mol) of methyl anthranilate and 135 g of an aqueous sodium nitrite solution (containing 0.5 mol of $NaNO_2$) were added gradually in the course of 30 min to 196 g of 30% strength by weight aqueous hydrochloric acid (=1.61 mol of HCl) from −5 to +10 °C. The diazonium salt solution was then brought into contact with a solution of 200 mL of 1,2-dichloroethane and 42 g (0.65 mol) of SO_2 for 10 min at 20 °C, with vigorous stirring.

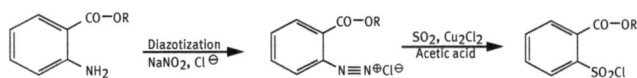

Fig. 5.10: Partial chemistry of Example 1 of USP '537.

About 5 g of an aqueous $CuCl_2$ solution (containing 0.9 g of Cu = 0.014 mol) was added to the reaction mixture, which was then heated to 50 °C, in the course of which a vigorous stream of N_2 was evolved. This process step required 80 min, after which the aqueous phase was separated off.

Tab. 5.9: Stoichiometry of step 1 of USP 4464537.

Chemical	Chemical formula	Mol. wt	Grams	Moles	Mole ratio	Theoretical
Methyl anthranilate	$C_8H_9NO_2$	151	76	0.5	1.0	1.0
Hydrochloric acid, 37%	HCl	36.5	196@30%	1.61	3.2	2.0
Sodium nitrite, 40%	$NaNO_2$	69	1.2	0.5	1.0	1.0
Sulfur dioxide	SO_2	64	42	0.65	1.3	1.0
Copper II chloride	$CuCl_2$	134.5	1.3	0.014	0.028	??

Tab. 5.9 lists the stoichiometry of USP '537 is compares it against its theoretical use. Chemistry of the two patents '203 and '537 has been compared, and diazotization and sulfation process of '203 could have been significantly simplified and improved. Use of acetic acid and excessive use of hydrochloric acid could have been curtailed or eliminated and the process significantly simplified. Productivity of '203 process could be further simplified through other methods such slurring the chloroaniline in water or a solvent rather than acetic acid as suggested which can be recovered and reused.

The purpose of such literature and patent [47, 48] search and their review not only facilitates process development but teaches how processes can be simplified.

5.6 Metformin hydrochloride

Metformin hydrochloride is an important drug widely used by significant number of global diabetic population for type 2 diabetes. History of metformin is an interesting read [49]. With rising global diabetic population, 700 million by 2045 [50] demand for this drug will increase. Table 5.10 estimates global metformin API annual demand.

Tab. 5.10: Projected demand for metformin [50].

Projected diabetic population 2045, million	700
50% type 2 million	350
Only 50% of type 2 take metformin, million	175
Daily dose 1,000 mg	
Total API needed, MT per year	63,875

Due to the high API volume need, metformin can be produced using a continuous process; however, the large number of plants, about 81 plants [51], suggests that it is produced using a batch process, a tradition in pharmaceutical manufacturing. If all of the global population with type 2 diabetes decides to take metformin, their demand cannot be met due to shortage of necessary raw materials and plant capacities. Pricing and selling price of formulated metformin hydrochloride is reviewed in Tab. 5.1.

Synthesis of metformin hydrochloride illustrated in Fig. 5.11 is a classical chemistry. Its melt synthesis process was first published in 1922 [52].

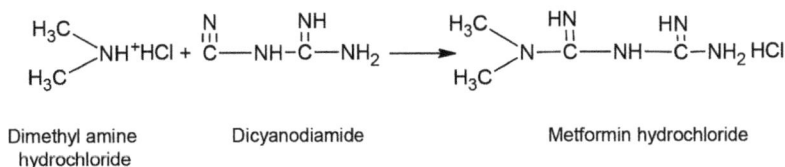

Dimethyl amine hydrochloride Dicyanodiamide Metformin hydrochloride

Fig. 5.11: Metformin synthesis.

Theoretically, synthesis of metformin hydrochloride is a molar addition of the two chemicals. Some stoichiometric excess of dimethylamine hydrochloride is necessary for the reaction. This has to be precisely managed as the residual concentration of dimethylamine in the final product is critical. Most of the patented processes relate to its purification [53–55]. Different solvents, for example, xylene/toluene or n-butanol are used in the reaction, purification, and crystallization.

There are similar fusion chemistries that are commercial using continuous process. An example is production of phthalimide by reacting liquid ammonia with molten phthalic anhydride. Besides reacting to produce the product, heat of the exotherm evaporates liquid ammonia, and the evaporation of gas with liquid phthalic anhydride improves the reaction rate [56–58]. Metformin could be similarly produced but not much effort has been made for manufacturing technology innovation. Reason for this could be lack of equipment in the laboratory or imagination or the need for developing efficient drug processes.

Supply of raw materials for certain synthesis can be an encumbrance for its commercial production. Companies have to investigate alternate sources. In the case of metformin [59, 60], use of dimethyl amine gas and hydrochloric acid solution to produce dimethyl amine hydrochloride is very feasible. They are reacted with 2-cyanoguanidine to produce metformin. In certain cases, producing onsite dimethyl amine hydrochloride could be economic than buying dimethyl amine hydrochloride. Each company has to do its own cost analysis. Such analysis are routinely done at fine/specialty chemical manufacturing companies for every product they develop and commercialize.

There are three options available to produce metformin manufacturing. Either route can be a continuous process. Since the global need for metformin is among highest API volumes, such an opportunity should be considered:

Tab. 5.11: Physical properties of chemicals for metformin hydrochloride.

Chemical	CAS	Chemical formula	Mol. wt	MP (°C)	BP (°C)
Dimethyl amine hydrochloride	506-59-2	C_2H_8ClN	81.5	160	
2-Cyanoguanidine	461-58-5	$C_2H_4N_4$	84	209.5	252
Metformin hydrochloride	1115-70-4	$C_4H_{11}N_5.HCl$	165.5	223–226	
Dimethyl amine	124-40-3	C_2H_7N	45	−93	7
Hydrochloric acid, 37%	7647-01-0	HCl	36.5		85

1. Use of traditional solvent process followed by crystallization has been discussed earlier. Use of wiped film dryer [61] is an excellent alternate to traditional drying. It can simplify the manufacturing process. Technologies are commercially available [62, 63]. Use of wiped film evaporators and other unit operations are not part of the laboratory process development exercise. They can come into play at any progressive companies who want to excel and are a manufacturing technology innovator. Development work would be needed. Properly designed wiped film evaporator is ideal for a continuous process and it fits the bill for metformin hydrochloride.
2. Production of dimethyl amine hydrochloride by reacting with dimethyl amine and commercially available hydrochloric acid to produce dimethylamine hydrochloride which reacted with 2-cyanoguanidine followed by traditional drying or using wiped film evaporator mentioned above. Such process would have to be developed and scaled up. Figure 5.12 is a schematic of this process.

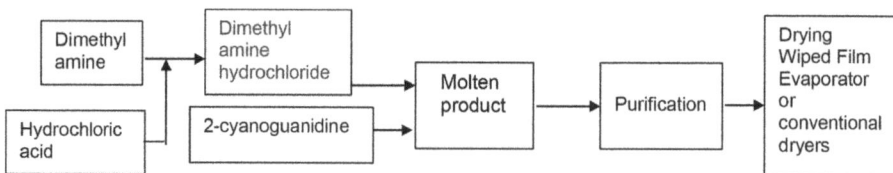

Fig. 5.12: Schematic of continuous metformin hydrochloride process.

In the reaction of molten dimethylamine hydrochloride and cyanoguanidine produced metformin that can be flaked and ground. This will eliminate use of any solvent. This would require complete control of the process stoichiometry. This particular process will pose a challenge of ready availability of the two raw materials. Molten delivery would be ideal but solid delivery would be acceptable and require melt production and the necessary handling equipment. Figure 5.13 is a schematic of such a process. Alternate process schemes starting with dimethyl amine are discussed in Chapter 7.

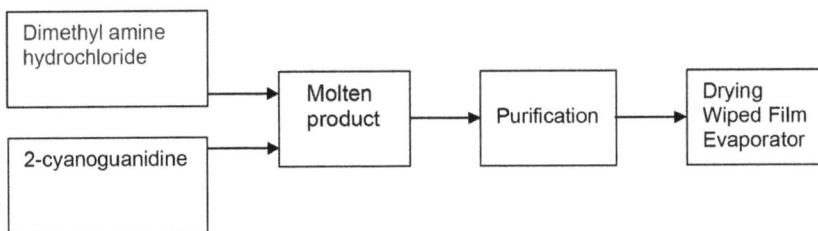

Fig. 5.13: Schematic of continuous metformin hydrochloride process.

Each of the above process schemes can be a continuous process. In either of the processes instead of using reactors, reactions can be conducted in pipes with static mixers or similar devices that can enhance mixing and have necessary provisions for heat control. Process control instrumentation that is reliable and will control it to produce quality products that has been available for over 50 years can be used. Each will be a dedicated production process and will have to be carefully tested and designed.

The biggest challenge for a continuous metformin hydrochloride process lies in the raw material supply chain. Production from each continuous manufacturing plant would necessitate multiple formulation plants which will have to operate continuously as multiple parallel lines producing different dosage tablets.

5.7 Omeprazole

Omeprazole is a drug used to alleviate acid reflux and is widely used. Over the years, it has made the top drug lists [64–66].

Omeprazole synthesis (Fig. 5.14) is a classical chemistry and the product can be manufactured easily in a properly designed plant.

A review of the related US patents presents interesting processing strategies [67–76].

A review of these patents brings out the efforts that have been made to improve the process. USP '024 [75] is for step 1 of the synthesis. It is an interesting patent as it simplifies and promotes the use of recovered wet toluene. Since toluene/water mixtures have an azeotrope, use of wet toluene simplifies solvent recovery and lowers the manufacturing cost. Such simplifications are of value for every synthesis. Properties of reactants and intermediates are of value in simplifying the process.

These [67–76] and additional patent [77] review chemistries can be used in step #3 to improve oxidation. An effort to optimize synthesis of omeprazole has continued [78, 79]. Tab. 5.12 catalogs physical properties of the raw materials, intermediates and the final product of omeprazole chemistry.

An estimate of 14% of the global population [80] in need of omeprazole is used to illustrate total API need (Tab. 5.13). About 90 plants [81] produce it. They most likely

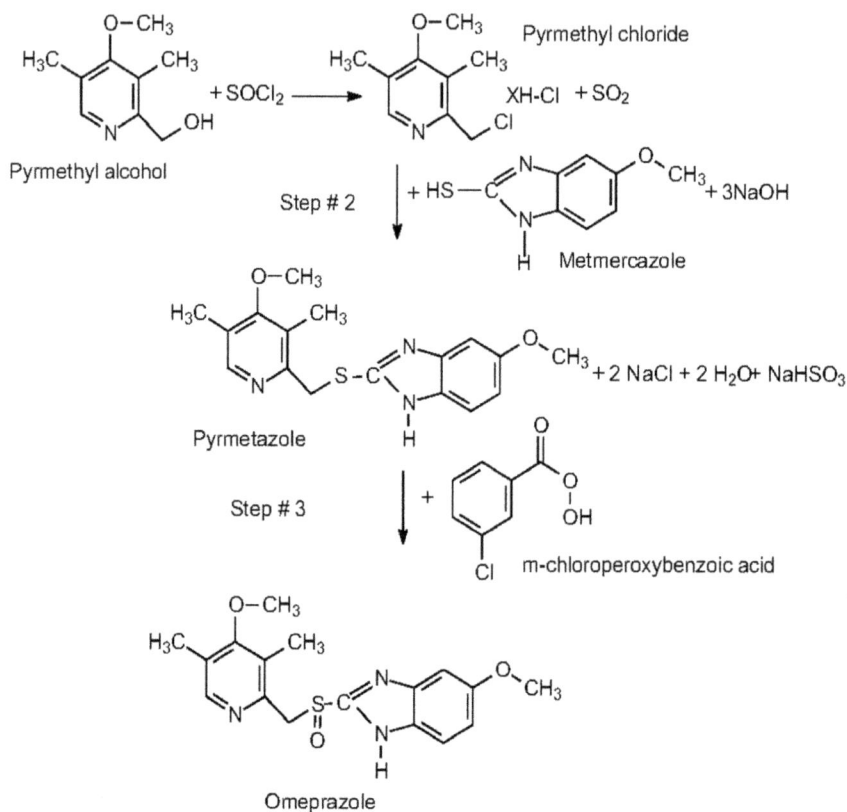

Fig. 5.14: Omeprazole synthesis.

Tab. 5.12: Properties of the chemicals involved in the production of omeprazole.

Chemical	CAS no.	Chemical formula	Mol. wt	MP (°C)	BP (°C)
Pyrmethyl alcohol	86604-78-6	$C_9H_{13}NO_2$	167.2	−54	69
Thionyl chloride	7719-09-7	$SOCl_2$	119	−104.5	74.6
Toluene	108-88-3	C_7H_8	92	−95	110.6
Pyrmethyl chloride	86604-75-3	$C_9H_{13}NOCl_2$	222	128–131	
Caustic 50%	1310-73-2	NaOH	40		140
Metmercazole	37052-78-1	$C_8H_8N_2OS$	180	261–263	309.4
Pyrmetazole	73590-85-9	$C_{17}H_{19}N_3O_2S$	329	Solid @ RT	
m-Chloroperoxybenzoic acid	937-14-4	$C_7H_5ClO_3$	172.5	92–94	305
Omeprazole	73590-58-6	$C_{17}H_{19}N_3O_3S$	345	120	

use batch processes even when the demand is high enough (Tab. 5.13) to produce it using a continuous process.

If the global demand for Nexium (esomeprazole sodium, the s-enantiomer of omeprazole) is included with omeprazole demand, the overall demand for omeprazole far exceeds the demand illustrated in Tab. 5.13. Concept and process scheme of a continuous omeprazole process are reviewed [82].

Tab. 5.13: Global annual omeprazole API demand.

Global population	7,800,000,000
Population in need of omeprazole	14%
Dose, 40 mg/day	43,680,000,000
@ 50 Days per patient, total milligrams	2,184,000,000,000
API, kg	2,184,000

A well-trained chemical engineer or chemist can take teachings of the referenced patents [67–76, 83] and physical and chemical properties to create an excellent continuous process. However, most likely they may use traditional design of new reactors ranging from 1,500 to 5,000 L and/or most likely use existing reactors to their process in the available equipment [15, 16]. There are alternate manufacturing technologies and strategies that are available but have not been considered for commercial production. They are discussed later in this chapter.

5.8 Metoprolol

Metoprolol is another drug that makes the top 20 drug list [37]. Using US demand, this drug's global demand is extrapolated. With one prescription per person about 20% of the US population [84] uses this drug. This number cannot be applied to the global population, and it is assumed that only 1% the global population uses this drug at 50 mg per day [Tab. 5.14].

API metoprolol's global demand [Tab. 5.14] is large enough to have a very efficient continuous process. Synthesis of metoprolol is a classical fine/specialty chemistry that is simple and can be easily executed. Figure 5.15 is the illustration of the chemical reaction.

Manufacturing process of metoprolol [85] is reviewed further. Tab. 5.15 lists physical properties of the chemicals involved in Metoprolol synthesis.

1-(2,3-Epoxypropoxy)-4-(2-methoxyethyl) benzene p-(2-methoxyethyl)phenol (A, ~ 6.6 mol), epichlorohydrin (B, 1.45 equiv.), and water (~2 kg) were combined and the mixture is heated to ~ 50 °C.

Tab. 5.14: Global metoprolol API demand.

Global population	7,800,000,000
% using metoprolol	1%
Dose, mg/day	50
Days/year	365
API needed kg/year	1,423,500

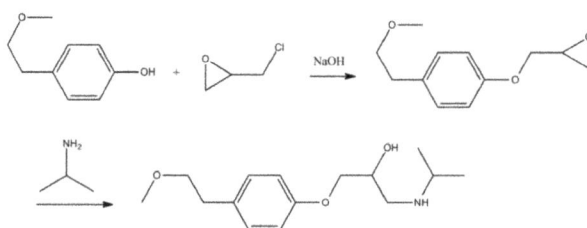

Fig. 5.15: Metoprolol synthesis.

Tab. 5.15: Physical and chemical properties of metoprolol chemicals.

Chemical	CAS no.	Chemical formula	Mol. wt	MP (°C)	BP (°C)
p-(2-Methoxyethyl) phenol	56,718-71-9	$C_9H_{12}O_2$	152	42	125@ 3 mm Hg
Epichlorohydrin	106-89-8	C_3H_5ClO	92.5	−25.6	118
Caustic, 50%	1310-73-2	NaOH	40	12	140
[4-(2-Methoxyethyl)phenoxy]methyl]oxirane	56,718-70-8	$C_{12}H_{16}O_3$	208	67	306
Isopropyl amine	75-31-0	C_3H_9N	59	−101	33
Metoprolol [1-[4-(2-methoxyethyl)-phenoxy]-3-(isopropylamino)-2-propanol]	51,384-51-1	$C_{15}H_{25}NO_3$	267	120	398

Sodium hydroxide solution (50%; 1.4 equiv.) was added during 3 h and the temperature was elevated to reach approximately 60 °C. during the addition. Formation of the title compound occurred during this period.

The batch was stirred for another hour at approximately 60 °C, then cooled to approximately 50 °C, and the phases were separated and the product washed with water.

The residue was distilled at ≤190 °C, and a pressure of ≤20 mm Hg and the distillate was collected. The yield of the title compound was 80% of theory and the purity was 98% according to GC analysis.

1-(2,3-Epoxypropoxy)-4-(2-methoxyethyl)benzene (1 kg, 4.8 mol), isopropyl alcohol (~0.9 kg), and isopropylamine (0.8–1.7 kg, 3–6 equiv.) were mixed and reacted for 2–5 h at reflux. Formation of metoprolol base occurred during this period.

The reaction mixture was then concentrated at atmospheric pressure until the inner temperature reached ~100 °C. Water was added to the batch and then distilled off in vacuo until the inner temperature reached ~100 °C to form a concentrate.

The resulting concentrate was diluted with isobutyl methyl ketone (~0.6 kg) and water (~2.2 kg), and concentrated sulfuric acid was added to adjust the pH to 4–6.

After separation, isobutyl methyl ketone (~1 kg) was added to the water layer, and concentrated sodium hydroxide solution was added to adjust the pH to 13.

The organic layer was concentrated in vacuo at ≤80 °C, until distillation ceased, and the concentrated batch was redissolved in acetone (~1.6 kg) and filtered, to yield metoprolol base solution. The assay of metoprolol base in the solution was determined by titration. Yield: ~1.2 kg metoprolol base (100%) ~95% of theory. The purity of the metoprolol base was 96%.

In this process, reactants are added sequentially per the reaction mechanism to create a good process. This has to be translated to a commercial process similarly. Additional optimization and improvements are possible. No suggestions are made as they can be best reviewed and implemented by the scale-up team. Table 5.16 reviews the theoretical and patent-suggested stoichiometry. In the patent compared to a molar need for isopropyl amine, a large excess is used.

Tab. 5.16: Theoretical and actual mole ratio comparison.

Patent US 2009/0247642 A1 [86]		
Chemical	Theoretical	Actual
p-(2-Methoxyethyl) phenol	1.0	1.0
Epichlorohydrin	1.0	1.03
KOH	1.0	1.0
Isopropyl amine	1.0	8.06

Theoretically, 1 mol of p-(2-methoxyethyl) phenol is reacted with a mole of alkali to produce the sodium salt which is then reacted with epichlorohydrin. The resulting oxirane compound is reacted with isopropyl amine to produce the desired product. Schematically, Fig. 5.16 is an illustration of the reaction and its mechanism.

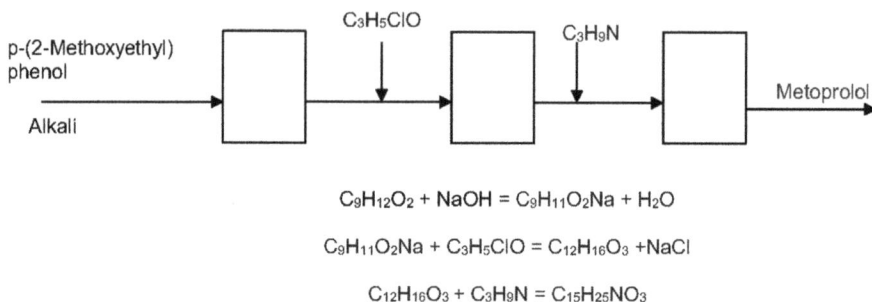

$$C_9H_{12}O_2 + NaOH = C_9H_{11}O_2Na + H_2O$$

$$C_9H_{11}O_2Na + C_3H_5ClO = C_{12}H_{16}O_3 + NaCl$$

$$C_{12}H_{16}O_3 + C_3H_9N = C_{15}H_{25}NO_3$$

Fig. 5.16: Metoprolol reaction mechanism.

There are other synthesis methods disclosed [86–88]. They can be reviewed and their teachings used to design and commercialize excellent process.

Traditionally, manufacturing of metoprolol like other fine/specialty chemicals is produced in commercially available equipment. Due to the simplicity and the volume of the needed product, this can also be produced using a continuous process. This is reviewed later in the chapter.

5.9 Modafinil

Modafinil synthesis is reviewed here. It was a blockbuster drug in around 2010–2012. It was categorized that way due to its yearly sales being over 1 billion dollar per year rather than high consumption. Five million patients using the drug are estimated at 0.06% of the global population (Tab. 5.17). This could be considered as a conservative number.

Tab. 5.17: Global modafinil API need, kg/year.

Global population	7,800,000,000
% Global patients	0.06%
Patients number	5,000,000
Dose, 100 mg/day	200
Days/year	365
Yearly API need, kg	365,000

Its process chemistry [89, 90] is a classical organic chemistry that is illustrated in Fig. 5.17. Many other patents have been filed for modafinil manufacturing processes [91–101]. Additional synthesis methods have also been published [102–105]. These teach different approaches and synthesis methods and are of value. Figure 5.17 illustrates the theoretical reaction chemistry.

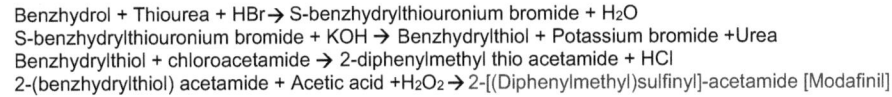

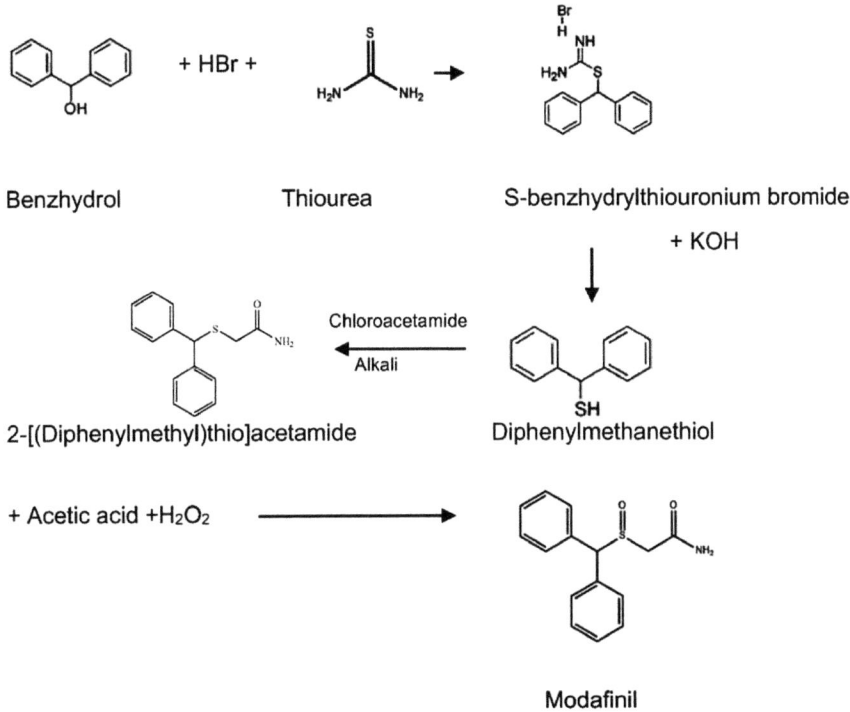

Benzhydrol + Thiourea + HBr→ S-benzhydrylthiouronium bromide + H₂O
S-benzhydrylthiouronium bromide + KOH → Benzhydrylthiol + Potassium bromide +Urea
Benzhydrylthiol + chloroacetamide → 2-diphenylmethyl thio acetamide + HCl
2-(benzhydrylthiol) acetamide + Acetic acid +H₂O₂ → 2-[(Diphenylmethyl)sulfinyl]-acetamide [Modafinil]

Benzhydrol   Thiourea   S-benzhydrylthiouronium bromide

+ KOH

Chloroacetamide

Alkali

2-[(Diphenylmethyl)thio]acetamide   Diphenylmethanethiol

+ Acetic acid +H₂O₂ ——————→

Modafinil

Fig. 5.17: Modafinil chemistry.

Purpose of this discussion is to review how the nuances of the reaction mechanism, chemistry and physical and chemical properties of a synthesis can be exploited to create a simple manufacturing process. It is difficult to say if these are exploited to the fullest with the current manufacturing practices. Since it is being produced at multiple sites [106], most likely they are not. Companies are also not able to capitalize on economies of scale and the product is being produced using existing process equipment [15, 16] that is too large for the process. This is generally the case when a process is fitted in the existing equipment.

    Table 5.18 illustrates some of the chemical and physical properties of the reaction components.

Tab. 5.18: Physical and chemical properties of modafinil reactants.

Chemical	Formula	CAS no.	Mol. wt	MP (°C)	BP (°C)
Benzhydrol	$C_{13}H_{12}O$	91-01-0	184	69	298
Hydrobromic acid	HBr	10,035-10-6	81	−11	122
Bromodiphenylmethane	$C_{13}H_{11}Br$	776-74-9	247	~35	184 °C (20 mm Hg)
Tetrahydrofuran	$(CH_2)_4O$	109-99-9	72	−108	65
Thiourea	CH_4N_2S	62-56-6	76	182	
KOH	KOH	1310-58-3	56	−29	132
Diphenylmethanethiol	$C_{13}H_{12}S$	4237-48-3	200		310
Chloroacetamide	C_2H_4ClNO	79-07-2	93.5	118	225
Acetic acid	CH_3COOH	1186-52-3	64	16	118
Hydrogen peroxide	H_2O_2	7722-84-1	34	−0.4	150.2
Modafinil	$C_{15}H_{15}NO_2S$	68,693-11-8	273	164	560

An example of laboratory modafinil synthesis outlined in USP '893 [89] is as follows and is analyzed further.

To a suspension of benzhydrol (35.00 g, 0.188 mol, 1 equiv.) and thiourea (17.40 g, 0.226 mol, 1.20 equiv.) in tetrahydrofuran (THF)/water (35.5 mL/52.5 mL), an aqueous 48% HBr solution (25.3 mL, 0.226 mol, 1.2 equiv.) over a 10 min period was added. During the addition, the reaction mixture is heated to 70 °C. After 3 h stirring at 70 °C, the uronium intermediate was hydrolyzed by addition of an aqueous 9.3 N potassium hydroxide solution (58 mL, 0.542 mol, 2.88 equiv.) over a 55 min period. After 1.5 h stirring at 70 °C, chloroacetamide (26.6 g, 0.282 mol, 1.5 equiv.) in a THF/water (80 mL/79 mL) solution was added over 15 min. After 1 h stirring at 70 °C, the reaction mixture was cooled down to 55 °C and the stirring was stopped. The lower aqueous phase was removed, and the reaction mixture was again stirred. Acetic acid (34.7 mL, 0.601 mol, 3.2 equiv.) was added. Hydrogen peroxide 30% (38.4 mL, 0.376 mol, 2 equiv.) was slowly added over 30 min. After 1 h stirring, the reaction mixture was cooled to 20 °C and water (263 mL) was added. The resultant suspension was stirred at 0 °C overnight. The suspension was then filtered, and the solid was washed with water and dried to yield modafinil (47.9 g, 80.4%). The crude modafinil was purified by recrystallization in methanol.

Table 5.19 compares the theoretical and laboratory mole ratios. Such comparison and review is of value and indicates the potential for process optimization. THF/water mix is used as a solvent. Sodium hydroxide can be an alternate for KOH [94].

Modafinil reaction chemistry and physical properties offer opportunities to exploit and simplify the process. Benzhydrol is solid at room temperature. By tradition,

Tab. 5.19: Modafinil stoichiometry comparison.

Chemical	Theoretical	Actual
Benzhydrol	1.0	1
Thio urea	1.0	1.2
KOH	1.0	2.88
Chloroacetamide	1.0	1.5
Acetic acid	1.0	2.0
H_2O_2	1.0	2.0

a solvent is used in the laboratory to facilitate the reaction. This is an acceptable practice. This would also be followed in scale-up in a pilot plant and on a commercial scale. Generally, between 25% and 40% solvent is used if a solid reactant is involved. This is also necessary for 1,500 L or above volume reactors. Solvent facilitates mixing and heat transfer but contributes to lower productivity. For higher productivity, their use has to be minimized.

In step 1 of Modafinil synthesis, a slurry of benzhydrol and urea in THF/water and hydrobromic acid are heated to 70 °C. An alternate [97] could be solubilizing thiourea in an appropriate solvent [35, 107, 108] with hydrobromic acid and reacting the mixture with benzhydrol to produce S-benzhydrylthiouronium bromide which is processed further to produce the modafinil.

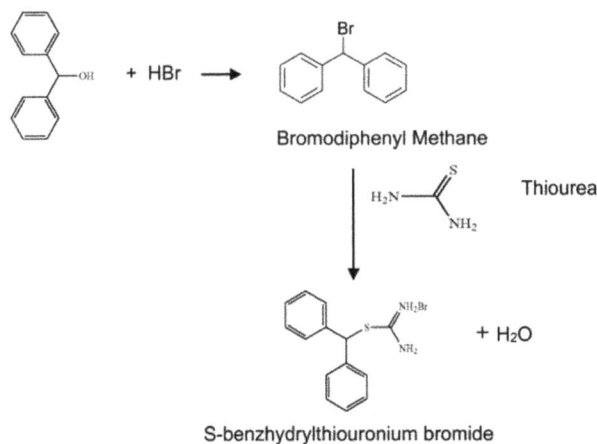

Bromodiphenyl Methane

Thiourea

S-benzhydrylthiouronium bromide

Fig. 5.18: Step 1 alternate process.

Another alternate for step 1 (Fig. 5.18) is reacting benzhydrol as a melt with hydrobromic acid to produce bromodiphenyl methane which is reacted with thiourea to produce

S-benzhydrylthiouronium bromide (Fig. 5.18). Since bromodiphenyl methane's melting point is about 35 °C, this can be used as a fluid in the process to reduce or minimize the use of THF/water or any other solvent in this step. Formation of bromodiphenyl methane also generates a mole of water. Thus, the solvent use can be reduced.

Reviewed alternates could significantly reduce the reactor size. The overall reaction time can be reduced if the reaction temperature is raised. In the annals of chemistry this is well known that raising the reaction temperature by 10 °C doubles the reaction rate [109].

5.10 Levothyroxine

Levothyroxine is an important drug for hypothyroidism. It is estimated that there are approximately 1 billion users globally. An average use of 112 µg/day per person would require about 41,000 kg to satisfy the global need. USP '295 [110] is reviewed.

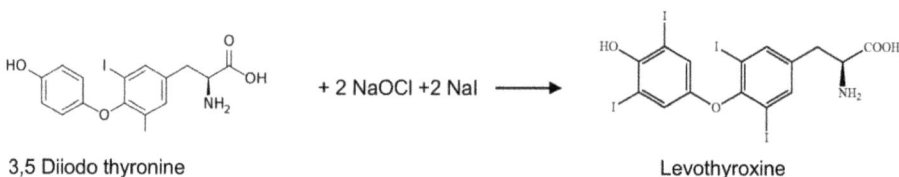

3,5 Diiodo thyronine   + 2 NaOCl +2 NaI ⟶   Levothyroxine

Fig. 5.19: Levothyroxine synthesis USP '295 [110].

This chemistry is an application of iodization processes developed using bleach and sodium iodide [111–113].

Step 1 of Example 1 of USP '295 [110] is as follows.

3,5-Diiodo thyronine (50 g, 0.095 mol) was added in methanolic methyl amine (250 mL) at 20–30 °C followed by addition of sodium iodide (49.99 g, 0.333 mol), and reaction mixture was stirred to get a clear solution. Aqueous solution of sodium hypochlorite [preparation: sodium hypochlorite (464 mL, 0.333 mol) in water (193 mL)] was slowly added in the reaction mixture and the mixture was stirred for 2–4 h. After completion of the reaction, the mixture was cooled at 5–10 °C and acidified to pH 4.0–5.0 using 50% hydrochloric acid (~22 mL). The reaction mixture was stirred for 30 min at 25–30 °C. The resultant solid was filtered and spray washed with methanol (50 mL). The obtained solid was dried at 50 °C under reduced pressure to get title compound (weight: 72 g, yield 97%).

In this step, it would be worth reviewing the sequence of addition of sodium iodide and bleach. It could improve the reaction and reduce the excessive use of iodide and bleach.

5.11 Status of the current manufacturing technologies

The history of pharmaceuticals discussed in Chapter 1 indicates that the development and commercialization of many small-molecule drugs had their origin in the late nineteenth and early twentieth centuries at fine/specialty chemical companies. Since there were/are chemistry and manufacturing similarities, equipment available at these companies was used to manufacture these APIs. The need to change manufacturing practices has not been considered necessary. As a result, manufacturing technologies between API and fine/specialty chemical manufacturing practices besides quality considerations have not changed much in over 100 years.

Another factor influencing lack of innovation is the price differential between API manufacturing cost and sale price illustrated in Tab. 5.20 and the drug selling price (Tab. 5.1). Innovation might be beneficial to the API manufacturers resulting in cost reduction: however, the average patient will never see any impact.

With many companies producing the same API at low volumes, there is lack of economies of scale. This prevents innovation [114].

The resulting lack of economies of scale benefits neither manufacturer nor user. Despite the lack of economies of scale and production inefficiencies, profits from API sales remain high compared to "average" fine/specialty chemical (~$2–10/kg). Indeed, so high that there is no driver for innovation in many/most cases.

Tab. 5.20: Factory cost and selling price of selected APIs.

API	Factory cost ($/kg)	Selling price ($/kg)
Mycophenolate mofetil	133 [115]	233
Dimethyl fumarate (Tecfidera)	2.23 (Tab. 5.4)	216 [29]
Modafinil	13.91 [24]	233 [106]

Under the current business scenario, brand API manufacturers have no price benchmark for their new molecules. Their drug selling prices are set to maximize their profits. The API price component of the drug selling remains low (Tab. 5.1). Thus, having efficient chemistry and excellent manufacturing technology does not matter to an average patient. It is up to each company to optimize the chemistry and commercialize the most economic manufacturing. Getting the product approved and to the market is the highest priority. Due to availability of the equipment and company's ability to fit the process in the existing equipment [15, 16], a need for innovation does not arise.

Generic API producers attempt to develop better chemistries and manufacturing technologies than the brand API. They have the time and they must compete against other generic companies. This is evident by many patents that are filed by the generic companies for the same drug. However, they still fit the chemistries in the existing equipment [15, 16], that is, lack of manufacturing technology innovation remains. This generally happens as it is difficult to justify investment for the new equipment for the better process. Production volumes also influence the justification [117].

Lack of manufacturing technology innovation begins from the onset of chemistry development. This happens at the brand as well as generic API producers as no one challenges the developed chemistry and the processing methods before the process is considered for scale-up and commercialization. Unchallenged processes result in translation of the developed chemistries to commercial processes even if they are inefficient and they will generally have high emissions for every kilo of product.

A chemist or chemical engineer on a scale-up team, after reviewing the chemistry and the process, assesses the available equipment in their organization and their capabilities so that the laboratory process after it has been scaled-up and tweaked in the pilot plant can be commercialized. Tradition of chemistry status quo and using the available process equipment continues.

The fine/specialty chemical sector from the start of a product commercialization uses batch processing as its preferred manufacturing practice until the product volume warrants a continuous process.

In the fine/specialty chemical industry, a robust process design is essential. This is due to competitive pressures. Producers cannot tolerate rework and disposal costs as they lower profitability. Every effort is made to minimize rework and waste. Process perfection is expected from the onset.

In pharma API, due to stringent product quality requirement, rework is avoided and the off-spec material is disposed. Repeat quality testing (quality by analysis, "QbA") even after a robust design due to batch processing is a normal practice. Testing, rework, and disposal costs are absorbed and passed on to the patients. This is obvious from the API factory cost and their selling price differentials (Tab. 5.20).

5.12 Innovation opportunities for API manufacturing

Reaction mechanism, physical properties, and chemical properties and their mutual behavior give us clues about how the reaction takes place. This information can be used for process design and simplification. Chemists and chemical engineers, as has been mentioned in earlier chapters, must understand the chemistry, value of physical and chemical properties of chemicals used and produced in the reaction, and their mutual behavior. Due to limitations of the available laboratory equipment, much of the aforementioned cannot be explored prior to pilot or production scale, although each has to be tested before it can be applied in a commercial process.

Their exploitation can result in using equipment of the appropriate scale and can improve the profitability of API producers. Such practices have been used in commercial operations by the fine/specialty chemical industry. With much lower profit margins, process optimization is mandatory to remain viable. Process optimization needs to be explored and incorporated in the API manufacturing sector also. All this has to be done before filing of NDA or abbreviated NDA documentation. Incorporation of any process changes can be a challenge especially after regulatory documents have been submitted.

Herein lies a major difference between specialty/fine and API manufacturing. Regulatory filing is required globally for production and sale of any chemical including an API. The cost of regulatory filing for a typical, non-API chemical is a small fraction of the cost for an API. Any process change in API manufacturing will require a reexamination of the drug's performance and efficacy with substantial added cost. This could also trigger the need to refile regulatory documentation with substantial added cost and time. Unless there is a large profit driver, API manufacturers will be reluctant to implement any process changes after an API is approved by government agencies. Without a profit-driver there is little motivation for innovation.

One of the biggest drawbacks or it could be called hesitation for inclusion of continuous improvement in API manufacturing is potential change in the performance and efficacy of the final product which can come about due to the process changes. Money needed to prove or assure that there is no change in the efficacy could negate any benefit. Since the API volume at each site is small, the need of process improvement is not considered a necessity. Thus, status quo is acceptable.

5.13 Nondestructive creation [118, 119] in API manufacturing

For innovation in pharma API manufacturing, it is necessary to look at the chemistries and their execution differently from the current practices. They get overlooked due to smallness of the needed API volume necessary to fulfill the demand (each kilogram API on 100% basis produces 1 million tablets at 1 mg). Currently available equipment gets used even when the equipment is inadequate for the process. Many companies producing the same API also suggests that inefficient processes are used for API manufacturing and they are most decidedly profitable.

This presents an opportunity to review the current practices and select the best alternate. A paradigm shift or some may call it "out of the box thinking" is needed. Nondestructive creation [118, 119] needs to be considered for API manufacturing. Pharma needs to shed its old ways [120]. If it happens, automatic consolidation of API manufacturing segment could result.

The concepts outlined here will have to be tested by the API manufacturers. Like any other change, considerations for alternate manufacturing methods and configurations could meet significant internal resistance [121]. "Not invented here syndrome" [122] can prevent any suggested change as the chemists and chemical engineers would be stepping out of their comfort zone. It is possible that companies, in the name of investment and without evaluating their benefits, may shrug from exploring alternate concepts.

Elements for nondestructive creation [118, 119] for the API manufacturing technology innovation exist but with the limitations of glassware, they are not recognized, reviewed, or tested. Speed to market and profitability of the current manufacturing methods also hinder. In the suggested methodologies, principles of science and engineering of mutual behavior of chemicals will still be used. They might be applied differently. Every change is science and engineering based. Ensuing results will be higher productivity, profitability, quality, and waste minimization.

A closer review of the reaction chemistry along with the physical properties of each reactant and intermediates suggest that one can capitalize on them and it is possible to simplify and create excellent manufacturing processes. Discussion here can be applied to every chemical synthesis. However, there are bottlenecks. They are the *traditions*, limitations of the labware to exploit physical and chemical properties and inability to manipulate the process unit operations. Speed to market and ready availability of the scale-up equipment are also a hindrance to innovation in API manufacturing and their formulations. With the exploitation of properties of the chemicals, it is possible to carry out reactions differently and thereby improve their productivity.

Unless there is a justification, no company will invest in any new equipment for every new or existing product. This is due to underutilization of the current assets [19–22] and as a result processes are fitted in the existing equipment [15, 16]. This is a norm in pharma API and fine/specialty chemical manufacturing.

Since the concepts outlined here are different from the current practices, they will need rigorous evaluation and they can be tested using the existing pilot plant equipment. Success would lead to adoption across other products and processes. Testing these alternates is not a waste of time but a learning opportunity for future use and ensuing value.

Pitfalls of the existing equipment need to be recognized:
- Too large for the process
- Same equipment is used for multiple products
- May not be the right equipment for the unit process and unit operation
- Auxiliary equipment and that includes utilities may not be the right size for the process

With the equipment that is available in the commercial operations, fitting the laboratory processes is a challenge. However, it is still used even when the equipment might not be adequate and not have adequate process controls and utilities. Processes musts be managed and can pose manufacturing challenges. Every such effort is not productive and can result in commercialization delays.

Commonly available equipment commercial equipment in API and fine/specialty for batch manufacturing can range from 1,500 to 30,000 L. Pilot plant equipment in API and fine/specialty chemical facilities varies between 100 and 300 L or smaller.

Some nondestructive creation [118, 119] ideas are presented. In the later chapter, how some of the discussed concepts can lead to continuous manufacturing of APIs are reviewed. They can be exploited and expanded to create simpler and economic processes. They can be easily tested in the current process development and manufacturing environment and applied to simplify commercial-scale manufacturing. Such exploitation might avoid any new investment, dreaded by most companies.

For "nondestructive creation" [118, 119], it is necessary to emphasize the fact that 1 kg of API can produce 1 million tablets, thus limited quantities of the product can serve the needs of many. In addition, the current practices besides having the drawbacks mentioned earlier have additional drawbacks. They are:

- Too much waste is generated and the long-term toxicological impact of the waste is not known [123, 124]
- Low asset utilization [19–22]
- Lack of manufacturing technology innovation

5.14 Modular plants

Productivity and efficacy of any product can be improved by carrying out the reaction differently. Modular plants offer that "out of the box" opportunity for pharmaceutical API manufacturing instead of the current sized commercial fixed equipment needs to be considered [125–129]. Their design is not different than the conventional plants. These plants would be like the erector sets [130] which can be assembled for the given process. Companies offer preassembled plants but the ideal would be that the companies assemble their own from the components. Principles of process equipment design that apply to conventional chemical plants will still apply, except the equipment would be smaller in size and can be conveniently used in multiple processing situations. Such plants are used in food processing industry. Modular processes are used in huge scale such as milk collection from dairy cows: 1 cow + 1 milking fixture = profit; 100 cows + 100 milking fixtures = more profit; 1 cow + 100 milking fixtures = less profit.

Since the modular plants are not in pharmaceutical companies' routine vocabulary, pilot plant equipment could be used initially to emulate "modular plants." An inventory of the existing pilot plant will assist in initial application and testing of "modular plant" concept in pharma. Once successful, the concept could be expanded. Every company will have to assess their own needs. Modular plants and their considerations are discussed in Chapter 6. Furthermore, there is an array of available simulation software to test modular concept in silico.

In this alternate process design, the equipment size would be between labware and an actual commercial plant currently being used. It may be very much like the size of pilot plant equipment. Actual size and process will depend on product demand and designer's creativity and imagination and how they exploit reaction mechanism and physical and chemical properties. Chemicals would move very quickly between the reactors and associated equipment. The actual process would become similar to continuous flow of materials, a continuous process. Schematically, the process could look like each block of Fig. 5.20.

For such a process to happen, one needs to consider that the chemistry of each reaction step is being executed separately, each reaction step product is the final product and that will be used immediately for the next reaction step. There will be no separation or purification of each product after each reaction.

Solvents used in various patents suggest that there are possibilities to simplify the process. When the process is scaled up, each reaction step must be reviewed individually but the solvent selection must be done collectively, limiting to one additional solvent besides water. Any solvent, other than water, will have regulatory, environmental, storage, and disposal concerns and associated costs. Recycling of solvent must be a necessary selection criterion.

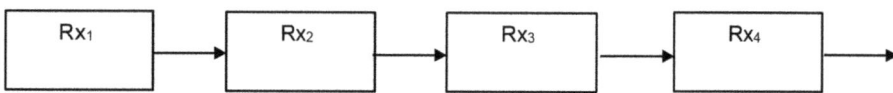

Fig. 5.20: Schematic of a simplified sequential process.

The range of reactor and accessory tank size determination will depend on the process designers. Such plants and their quick assembly can give any operation the needed flexibility and efficiency. Such modular equipment would have ways so that they can be assembled for flexible manufacturing. Instead of being the fixed equipment dedicated to certain processes, as is the case with the current equipment, the biggest advantage of such equipment and configuration would be the "right sizing" of equipment for different products.

The choice of modular reactors of optimum volume and size would have better mixing along with proper concentration of reactants and could increase the reaction rates. Better temperature control (heating/cooling) will also accelerate processing. Right sized equipment would also lead to reduction in solvent use. Since the reactants will move quickly between the reactors, under the right circumstances, this could lead to continuous manufacturing of APIs. Creativity and imagination of the chemical engineers and chemists would be put to test to create excellent processes. All this could improve throughput.

Static mixers [131] and plate and frame heat exchangers [132] have been used in the chemical industry to facilitate and improve mixing, augment process design, and improve reaction rates and mixing. These have been used for over 50 years in the fine/specialty chemical industry and can also be used in API manufacturing as the process equipment designs are same. Piping with static mixers can be part of the process design.

Microreactors [133] are also available. They are extremely high priced compared to static mixers and off-the-shelf available plate and frame heat exchangers or the electrically heated heat exchangers. Investment in these has to be justified on a cost/performance basis. In addition, compared to regular reactors their use may be limited making their justification difficult for across the board use.

The current "QbA" modus operandi to produce quality product could become history and "quality by design" could finally take hold for API manufacture. It is well known that continuous processes compared to batch processes have higher yield, lower cost, and better quality products. It is highly likely that continuous API manufacturing that has eluded pharma could finally take hold.

All of the above will give the companies advantages of economies of scale and using the right equipment to minimize excessive solvent use that is predominant in the current API manufacturing. Waste would also be reduced. Another benefit would be lower inventory of raw materials, in-process materials, and finished products. These result in significant financial gains from better inventory management, working capital, and cash flow.

5.14.1 Process design considerations

Modular plants will also provide the designers an opportunity to exploit the reaction mechanism and mutual behavior of chemicals to create excellent processes. In the current plants, their exploitation and incorporation can be a challenge. Each opportunity will have to be tested in the lab and pilot plant.

5.14.1.1 Solvent use/selection

Solvents act as a process facilitator. They function as thermal transfer fluids and by bringing reactants into reactive proximity. The optimum volume utilization of a typical reactor is 40–75% for proper mixing and thermal management. In a tradition inherited from fine/specialty manufacturing, API manufacturers will use existing reactors that are often too large based on reactant loading. With the 40–75% reactor filling requirement, this practice results in using more solvent than required.

At times, commercial processes also replicate laboratory experiments. Such situations, due to the time needed for heating and cooling the volume of liquid, can prolong the batch cycle time and generate more than necessary waste. If the solvent volume can be lowered for any reaction, process productivity will be improved.

Soluble and insoluble solvents along with their density difference and azeotropic behavior with water are of immense value. They have been reviewed in Chapter 3. It is necessary to understand the stability of the reaction intermediates. Many intermediates do not have long shelf life. Thus, their processing of the next reaction becomes timely. These result in short product cycles, that is, quick turnover of the equipment for better asset utilization. Every chemist and chemical engineer is familiar with these situations but their incorporation in a process still requires creativity and imagination.

In many of the API patents, intermediates are isolated at times, dried, and processed for the next step. Intermediate isolation may be acceptable for the laboratory but need to be totally eliminated in scale-up and definitely in a commercial process. Solvent selection should be such that the solid is soluble in the selected solvent. This will significantly lower the processing time if it is a batch process. For a continuous process, solid intermediate suspended in a solvent would be used immediately in the next step. Emphasis is to review and keep in mind how the reaction time can be lowered. The purpose of every process is to produce a saleable quality product versus have a process where intermediates become part of the in-process inventory.

5.14.1.2 Reaction mechanism and order of addition

Attention has to be paid to the reaction mechanisms. Every chemist knows that each reaction tells us how the reaction proceeds. If the reactants are not added to the equipment in the sequence suggested by the mechanism, the expected yields/results will not be achieved. A simple example is: acid added to water versus water added to acid. Results of addition, acid to water, can be controlled and temperature maintained versus water added to acid can result in dangerous eruption of the mixture due to an uncontrolled exotherm.

Some of the chemistries are reviewed. No process reviewed uses reaction vessels that are larger than 500 L. In each process step, physical and chemical properties of reactants and intermediates are incorporated. Stoichiometry and temperatures are precisely controlled. Chemical engineers, who are well versed in process design, will totally understand the value of the processes described.

In diazo reactions, an amine reacts with hydrochloric acid to produce the amine hydrochloride that reacts with sodium nitrite to produce the diazonium salt which is reacted further (Fig. 5.9). In USP '203 [45] and USP '537 [46], the diazo salt reacted with sulfur dioxide to produce chloro sulfonyl. This is insoluble in water. Generally, the suggested practice is to isolate it as solid for further reaction.

Alternate consideration would be to use a suitable solvent that will dissolve the sulfonyl chloride. The selected solvent should be such that it is water insoluble and its density is sufficiently different from water. The water phase would be decanted to reduce the volume of the reaction mass for the subsequent reaction. This works as it is used in commercial processes. Decantation might look like an extra step and it is, but is definitely faster than filtration, isolating the solid and subsequent dissolution.

Solvent can be added to the sulfation reactor and the mix pumped to the decanter. This significantly lowers the processing time. Figure 5.21 is the process schematic of a sulfonyl chloride process described above.

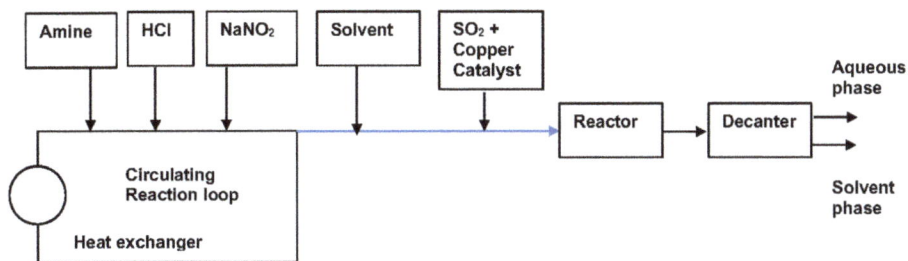

Fig. 5.21: Production of sulfonyl chloride of amine .

The chemistry described above would always be a challenge in conventional batch reactors due to many limitations, for example, large volume, residence time, and heat exchange needs. A batch process will also lower the conversion yield. Since the amine hydrochlorides are water soluble, amines can be slurried in water if they are not liquid at room temperature. The diazo exotherm can be controlled with proper heat exchanger design.

Equipment configuration similar to Fig. 5.21 can also be used for the synthesis of levothyroxine (Fig. 5.19), where 3,5-diiodothyronine is iodized using bleach and sodium iodide. Bleach can be prepared in line. This would be followed by addition of sodium iodide to produce levothyroxine. Since the global demand for the product is not large, a single continuous modular production site could be sufficient for the global need.

There are alternate methods available but the pharmaceutical industry, being bound by tradition, has not explored and/or practiced them. Some are reviewed. These processes do not use reaction vessels that are larger than 500 L. In each process step, physical and chemical properties of reactants and intermediates are exploited and incorporated. Stoichiometry and temperatures are precisely controlled. Chemical engineers who are well versed in process design will totally understand the value of the processes described.

5.14.2 Dimethyl fumarate

It is necessary to take advantage of methanol, thiourea [36], and maleic anhydride solubility [134] and temperature relationships (appropriate concentration solutions can be prepared continuously in a 100 L or smaller feed tank). No specific chemistry and stoichiometry is suggested as chemists and chemical engineers well versed in scale-up will understand. Reactors mentioned in Fig. 5.7 can be sized for sufficient residence time.

Chemical engineers well versed in process design can incorporate process controls to precisely manage the stoichiometry and temperatures. Global demand (Tab. 5.2) can be produced at ~ 200 kg/h in a single plant.

5.14.3 Metformin hydrochloride

Reactions like formation of metformin are an excellent opportunity to exercise and practice chemistry and chemical engineering. They test chemists' and chemical engineers' creativity and imagination. Based on review of the various patents related to metformin hydrochloride, one can conclude that the laboratory processes are used for commercial manufacturing and most are batch processes. With the needed volume, it can be produced using a continuous process.

Translation of the two routes (Figs. 5.12 and 5.13) presents us an opportunity to innovate. Route in Fig. 5.12 can be the route if dimethylamine hydrochloride is not available as a raw material and has to be produced for metformin synthesis. Raw material economics and logistics dictate this route. Xylene can assist in the removal of water using azeotropic property and density difference for the production of dimethylamine hydrochloride. Xylene can also act as a solvent for the fusion to produce metformin hydrochloride. Once the product is produced, it can be purified and a wiped film evaporator can be used to produce the final product. Figure 5.22 is a process flow diagram for this route.

An all melt fusion process (Fig. 5.13) can also be practiced. In this process, dimethylamine hydrochloride and cyanoguanidine in stoichiometric ratio can be reacted as a melt and processed with xylene as the purification solvent. Each of the processes would have to be tested, optimized, and scaled-up. One of the most critical parts of the design would be stoichiometry and temperature controls. Similar processes are commercially practiced [135, 136].

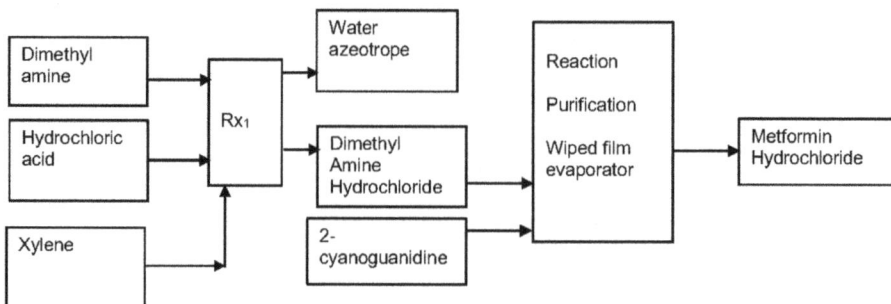

Fig. 5.22: Metformin route from dimethyl amine and HCl .

Either of the processes can be commercialized using modular plants. Based on the global need, continuous operation (~7,200 h/year) is definitely possible. It is very likely that a single plant would require multiple formulation lines to produce different dosage tablets. Technology for their formulations and their subsequent tableting exists but has not been used [137–139].

Process designers will have to evaluate the technologies themselves and apply their own creativity and imagination to design formulation and a tableting production train.

5.14.4 Omeprazole

Synthesis of omeprazole outlined in Fig. 5.14 is an ideal process design chemistry for most chemists and chemical engineers. This process design requires excellent application of process controls to feed various raw materials in precise ratios. The process could have inline mixing with back mix plug flow or small reactors with recirculating loops for the required residence time. Chemical engineers who have designed such processes would be familiar with the suggested or similar designs [140].

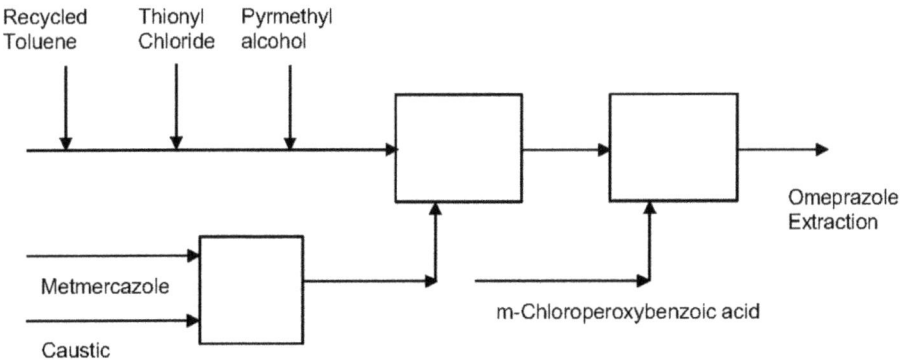

Fig. 5.23: Simplified flow for omeprazole .

Many alternatives [67–76] are available for the oxidation step. Chemists and chemical engineers can easily select one that would fit well in their process scheme and is economic and safe. With advancements in chemistry, they might find other alternate methods. Toluene from the last step would be decanted and azeotroped to yield wet toluene which facilitates the process [75]. Appropriate and precise process controls can be incorporated to control the stoichiometry. With the estimated yearly product demand (Tab. 5.11), a continuous process can produce global needs of omeprazole at a single site. Currently, it is being produced at multiple sites [81]. Figure 5.23 is a simple schematic flow diagram for omeprazole.

AstraZeneca in the development of omeprazole [67–76] did extensive work, reviewing many alternates. They can be used by the developers to explore simpler ways.

Since Nexium is the S-enantiomer of omeprazole, similar continuous processes can be commercialized. It will take multiple formulation and tableting parallel chains to convert the API into saleable tablets.

5.14.5 Metoprolol

This beta blocker drug is used worldwide. Its synthesis is another classical chemistry example. Based on its reaction mechanism, it is most suited for a continuous process. This observation is based on the fact that phenol reacts with caustic soda to produce the phenolate salt which is sequentially reacted with epichlorohydrin and the product is further reacted with isopropyl amine to produce metoprolol. Process engineers will have to test each step to use their operating conditions and parameters to design the optimum process.

In this chemistry, except for the final product, each of the raw materials and the intermediate are low melting solids or liquids. p-(2-Methoxyethyl) phenol would be slurried in sufficiently hot water in the prescribed stoichiometric ratio and produce an oil/water mix, (Tab. 5.15). This would be continuously fed to a reaction system (reactor or circulating loop with sufficient residence time) to produce the sodium phenolate which is reacted with epichlorohydrin. An excess of epichlorohydrin is suggested in patents [85, 141]. If the excess is necessary for the process, it can be removed due to its azeotropic behavior with water, recovered, and reused. Phase separation could also be used to facilitate processing.

Oxirane water slurry produced in reaction step 1 can be reacted with isopropyl amine in stoichiometric ratio to produce metoprolol which can be processed to produce the desired product. Figure 5.16 can be used as a simplified process flow diagram for metoprolol.

Use of modular plants or reactors that are 500 L could be ideal for processing. As stated earlier, they provide shorter processing time and with the control of stoichiometry quality product. Process control technologies and equipment to commercialize such processes are off-the-shelf items. They just have to be assembled to commercialize an efficient process. Operating conditions and stoichiometry will have to be optimized for an economic process.

5.14.6 Modafinil

As reviewed earlier, modafinil's overall annual demand volume (Tab. 5:17) is ideal for a single continuous production plant. Due to simplicity of its chemistry, it is most likely that it is being produced at multiple plants using existing equipment. With this being

the case, it presents an opportunity to further streamline the process. Production in a single modular plant would be beneficial due to the following benefits:
1. Relatively efficient process compared to normal plants of 1,500 L and higher reactor volume and the associated process equipment.
2. Reduced production time per kilogram. This will depend on how the process is organized.
3. Reduced solvent use due to using process equipment of appropriate volume.
4. Reduced waste from process and cleaning of the equipment.

Process chemistry and the reaction mechanism of Modafinil is ideal for a continuous process. Its reaction mechanism and the physical properties provide this opportunity. Short residence time will be of value. Process (reaction mechanism) could be broken down in steps and reviewed for the best processing options. A simplified process diagram for Modafinil synthesis (Fig. 5.17) is illustrated in Fig. 5.24.

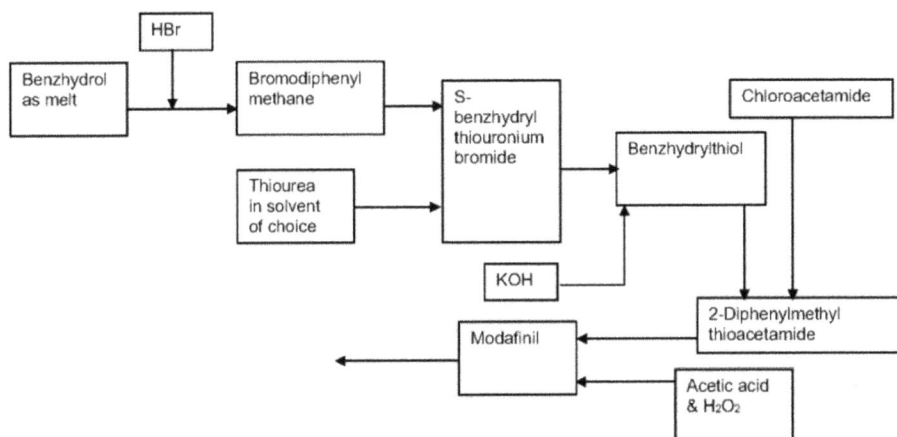

Fig. 5.24: Simplified Modafinil process flow diagram.

5.15 Future of technology innovation in API manufacturing

For the "REAL" innovation to happen in the generic API manufacturing, the landscape has to change. There are two alternate routes. In each case, outliers are needed.

"Nondestructive creation" [118, 119] would be an excellent alternate. As discussed earlier, modular plants through better processes and reduced waste will improve profits. Taste of improved profitability would lead generic companies to capitalize on economies of scale and higher market share. This could lead to additional innovations. This route presents excellent possibilities of waste reduction that is direly needed for the manufacture of API and has been ignored.

The other route "creative destruction" [142] would result in higher profitability for the API manufacturers and their formulators. It would involve selling FDA and other regulatory body approved drugs directly to patients using authorized channels [143]. This route is more disruptive and would face significant resistance from many camps such as pharmacy benefit managers, insurance providers, and even the governments and the regulators. This route should possibly bring manufacturing home to many countries and alleviate some of the resistance.

It is necessary to recognize that innovation comes from the companies who want to be in the market to serve a need. Innovators need to "Think Different" [144]. Outsiders, for example, regulators, who have NO experience of operating a profit-making organization should not be telling profit-making companies how to innovate and/or operate. It is a well-established and known phenomenon of every profit-making company that they develop, design, commercialize, and operate profit-making manufacturing plants. It is time for the profit-making companies take the reins in their own hands [step up to the plate (in baseball) or to the wicket (in cricket)] and take the lead to improve the practices of the last 75+ years and change the landscape.

Brand companies whose major focus is to get innovative drugs to the market could start with "nondestructive creation" [118, 119] to innovate API manufacturing and formulations.

Hesitation [145] to change is and will be there. Overcoming it is not only necessary, but it is critical to reduce the longer term impact on our ecosystem. Hence, further use of skunk works [146] might be very helpful. It may be a different territory for pharma but is worth exploring as the benefits are beyond imagination.

Abbreviations

cGMP	Current good manufacturing practice
HCTZ	Hydrochlorothiazide
THF	Tetrahydrofuran
mL	Milliliter
QbA	Quality by analysis

References

[1]   Unit Processes and Unit Operations: https://encyclopedia2.thefreedictionary.com/Unit+pro
      cesses Accessed November 5, 2020.
[2]   Perry, J. H., et al. Chemical Engineers' Handbook, McGraw-Hill Chemical Engineering Series,
      Fourth Edition, McGraw-Hill Book Company 1974 Accessed January 29, 2020.
[3]   Kirk-Othmer Encyclopedia of Chemical Technology, John Wiley & Sons, Inc. December 4,
      2000 Accessed January 15, 20202.

[4] Gadamasetti et al. Process Chemistry in the Pharmaceutical Industry, Taylor and Francis Group 2008 Accessed September 6, 2020.

[5] Malhotra, G.: Systematic Demystification of Drug Price Mystique and the Needed Creative Destruction, Profitability through Simplicity, https://pharmachemicalscoatings.blogspot.com/2019/10/systematic-demystification-of-drug.html, October, 2, 2019 Accessed November 19, 2020.

[6] Walker, D.: The Management of Chemical Process Development in the Pharmaceutical Industry, John Wiley & Sons. Inc. 2008.

[7] Malhotra, G.: The Good, the Bad, the Ugly (1)complexities of Pharmaceutical Manufacturing, Profitability through Simplicity, https://pharmachemicalscoatings.blogspot.com/2018/04/the-good-bad-ugly-1-complexities-of.html, April 9, 2018 Accessed November 1, 2020.

[8] Malhotra, G.: Chapter 4 "Simplified Process Development and Commercialization" in " Quality by Design-Putting Theory into Practice" co-published by Parenteral Drug Association and DHI Publishing© https://store.pda.org/TableOfContents/17296_TOC.pdf, February 2011.

[9] Malhotra, G.: Quick Review of Chemicals Related Process Development, Design and Scale up Considerations, Profitability through Simplicity, https://pharmachemicalscoatings.blogspot.com/2018/11/quick-review-of-chemicals-related.html, November 7, 2018 Accessed November 6, 2020.

[10] Malhotra, G.: Process Centricity is the Key to Quality by Design, Profitability through Simplicity, https://pharmachemicalscoatings.blogspot.com/2010/04/process-centricity-is-key-to-quality-by.html April 6, 2010 Accessed November 6, 2020.

[11] Malhotra, G.: Toward a Process-Centered Pharma, Contract Pharma, http://www.contractpharma.com/issues/2014-06-01/view_pharma-beat/toward-a-process-centered-pharma/ 16, 5, Pg. 46–48, June 2014.

[12] cGMP Regulations, https://www.fda.gov/files/drugs/published/Q7-Good-Manufacturing-Practice-Guidance-for-Active-Pharmaceutical-Ingredients-Guidance-for-Industry.pdf April, 2016 Accessed November 5, 2020.

[13] Anderson, N. G.: Practical Process Research & Development, Academic Press, April 9, 2012 Accessed September 5, 2020.

[14] Blacker et al.: Pharmaceutical Process Development: Current Chemical and Engineering Challenges (Drug Discovery, Volume 9) Royal Society of Chemistry; 1st edition (August 30, 2011).

[15] Malhotra, G.: Square Plug in A Round Hole: Does This Scenario Exist in Pharmaceuticals?, Profitability through Simplicity, https://pharmachemicalscoatings.blogspot.com/2010/08/square-peg-in-round-hole-does-this.html August 17, 2010.

[16] Malhotra, G.: Why Fitting a Square Plug in a Round Hole Is Profitable for Pharma and Most Likely Will Stay?, Profitability through Simplicity, https://pharmachemicalscoatings.blogspot.com/2014/08/why-fitting-square-plug-in-round-hole.html August 1, 2014.

[17] Malhotra, G.: Impact of Regulations, Manufacturing and Pharmaceutical Supply Chain (PBMs) on Drug Shortages and Affordability Part 2, Profitability through Simplicity, https://pharmachemicalscoatings.blogspot.com/2019/04/impact-of-regulations-manufacturing-and.html April 3, 2019.

[18] Berndt, E. R., Conti, R. M., Murphy, S. J.: The Generic User Fee Amendments: An Economic Perspective, NBER Working Paper 23642, August 2017.

[19] OEE Benefits for Pharmaceutical, BioTech and Medical Device Manufacturers in Life Sciences, https://www.pharmaceutical-technology.com/contractors/process_automation/zenith%20technologies/pressreleases/pressoee-benefits-life-sciences/ August 27, 2015 Accessed May 19, 2020.

[20] Benchmarking Shows Need to Improve Uptime, Capacity Utilization, https://www.pharmama nufacturing.com/articles/2007/144/ Sep 20, 2007, Accessed May 19, 2020.

[21] Malhotra, G.: Continuous Pharmaceutical Processes and Their Demands, Contract Pharma, https://www.contractpharma.com/issues/2016-04-01/view_features/continuous-pharmaceutical-processes-and-their-demands/?userloggedin=true, April 5, 2016, Accessed May 15, 2020.

[22] Tyson, T.: Solving the Industry's Capacity Utilization Problem, https://www.pharmasalma nac.com/articles/solving-the-industrys-capacity-utilization-problem, October 28, 2019.

[23] Malhotra, G.: Opportunities for Generic Pharma to Clear the Quality Stigma, https://pharma chemicalscoatings.blogspot.com/2019/05/opportunities-for-generic-pharma-to.html, May 23, 2019, Accessed April 25, 2020.

[24] Malhotra, G.: Chemical Process Simplification, John Wiley & Sons, Inc. 2011.

[25] Malhotra, G.: ONE PAGE Road Map to Reduce Drug Shortages, Assure Quality and Improve Affordability, Profitability through Simplicity, https://pharmachemicalscoatings.blogspot. com/2019/12/one-page-road-map-to-reduce-drug.html, December 6, 2019 Accessed November 11, 2020.

[26] Malhotra, G.: Pharmaceutical Quality: Concepts, Misconceptions, Realities and Remedies, Profitability through Simplicity, https://pharmachemicalscoatings.blogspot.com/2019/11/ pharmaceutical-quality-concepts.html, November 4, 2019 Accessed November 25, 2020.

[27] Narayan, S., et al. "On Water": Unique Reactivity of Organic Compounds in Aqueous Suspension, Angewandte Chemie International Edition, 2005, 44, 21, pgs. 3275–3279 Accessed October 23, 2020.

[28] National Psoriasis Foundation, https://www.psoriasis.org/psoriasis-statistics/ Update October 8, 2020 Accessed November 13, 2020 Accessed November 19, 2020.

[29] Dimethyl fumarate API Sites https://www.pharmacompass.com/active-pharmaceutical-ingredients/dimethyl-fumarate Accessed November 30, 2020.

[30] Guzowski et al. Biogen MA Inc. USP 9422226 B2, Process for preparing high purity and crystalline dimethyl fumarate, August 23, 2016 accessed June 10, 2017.

[31] Pullaguria et al. Biophore India Pharmaceuticals Inc. WO 2015/140811 A2, US 2017/009638 A1, USP 10636076 B2, An improved process for the synthesis of dimethyl fumarate April 20, 2020 accessed June 20 2020.

[32] Bhirud et al. Glenmark Pharmaceuticals USP 9840456 B2, Process for preparation of dimethyl fumarate December 12, 2017 accessed March 10, 2018.

[33] Raillard et al. XenoPort Inc. USP 9302977, Method of making monomethylfumarate, April 5, 2016 accessed June 10, 2017.

[34] de Souza, R. O. M. A., et al. Continuous-flow Synthesis of Dimethyl Fumarate: A Powerful Small Molecule for the Treatment of Psoriasis and Multiple Sclerosis, RSC. Advances 2020,10, 2490 Accessed January 16, 2020.

[35] Chapter 2, Table 2.4 Comparison of Selling Price as a Fine Chemical vs. an API and Number formulation plants needed.

[36] Shnidman, L.: The Solubility of Thiourea in Water, Methanol, and Ethanol, Journal of Physical Chemistry 1933, 37, 6, 693–700 Accessed December 13, 2020.

[37] Top 20 generic molecules worldwide, Fiercepharma.com, https://www.fiercepharma.com/ special-report/top-20-generic-molecules-worldwide, Nov 20, 2012, Accessed March 15, 2017.

[38] Nasim, A.: http://medicinal-chemistry-notes.blogspot.com/2015/12/synthesis-of-hydrochlorthiazide.html Accessed July 4, 2017.

[39] Novello, F. C. Merck & Co. USP 2965675 Aniline Compounds, Dec. 20, 1960 Accessed November 10, 2020.

[40] Schoenewaldt et al.: Merck & Co. Inc. USP 3161675, Process for preparing sulfonamide compounds, December 15, 1964 accessed June 10, 2017.

[41] Topliss et al. Schering Corp. USP 3326908 Novel Process for the Preparation of Dihydrobenothiadiazines June 20, 1967 accessed December 17, 2020.

[42] Ramakrishnan, A. et.al. Unichem Laboratories, WO 2007/026376 A2, A novel process for preparation of highly pure crystalline hydrochlorothiazide Accessed July 5, 2017.

[43] Deo et. al. Alembic Limited, WO 2009/150497 A1, A process for the preparation of highly pure hydrochlorothiazide Accessed July 5, 2017.

[44] Hydrochlorothiazide, https://www.pharmacompass.com/active-pharmaceutical-ingredients /hydrochlorothiazide Accessed December 17, 2020.

[45] Hart et al. Novartis AG, USP 7109203 B2 Sulfonamide Derivatives, September 19, 2006 Accessed March 12, 2013.

[46] Tonne et al. BASF Aktiengesellschaft, USP 4464537, Preparation of saccharin, August 7, 1984 Accessed January 2, 2016.

[47] United States Patent and Trademark Office, http://patft.uspto.gov Accessed January 5, 2021.

[48] Espacenet Patent Search, https://worldwide.espacenet.com/patent/ Accessed January 5, 2021.

[49] Bailey, C. J. Metformin: Historical Overview. Diabetologia 60, 1566–1576 (2017) Accessed December 10, 2020.

[50] International Diabetes Foundation, 9th Edition, https://www.diabetesatlas.org/en/sections/ worldwide-toll-of-diabetes.html Accessed December 20, 2020.

[51] API Plants: https://www.pharmacompass.com/active-pharmaceutical-ingredients/metfor min-hydrochloride Accessed December 20, 2020.

[52] Werner, E. A., Bell, J. (1922) The Preparation of Methylguanidine, and of ββ-dimethylguanidine by the Interaction of Dicyandiamide, and Methylammonium and Dimethylammonium Chlorides Respectively. Journal of the Chemical Society Transactions 121: 1790–1794, Accessed March 3, 2020.

[53] Timmins et al. USP 6031004, Salts of Metformin and methods Bristol-Meyers Squibb, Feb. 29, 2000 Accessed December 15, 2020.

[54] Patel et al. WO 2010/1466604 A2, Processes for Preparing Metformin Hydrochloride, Exmed Pharmaceuticals, December 23, 2010 Accessed December 15, 2020.

[55] KAMAVARAPU et al. WO 2016/059507 A1, IMPROVED PROCESS FOR THE PREPARATION OF HIGH PURE METFORMINE, April 21, 2016 Accessed December 15, 2020.

[56] Cain, J. C., Carboxylic Acids and Their Derivatives, T Pg. 147, He Manufacture of Intermediate Products for Dyes, Macmillan and Co. Ltd, 1919, https://books.google.com/books?id=SEFth-Xium8C&pg=PA147&source=gbs_toc_r&cad=3#v=onepage&q&f=false, Accessed June 10, 2019.

[57] Schlaudecker, G. F. USP 2668326, Process for the Preparation of Phthalimide, August 3, 1951, Accessed June 10, 2019.

[58] Vogel et al. USP 4001273 BASF Aktiengesellschaft, Continuous Manufacture of Phthalimide June 4, 1977 Accessed June 21, 2020.

[59] CN 100391939C https://patents.google.com/patent/CN100391939C/en, Accessed July 10 2020.

[60] Aarti Chemicals http://environmentclearance.nic.in/writereaddata/form-1A/2013_8_06_ Aug_2013_1035311301214_13.pdf Accessed August 19, 2020.

[61] Chava et al., WO 2014/041566 A2, Laurus Labs Pvt. Ltd. AN IMPROVED PROCESS FOR THE PREPARATION OF METFORMIN HYDROCHLORIDE, March 20, 1994 Accessed January 10 2020.

[62] https://www.3v-tech.com/en/process-equipment/21/evaporators/42/vertical-agitated-thin-film-evaporator Accessed December 2, 2020.

[63] https://www.technoforce.net/technology/agitated-thin-film-evaporator/ Accessed December 2, 2020.

[64] https://www.drugreport.com/50-commonly-prescribed-drugs-in-america/, March 23, 2020 Accessed December 26, 2020.

[65] https://www.statista.com/statistics/233986/top-us-pharma-products-by-prescriptions/, August 2020 Accessed January 10, 2021.

[66] Top 20 Generic molecules worldwide https://www.fiercepharma.com/special-report/top-20-generic-molecules-worldwide Nov 20, 2012, Accessed March 15, 2017.

[67] Junggren et al. USP 4255431 Aktiebolaget Hassle, Gastric acid secretion inhibiting substituted 2-(2-benzimidazolyl)-pyridines, pharmaceutical preparations containing same, and method for inhibiting gastric acid secretion, March 10, 1981 Accessed June 10, 2008.

[68] Brändstöm et al. USP 4620008 Aktiebolaget Hassle, Process for the Preparation of Omeprazole and Intermediates, October 28, 1986, Accessed June 10, 2008.

[69] Brändstöm, A., USP 5386032 Aktiebolaget Astra, Method of synthesis of 5-methoxy-2-[(4-methoxy-3,5-dimethyl-2-pyridinyl)-methyl]sulfinyl-1H-benzimidazole (omeprazole) January 31, 1995 Accessed June 10, 2008.

[70] Hoerrner et al., USP 5391752 Merck & Co, Inc. Process for the Preparation of Antiulcer Agents, February 21, 1995 Accessed June 10, 2008.

[71] Gustavsson et al. Astra Aktiebolag, USP 5958955 Method for the Synthesis of A Benzimidazole Compound, September 28, 1999 Accessed June 11, 2008.

[72] Milač et al., USP 6268502 LEK, Tovarna Farmacevtskih in Kemicnih Izdelkov. D.D., Process of synthesis of 5-methoxy-2-[(4-methoxy-3,5-dimethy-2-pyridyl)methyl]sulfiny-1h-benzimidazole, July 31, 2001 Accessed June 11, 2008.

[73] Prasad, K. B. Natco Pharma Limited, USP 6303787 Intermediates and an improved process for the preparation of Omeprazole employing the said intermediates October 16, 2001 Accessed June 11, 2008.

[74] Cotton et al. USP 6303788 B1, AstraZeneca AB, Process for Preparing Omeprazole, October 17, 2001 Accessed June 11, 2008.

[75] Anders, G., AstraZeneca, A. B., USP 7227024, Method for the Preparation of A Benzimidazole Compound, June 5, 2007 Accessed June 11, 2008.

[76] Anders, G., AstraZeneca, A. B., USP 7683178 B2, Method for the Synthesis of A Benzimidazole Compound, March 23, 2010 Accessed May 22, 2017.

[77] Broeckx et al. Janssen Pharmaceutica N.V. USP 6919459, Process for Preparing Benzimidazole-type Compounds June 19, 2005 Accessed May 22.

[78] Bhalerao et al. Novel Approach to the Synthesis of Omeprazole: An Antipeptic Ulcer Agent, Synthetic Communications, 40, 2010, Pgs. 2983–2987 Accessed June 11, 2012.

[79] Vojčić et al. Optimization of Omeprazole Synthesis: Physico-Chemical Steering Towards Greener Processes, ChemistrySelect 2017, 2, 4899–4905 Accessed June 20, 2020.

[80] Nirwan et al. Global Prevalence and Risk Factors of Gastro oesophageal Reflux Disease (GORD): Systematic Review with Meta-analysis https://doi.org/10.1038/s41598-020-62795-1 April 2, 2020 Accessed January 11, 2021).

[81] Omeprazole https://www.pharmacompass.com/active-pharmaceutical-ingredients/omeprazole-4594 accessed January 11, 2021.

[82] Malhotra, G.: Alphabet Shuffle: Moving From QbA to QbD – An Example of Continuous Processing, Pharmaceutical Processing, https://www.pharmaceuticalprocessingworld.com/alphabet-shuffle-moving-from-qba-to-qbd/ February 2009 pgs. 12–13, Accessed January 11, 2021.

[83] Anders, G., AstraZeneca, A. B., USP 7683178 B2, Method for the Synthesis of A Benzimidazole Compound, March 23, 2010 Accessed May 22, 2017.

[84] Metoprolol, Drug Usage Statistics, United States, 2008–2018, https://clincalc.com/Drug Stats/Drugs/Metoprolol, Accessed January 12, 2021.
[85] Palmer et al. AstraZeneca AB, Manufacturing Process for Metoprolol, USP 6252113 June 26, 2001 accessed August 1, 2013.
[86] Aguilar et al. Synthesis and Preparations of Metoprolol and its Salts, US 2009/0247642 A1 October 1, 2009 Accessed May 1, 2017.
[87] Gurjar et al. A New Route to Metoprolol Synthetic Communications, 20(22),3489–3496 1990 Accessed May 30, 2020.
[88] Mehra et al. IPCA Labs Ltd. Metoprolol Manufacturing Process, US 2005/0107635 May 19, 2005 Accessed March 14, 2017.
[89] Largeau, D., USP 6875893, Preparation of a Sulfinyl Acetamide, Cephalon Inc. April 5, 2005 Accessed January 23, 2006.
[90] Largeau, D., USP 7057069 B2, Preparations of a Sulfinyl Acetamide, Cephalon Inc. June 6, 2006 Accessed September 23, 2006.
[91] Lafon, L., USP 4066686 New Benzhydrylsulphinyl Derivatives, Laboratoire L. Lafon Janaury 4,1978 Accessed June 14, 2012.
[92] Lafon, L., USP 4098824 Benzhydrylsulphinyl Derivatives, Laboratoire L. Lafon July 4,1978 Accessed June 14, 2012.
[93] Lafon, L., USP 4177290, Acetamide Derivatives, Laboratoire L. Lafon December 4,1979 Accessed June 14, 2012.
[94] Naddaka et al., USP 6649796 B2 Process for the preparation of acetamide derivatives, Chemagis, Ltd., December 5, 2002 Accessed November 18, 2003.
[95] Baco et al. USP 6492396 B2 Substituted Thioacetamides, Cephalon Inc., December 10, 2002 Accessed November 18, 2003.
[96] Fornaroli et al. US 2004/0106829 A1 Process for the Synthesis of Modafinil, Procos SpA June 3, 2004 Accesses January 23, 2006.
[97] Liang,, Mallinckrodt, S., US 2006/0160903 A1, Process for preparing benzhydrylthioacetamide, July 20 2006 Accessed July 23, 2019.
[98] Bhatt et al. USP 7186860 Alembic Limited, Process for the preparation of 2-[(diphenylmethyl) thio] acetamide, March 6, 2007 Accessed June 14, 2012.
[99] Rose, S. US 2007/0015836 A1 Modafinil Synthesis Process, Organisation De Synthese Mondiale Orsymonde, January 18, 2007 Accessed July 23, 2019.
[100] Liang, S., USP 7244865 B2 Process for Preparing Benzhydylthioacetamide, Mallinckrodt Inc. July 17, 2007 Accessed July 23, 2019.
[101] Liang, S., USP 7345188 B2 Process for Preparing Benzhydylthioacetamide, Mallinckrodt Inc. January 31, 2008 Accessed July 23, 2019.
[102] Johnson et al., A New Method for the Preparation of Alkyl Sulfonyl Chlorides, Journal of the American Chemical Society 1936, 58, 8, 1348–1352, Accessed July 1, 2009.
[103] Sprague, J. M., et al., A Preparation of Alkyl Sulfonyl Chlorides from Isothioureas, II, Journal of the American Chemical Society 1937, 59, 10, 1837–1840, Accessed July 1, 2009.
[104] Kofod, H., Furfuryl Mercaptan An Improved Preparation Method, Acta Chemica Scandinavica 7, 1953, 1320–1306 Accessed July 1, 2009.
[105] Chatterjie et al. Anti-narcoleptic Agent Modafinil and Its Sulfone: A Novel Facile Synthesis and Potential Anti-epileptic Activity, Neurochemical Research, 2004 Aug;29(8):1481–1486, Accessed July 1, 2009.
[106] Modafinil https://www.pharmacompass.com/active-pharmaceutical-ingredients/modafinil, Accessed December 20, 2020.
[107] Kim et al., Solubility of Thiourea in C1 to C5 Alcohols, Journal of Chemical & Engineering Data, 1994, 39, pg. 228–230 Accessed July 1, 2009.

[108] Frank et al. The Preparation of Mercaptans from Alcohols, Journal of the American Chemical Society. 1946, 68, 10, 2103–2104 Accessed July 1, 2009.

[109] The Arrhenius Law https://www.chem.tamu.edu/rgroup/hughbanks/courses/102/slides/slides17_2.pdf, Accessed June 10, 2020.

[110] Deshmukh et al., USP 9932295 Process for preparation of levothyroxine and salts thereof, Lupin Limited, April 2, 2018 Accessed March 31, 2019.

[111] Kometani et al. An Improved Procedure for the Iodization of Phenols Using Sodium Iodide and Tert-butyl Hypochlorite, The Journal of Organic Chemistry 1985, 50, 25, pg. 5384–5387 Accessed October 16, 2020.

[112] Edgar et al. An Efficient and Selective Method for the Preparation of Iodophenols, Journal of Organic Chemistry. 1990, 55, 18, pg. 5287–5291 Accessed October 28, 2020.

[113] Eby et al. A Green, Guided-Inquiry Base Electrophilic Aromatic Substitution for the Organic Laboratory, Journal of Chemical Education 85 No 10 October 2008 Accessed October 15, 2020.

[114] Malhotra, G.: Pharmaceutical Quality: Concepts, Misconceptions, Realities and Remedies https://pharmachemicalscoatings.blogspot.com/2019/11/pharmaceutical-quality-concepts.html, November 4, 2019, Accessed June 6, 2020.

[115] Chapter Four Table 4.9.

[116] Mycophenolate Mofetil www.Pharmacompass.com,https://www.pharmacompass.com/active-pharmaceutical-ingredients/mycophenolate-mofetil Accessed October 12, 2020.

[117] Chapter Two Table 2.1, 2.2, 2.3.

[118] Kim et al. Nondisruptive Creation: Rethinking Innovation and Growth, MIT Sloan Review, February 21, 2019 Accessed March 6, 2019.

[119] Hubbard, G. Nondestructive Construction, TECH & INNOVATION https://www.strategy-business.com/article/07203?gko=dad6d, May 29, 2007 Accessed January 26, 2021.

[120] Malhotra, G., Pharmaceutical Manufacturing: Is It the Antithesis of Creative Destruction? https://www.pharmamanufacturing.com/assets/wp_downloads/pdf/Creative_Destruction_Malhotra.pdf, July 2008 Accessed June 20, 2020.

[121] Lawrence, P. R., How to Deal with Resistance to Change, Harvard Business Review, https://hbr.org/1969/01/how-to-deal-with-resistance-to-change, January 1969 Accessed January 27, 2021.

[122] Kenehan, S., Not Invented Here Syndrome explained, https://learnosity.com/not-invented-here-syndrome-explained/, Accessed January 29, 2021.

[123] Larsson, D. G. J. 2014 Pollution from Drug Manufacturing: Review and Perspectives. Philosophical Transactions of the Royal Society B 369: 20130571, http://dx.doi.org/10.1098/rstb.2013.0571, accessed February 4, 2021.

[124] Malhotra, G.: Pharmaceuticals, Their Manufacturing Methods, Ecotoxicology, and Human Life Relationship, Pharmaceutical Processing, November 2007 Pgs. 18–23 Accessed June 1, 2020.

[125] Klose,, et al. Chemical Engineering and Technology. 2019, 42, 11, pgs. 2282–2291 Accessed January 29, 2021.

[126] Is Modular Right for Your Project?, Chemical Engineering January 1, 2016 https://www.chemengonline.com/modular-right-project/ accessed December 18, 2020.

[127] Modular Plants Flexible chemical production by modularization and standardization- Status Quo and Future Trends https://dechema.de/dechema_media/modularplants.pdf Accessed January 28, 2021.

[128] Malhotra, G.: A Radical Approach to Fine/Specialty API Manufacturing, Profitability through Simplicity, https://pharmachemicalscoatings.blogspot.com/2010/01/radical-approach-to-finespecialty-api.html, January 20, 2010, Accessed December 10, 2020.

[129] Malhotra, G.: Considerations to Simplify Organic Molecule (API) Manufacturing Processes: My Perspective, Profitability through Simplicity,https://pharmachemicalscoatings.blogspot. com/2019/04/considerations-to-simplify-organic_77.html, April 20, 2019 Accessed December 10, 2020.

[130] Erector [http://www.meccano.com/products], Accessed December 10, 2020.

[131] Static Mixers https://komax.com/product-category/static-mixers/ Accessed January 29, 2021.

[132] Plate and Frame Heat exchangers, https://www.alfalaval.my/products/heat-transfer/plate-heat-exchangers/gasketed-plate-and-frame-heat-exchangers/heat-exchanger/how-plate-heat-exchanger-work/ Accessed January 29, 2021.

[133] Micro reactors, https://www.corning.com/worldwide/en/innovation/corning-emerging-innovations/advanced-flow-reactors.html Accessed January 29, 2021.

[134] Felthouse et al. Maleic Anhydride, Maleic Acid, and Fumaric Acid, Huntsman Petrochemical Corporation, April 26, 2001 Accessed October 1, 2013.

[135] Cain, J. C. The Manufacture of Intermediate Products for Dyes, https://www.google.com/ books/edition/The_Manufacture_of_Intermediate_Products/SEFth-Xium8C?hl=en&gbpv=1.

[136] Schlaudecker, G. USP 2668326 Process for the Preparation of Phthalimide, August 3, 1951 Accessed October 19, 2020.

[137] Continuous Manufacturing https://lbbohle.com/machines-processes/continuous-manufacturing/ Accessed February 4, 2021.

[138] https://www.gerickegroup.com/continuousmanufacturing Accessed February 4, 2021.

[139] Xelum Continuous Manufacturing https://www.syntegon.com/products/xelum-continuous-manufacturing Accessed February 4, 2021.

[140] Malhotra, G.: Alphabet Shuffle: Moving from QbA to QbD – an Example of Continuous Processing, Pharmaceutical Processing, February 2009 pg. 12–13.

[141] Baro et al. US 5,082,969, Industrial Process for Obtaining an Aryloxypropanolamine, Esteve Quimica S.A., January 21, 1991 Accessed June 20, 2020.

[142] Creative Destruction, https://en.wikipedia.org/wiki/Creative_destruction, Accessed April 26, 2018.

[143] Malhotra, G.: Improving Drug Affordability for the United States Populous through Alternate Business Models, https://pharmachemicalscoatings.blogspot.com/2018/05/improving-drug-affordability-for-united.html, May 4, 2018 Accessed February 10, 2021].

[144] Think Different, https://www.youtube.com/watch?v=5sMBhDv4sik Sep 30, 2013 Accessed February 10, 2019.

[145] Malhotra, G.: Hesitation In The Drive To A Continuous Pharmaceutical Manufacturing Process: Real or Imaginary? https://www.pharmaceuticalprocessingworld.com/hesitation-in-the-drive-to-a-continuous-pharmaceutical-manufacturing-process-real-or-imaginary/, July 7, 2009.

[146] Skunk works https://en.wikipedia.org/wiki/Skunkworks_project, Accessed February 12, 2021.

Chapter 6
Active pharmaceutical ingredients (API): innovation, design considerations, and waste reduction

> There is but one Earth, tiny and fragile, and one must get 100,000 miles away to appreciate fully
> one's good fortune in living on it.   – Michael Collins, NASA astronaut, command module pilot
> for the Apollo 11 Mission to the moon

Active ingredients of most drugs are small-molecule [1] chemicals. The chemical structures of these are similar to that of fine/specialty chemicals. Their manufacturing chemistries, processes [2], and equipment [3] are the same or similar and have been practiced and improved over the last 100 years. Their patent status classifies them as brand (under patent) or generic (outside patent).

Compared to fine/specialty chemicals, APIs are highly regulated. Either of the molecule, brand or generic, can be produced by any company that has the necessary processes, equipment, and staff to produce them.

Drug molecules that are under patent are generally manufactured or outsourced by the brand (inventor) company whereas the generic molecule could be produced by any company, and that includes outsourced manufacturing. Each manufacturer has to meet FDA's cGMP [4] regulations and requirements.

The process of continuous improvement and constant innovation are an integral part of human life. We have created products and their manufacturing processes that improve our quality of life. Like most animals, we do nothing when nothing is required. Unless provoked, motivated, or stimulated we do not create or innovate.

In the chemical/petrochemical landscape, pharmaceuticals having the highest "environmental factor" (E-factor) [5, 6] [Tab. 6.1] is a provocation. The E-factor is a ratio of the mass of waste produced/mass of product. A lower "E-factor" reflects a more environment-friendly or "greener" process. Greenness of API manufacturing processes has been ignored or valued less. It is on us and it is timely that we address environmental concerns, and do what we can and should do to lower/minimize the "E-factor" in pharmaceutical manufacturing.

Knowledge, methods, and processes that can simplify the API manufacturing and their formulation methods so that we can minimize the resulting emissions and continue to appreciate and maintain our good fortunes are well understood. They are known to us and, but are not fully practiced in pharmaceuticals. There are many reasons that include:
1. Profitability of the plethora of companies involved
2. Lack of our desire to leave a good legacy for our coming generations
3. Shrugging our responsibility to preserve and improve our environment

https://doi.org/10.1515/9783110702842-006

4. Regulations that have become an hindrance for continued improvements
5. Policies that interfere progress

It is well recognized that among the waste produced in different chemical-related industries, pharmaceutical industry stands out as the largest producer of waste. This waste is categorically defined as "E-factor." Pharma's "E-factor" [5, 6] has been anywhere between 25 and 100 kg waste per 1 kg of product. Tab. 6.1 Illustrates and compares the waste generated by different chemical-based industries.

Tab. 6.1: "E (kg waste/kg product)" factor [5, 6].

	Tonnes per year	E-factor (kg waste/kg product)	Total annual waste tonnage
Oil refining	10^6–10^8	<0.1	10,000,000
Bulk chemicals	10^4–10^6	<1–5	5,000,000
Fine chemicals	10^2–10^4	5–50	500,000
Pharmaceuticals	10–10^3	25 to >100	100,000

Of late, there has been significant discussion about "net zero" [emissions produced = emissions removed] [7]. In order to get to this goal, pharma needs to review its API manufacturing practices (active pharmaceutical ingredients) and their formulations.

It is noteworthy that in the chemical-based product category (Tab. 6.1), the one with the lowest yearly volume has the highest waste. Pharmaceutical manufacturing's waste has been a perennial problem [8, 9]. In reality, the pharmaceutical waste numbers would be much higher if the solvents were not recycled. In solvent recovery processes, distillation residue is generated. Its chemical composition and long-term impact on aquatic life or soil, even if properly disposed off, is not known. "Green chemistry" [10], of late, has become the vogue. Instead, if "good chemistry" is practiced, API manufacturing [11] and their formulations will transition to "green chemistry" [10] and move toward NET ZERO (carbon neutrality) [7] emissions. It can be achieved, but a concerted effort would be needed.

Inefficiencies in pharmaceutical manufacturing are well recognized [12, 13]. FDA's PAT Team and the Manufacturing Science Working Group Report [13] clearly states:

> Pharmaceutical manufacturing operations are inefficient and costly. Compared to other industrial sectors, the rate of introduction of modern engineering process design principles, new measurement and control technologies, and knowledge management systems is low. Opportunities for improving efficiency and quality assurance through an improved focus on design and control, from an engineering perspective, are not generally well recognized. For example, when discussions at the FDA Science Board and Advisory Committee for Pharmaceutical Science shed light on the current low efficiency and its cost implications (e.g., costs associated

with manufacturing can far exceed those for research and development operations in innovator pharmaceutical firms) many at FDA had difficulty understanding this and common reactions were "how could this be possible?" or "this can't be true." Regulators and many in manufacturing operations express their frustration by suggesting that manufacturing is a "step-child" in this industry, and that there is no economic motivation (e.g., cost and price difference) for improvement. Other suggestions include a general lack of systems perspective, organizational barriers that inhibit exchange of knowledge, and the attitude that much of pharmaceutical formulation and process development is an "art." Some in pharmaceutical development suggest that there are very limited opportunities ("development time crunch") to realize and/or demonstrate the level of science underlying current formulation and process development efforts.

FDA's 2004 statement still holds true. Reduction of waste in the manufacture of small-molecule API (active pharmaceutical ingredients) is extremely important, as the long-term effects of chemicals that end up in aquatic, soil, human, bird, and plant life are not totally understood. Since the short and long-term effects of chemicals that are produced as product and byproduct are not known and understood, it is best to minimize/eliminate their presence in every effluent [8, 9].

With inefficiencies (lower than optimum yield and waste) in pharma manufacturing being well-recognized [5, 6, 12–15], the issue needs to be addressed. A concerted effort is needed to minimize the discharge of organic chemicals in the environment, as a whole. By having an understanding of the reasons for this high level of waste, it is possible for the pharmaceutical industry to address the issues, and corrective measures can be taken. If done right, it could be possible to lower the levels of pharmaceutical waste to that of fine/specialty chemicals levels [5–50 kg/kg product] [5, 6]. This, or lower, waste could be set as a target. Some of the causes and methods to reduce waste are reviewed.

6.1 Why low yield and waste happen

These happen due to a combination of overlooking what the chemicals, their chemistry, their properties, their mutual behavior and reaction conditions, the reaction rate, and the equipment are telling us. Details of why the waste and why lower yields result have always been in front of us. Some recognize them and capitalize on them to take advantage; thereby improving profits and product quality.

Reaction chemistry and equipment are the two primary elements of each process. Generally, use of equipment that is available at a site is taken for granted. Every process has its own unit operation [3] (agitation, heat addition/removal, etc.) and unit process [2] requirements. Since most of the chemistries and unit operations can be fitted in the existing equipment [16, 17], unless it is critical for a product chemistry to have a hand in glove fitting, no one will invest in new processing equipment. This has been practiced for the last century and is the largest cause of waste generation and less-than-optimum yields. As explained later in this chapter, the selection/configuration of the existing equipment contributes to high solvent waste.

This is more true for active pharmaceutical ingredients, as a small quantity produces a large number of tablets (in a typical formulation, 1 kg API produces 1 million of 1 mg tablets] and serves the needs of many. Since processes can be modified to fit in the existing equipment [16, 17], no one will invest in new equipment for each API.

Process development starts at the initial evaluation of the chemistry. As has been discussed earlier and in Chapter 5, a village [18] needs to be involved from the onset of the chemistry development and its commercialization. If this does not happen from the start, low yields and high waste will occur. Another reason for the continued waste in manufacturing a product is the lack of financial return due to the limited volume of API per manufacturing site [19]. This has been discussed in earlier chapters.

6.2 Sources of pharma's waste

API manufacturing technology, unlike its older cousin, fine/specialty chemicals, has been stuck, for more than 50 years, in what can be labeled as "laboratory glassware technology/methodologies." These historical methodologies result in high solvent use and less-than-optimum yields. They are the acceptable norms of API manufacturing. The issues are recognized but not addressed. On the other hand, fine/specialty chemicals live with the process of continuous improvement.

Regulatory restrictions are a hindrance to the continued optimization of pharma manufacturing processes. Pharma companies, instead of tackling the issue directly at the inception, have been kicking the "manufacturing technology innovation can" to a later/future date. Brand companies, due to the limited patent life and the potential of altering of the product's efficacy and/or the performance, seldom venture in this area. As stated earlier, generic manufacturers do improve the synthesis and the manufacturing processes before they commercialize their products, but not enough to make a significant impact on waste reduction. This also happens due to the lack of economies of scale because too many [20] companies are manufacturing the same product. In addition, by kicking the can, pharma is also ignoring the increasing ecological impact it leaves behind, with every passing day.

6.3 Can pharma's yields and solvent use be improved to achieve net zero emissions?

In recent years, the discussion volume about low process yields and excessive solvent use has increased [8–11, 15]. Effort is needed to improve the yield and minimize the solvent use. It requires not only a different thinking but also "creative/imaginative" application of the fundamentals of science. Most of these have been discussed in earlier chapters but it would be beneficial to review them. API manufacturing, being the largest contributor to environmental pollution in the chemical industry, has to transition to *net zero* (carbon neutrality) [7] emissions. It can be done, but a concerted effort would be needed. Others can add their own criterion that benefit and simplify their process transition from bench to commercial scale [21].

6.4 Why has the pharma lagged in manufacturing technology innovation?

It would be beneficial to understand how and why the pharmaceutical API manufacturing industry has not made much effort to minimize its waste. It is equally important to understand the reason/s for why the issues still exist and why even after recognition of the problem, not much effort has been spent to lower/alleviate the problem. Major reasons are:

1. Laboratory processes are not optimum. If they are not optimized before NDA (new drug application, brand drug) filings, they rarely get optimized after the product is commercial. This is due to the regulatory hurdles and the associated costs.

2. After the patent expiration, the brand molecule becomes generic. Generic companies do make improvements in the brand molecule synthesis processes, but they are still not enough to make an impact on the total waste discharge. In addition, since many companies produce the API [20], values that could be generated from economies of scale are never achieved as their processes are not optimized.

3. Table [22] is an illustration of yield versus the number of steps for a process. For the brand drug producer, the priority is to commercialize the product, even if the manufacturing yield is 12% and the process has six steps. To a generic API producer, low yield matters, if their competitors for the same process have a higher yield. If they want to stay in the business, they have to put the necessary effort to improve their process yield.

4. To a brand or a generic API producer, in the final cost analysis, low yields are irrelevant. Their mission is to maximize their profits. Thus, the drugs are priced accordingly. Low yields or waste-related costs are a slight perturbation in the total financial analysis. Table 6.3 illustrates the factory cost and the selling price of the API for the drug, Tecfidera. Even if the factory cost of dimethyl fumarate is doubled or quadrupled, the cost impact would be about 5.00% reduction in the profit to the API seller, which in the overall analysis, may not be worth worrying about.

Tab. 6.2: Correlation of the yield and the number of reaction steps [22].

	Overall yield				
	Steps				
Yield per step	2	3	4	5	6
95%	90%	86%	81%	77%	74%
90%	81%	73%	66%	59%	53%
80%	64%	51%	41%	33%	26%
70%	49%	34%	24%	17%	12%

Tab. 6.3: Sample example of an API, its factory cost, selling price, and drug sell price.

	API cost ($/kg)	API selling price ($/kg)	Drug selling price, USA		
Dimethyl fumarate (Tecfidera)	1.71 [23]	216 [24]	Biogen (brand) List price	Mylan (generic) List price	14-Day supply is available from drug stores if there is a prescription.
30-Day supply			$5,463.00	$4,911.00	Price range $359.00–$1,579.00
API selling price contribution per 120 mg tablet = <3.00 cents					

If the API price goes up, the formulator will pass on their increased costs to the patients. The manner in which the middlemen in the supply chain react to the increased costs can influence the selling price to the patient, through increased copay or other means.

If there is a cost reduction due to manufacturing technology innovation, patients do not see the impact of such reduction. This is due to patients not knowing the real costs. Thus, the manufacturing technology innovation/s does/do not have any relevance to patients.

5. Active pharmaceutical ingredient (API) manufacturers, due to their chemistry and synthesis similarities to fine/specialty chemicals, in the last century, found an easy way to produce their products in similar equipment. Since the equipment, methods, and processes of the fine/specialty chemicals industry worked extremely well, processes were fitted in the existing equipment [16, 17]. No attempt has been made to develop or commercialize alternate equipment or processes methodologies to minimize waste. Traditions of the over hundred years are the standard modus operandi for the manufacture of most APIs and have been difficult to shed.

6. In most APIs, multiple and excessive amounts of solvents are used to achieve the needed product purity. Once the brand drug gets to phase III clinical trial, it

is difficult to alter/change the chemistry/synthesis process. It is possible that these chemistries may not be optimum.

7. Speed to get the product to the market is considered more important than the need to optimize the chemistries and/or the production processes. Optimizations are not considered necessary. This is due to the limited patent life after the brand drug approval. Revenue generated in the limited patent life far exceeds the monies that could be saved by process optimization or waste reduction. Since patients are most interested in getting the appropriate treatment, the price of the drug at times does not matter. This is especially true when patients are part of the mutually subsidized healthcare systems.

8. Adoption of the Hatch-Waxman Act [25], the WTO TRIPS agreements [26], and the lax environmental laws [27] of the developing countries [28] have also resulted in moving generic pharma manufacturing [29] from the developed countries.

All of the above and the regulatory restrictions to manufacturing process changes prevent continuous improvement of the API processes. With the API producer companies being profitable and due to the ever increasing global drug need, the need to innovate pharmaceutical manufacturing processes and technologies is not considered a priority. Once the product is commercial, except for following the cGMP [4] practices, not much improvement in most manufacturing technologies takes place.

6.5 Process steps: laboratory to commercialization

Steps needed from laboratory to the commercialization of a product have been reviewed in earlier chapters, but it is necessary to emphasize their value. These are also used to propose alternate processes that are simpler, produce quality products, and can quickly fill the drug demand in times of shortages. Schematics of the current process are illustrated in Fig. 6.1.

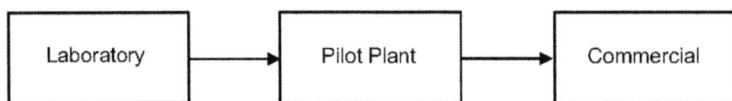

Fig. 6.1: Process commercialization process.

6.5.1 Laboratory development

1. Chemical equation: what are the reactants and products in each step.
2. Stoichiometric balance.
3. Definition of chemistry and the reaction parameters.

4. Proper and multiple commercial supplies for raw materials and solvents.
5. Solvent selection has to be such that it facilitates the reaction and needs to be limited to one additional solvent, in addition to water. More than two solvents add unnecessary complexity to the process.
6. Compilation of physical and chemical properties of the reactants used and produced. How can they be used in developing a commercially viable process?
7. Toxicity and safety of the chemicals used, produced, and the generated waste.
8. Cost comparison of the various synthesis routes and selection.
9. Proof and viability of the selected synthesis route.
10. Re-costing of the selected chemical synthesis route.
11. Laboratory equipment shows the pathway, but does not tell us how to translate the process in a commercially viable economic route.

All of the above (Items 1–11) are part of the exercise for the development of every fine/specialty chemical. Since the small-molecule active pharmaceutical ingredients are part of the fine/specialty chemical family, the above should be practiced for their commercialization also.

In the initial review, raw materials may be acquired from laboratory chemical supply houses to benchmark their prices and specifications. However, they should not be part of the lab development work as their prices and quality are far higher than the commercially available raw materials. Raw materials used for the laboratory development should be industrial grade, rather than pharmaceutical grade, as they do not represent reality. A word of caution about suppliers – they would like to know how and where their materials will be used, as use in pharma synthesis will increase prices. If the suppliers come to know about the use in API synthesis, rather than in fine/specialty chemical synthesis, prices go up. Suppliers have to meet customer specifications rather than where the product would be used. As has been suggested in earlier chapters, the knowledge developed and realized in the laboratory is helpful in process scale-up, commercialization, and trouble-shooting of processes.

6.5.2 Scale-up considerations

High "E-factor" [5, 6] and poor asset utilization [30] clearly suggest that pharma API needs to consider different operating philosophies to manufacture the APIs. Production volume at sites influences equipment and process selection. Chemists and chemical engineers practice this day in, day out for every product scale-up.
1. Can the solid reactants be solubilized, or slurried, or fed as a melt to the reaction vessel?
2. How can the physical properties of the reactants and reaction products be used to facilitate and/or simplify the reaction?

3. Can the reaction be carried out, following the reaction mechanism steps, thereby minimizing the solvent use?
4. Can the reaction kinetics be improved so that each reaction becomes a zero-order reaction [31]?
5. Can the alternate process operating parameters be tested in the laboratory or the pilot plant?
6. Can the whole process be carried out using reactors, or similar apparatus, or associated equipment that have ~ 400 liters or less volume per vessel?
7. Is the equipment available for commercial process, the right equipment that will minimize solvent use and maximize reaction yield?
8. Process centricity [32] has to be the key for commercialization of every process.

These considerations are used by most chemists and chemical engineers, but since the equipment between the lab bench and the manufacturing floor is standard, they are forced to use what is available. Lack of user flexibility of the available equipment is one of the major reasons for less-than-optimum processes. Their use is considered as an acceptable standard practice, but can also be the cause for high waste. We need to recognize that the large reactors force processes to use significantly large volume of solvent for sufficient agitation and heat transfer. Such inflexibility forces a high "E-factor" for most batch processes that are the mainstay of API manufacturing.

It may be time that pharma's API manufacturing needs a review of its practices. Three possible manufacturing routes are illustrated in Tab. 6.4 for every API.

Tab. 6.4: Possible API synthesis routes.

1.	Batch, as they are practiced
2.	Modular plant [33–36]: an alternate option for a batch or a continuous process discussed later
3.	Continuous

6.5.3 Equipment options

Selection of processing equipment and route depends on each company's business model for the involved products. As stated earlier, yearly production volumes dictate the routes.

Using laboratory glassware and process concepts, the operating conditions to be used in a commercial process can be defined. This is where chemical engineers, who are part of the village [18], would step in to create simple processes [37] that use minimum solvent (lower the "E-factor") and can be optimized for excellent yield.

Most companies are familiar with batch and continuous processes. Continuous processes are designed for specific products to produce a single product, about

8,400 h/year, with predefined (minimal) downtime for routine maintenance and unexpected downtime. This is an established definition and has been reviewed in earlier chapters.

For batch processing, processes are fitted in the existing equipment [16, 17]. Most of the fine/specialty chemicals, especially active pharmaceutical ingredients, are produced using batch processes in a tradition of more than 80 years. Fitting the chemistries in the existing equipment also necessitates the use of a large excess of solvents. Though these routes provide convenience, minimize the investment, and provide operating flexibility, they are the largest causes of high "E Factor" in the manufacture of APIs and equipment underutilization [30].

Use of modular plants for fine/specialty-chemical manufacturing is discussed occasionally. They are equipment assemblies for specific products or selected unit operations that can be used in a process. Unless there is an exact fit of unit operations in processes, their use would be no different than any fitting in a unit operation in an existing batch equipment assembly, leading to high "E-factor."

Currently, assembly of most of the modular plants (process and equipment design) is outsourced. This could be due to the lack of in-house expertise. If such assemblies do not have year-round use (8,400 h/year), they would be no different from any other batch process assemblies being used.

In order to have an excellent and economic process and consistently take advantage of in-house modular equipment assemblies, a total command and understanding of unit processes, unit operations, physical properties, and their mutual behavior are necessary. All these criteria are taught extremely well in our curricula, but their inclusion in process design, especially modular plants, happens through experience and application in actual processes.

In-house modular process equipment [33–35] would be assembled like an erector set [36] to create a viable process train to have an excellent process for the production of different products. However, the concept of modular assembly of chemical processing equipment is not very prevalent. This could be due to the fact that companies will have to have necessary manpower to assist in design and assembly of such processes. Most pharma API producing companies in today's environment and operating strategies may not have such prowess.

Traditions, easy availability of existing equipment where different processes can be easily fitted in the existing equipment, and uncertainty of product demand could be the additional reasons. As stated earlier, achieving low "E-factor" has never been a process consideration in API manufacture.

Using smaller reactor equipment, compared to conventional equipment, can lead to significant reduction in solvents needed. However, due to the low asset utilization [30] in the current plants, investment in additional equipment that could lower costs and improve product quality due to regulatory concerns is generally avoided. As has been discussed in different chapters, API costs have minimal impact on the selling price of the drug; thus manufacturing technology innovation is not a prime driver. Lowering the environmental impact can be the driver. Concepts discussed here can be used to review their application for the production of APIs.

Modular equipment would be used as plug-in devices to improve manufacturing processes. It is critical that chemists and chemical engineers have a total understanding of the physical properties [38] and their mutual behavior. They can be exploited to create simpler processes. It is best to review what could be the components of modular plants.

If modular equipment is not available, some of the pilot plant equipment, especially the reactors, due to their small size, can also be used and tested as modular devices. The same holds for other processing equipment. They provide a convenient place to test some of the concepts and also avoid any investment. Schematics of some of the concepts are illustrated. Chemical engineers, who are well versed in process design, will understand. The outlined concepts can be extended to any reaction.

In recent times, contract manufacturing companies have stepped in to produce many APIs. These plants are no different from the in-house plants. Manufacturing at these outsourced sites would be using batch processes, unless they have a long-term plant to produce the product using a continuous process that will operate about 7–8,000 h/year.

For modular plants, creativity and imagination would have to be unleashed, and it can lead to simplification of chemical synthesis processes. A critical element for the use and success of such plants is the internal intellectual capability and the expertise of each company, as the needs are very specific to each chemistry. Chemists and chemical engineers would have to be the outliers. "Can do" attitude has to be the modus operandi. They have all the necessary training and knowledge to reduce the "E-factor" [5, 6] of each reaction, and drive it to "net zero" [7]. In-house capabilities to assemble and disassemble equipment would be a must.

Some the applications are discussed in detail.

Following are some unit operation concepts where the available equipment can be used to improve chemical manufacturing processes. These concepts should not be limited to these equipment only.

– **Heat exchangers as reactors:** Reactor use in the fine/specialty chemical/pharmaceutical industry is an age-old tradition. Table 6.5 [39] illustrates the heat transfer area available for jacketed reactors. It is noteworthy that the surface area available decreases as the reactor size increases. Thus, to make sure that every advantage of the available surface is gained, excess solvent has to be used. This is a well-accepted practice. Generally, between 40–60% excess solvent is used for a good heat transfer and proper agitation.

It is easily possible to use heat exchangers as a substitute for reactors. They have been and are being used as reactors, as needed. Based on the physical properties and chemistry, it is highly possible to minimize the solvent use in reactions. Reduced solvent use would have to be included in the manufacturing process, and that can, in turn, significantly lower the "E-factor" [1, 2] for most pharmaceuticals.

Tab. 6.5: Available heat transfer area for different size reactors [39].

Liters	Surface area/volume (m^2/kL)
1,135.5	5.76
1,892.5	4.78
2,838.8	4.03
3,785.0	3.66
5,677.5	3.33
7,570.0	2.91
11,355.0	2.50
15,140.0	2.27

In general, shell and tube, and plate and frame heat exchangers are the external unit operations used to add or remove heat in any chemical reaction. Due to their limited space compared to a chemical reactor for the same function, they can provide distinct advantages in improving reaction rates and reducing the needed solvent.

Heat transfer is carried out using a cooling or a heating media. In both cases, the heating/cooling fluid supply link becomes an umbilical cord. This cord significantly lowers the portability of the equipment. However, electrically heated tubular heat exchangers used in semiconductor and electroplating industry [40] are available and can be used for many reactions in the chemical industry. They can, depending on the need, provide as much as 40 kW/sq. m. heat to the fluids [Table 6.6]. Compared to the conventional heat exchangers, they can be used as modular heaters and can facilitate reactions.

Plate and frame heat exchangers [41–43] can also be used. An integrated modular cooler/chiller can be used as a heat exchanger. These can simplify chemical processing. Tab 6.6 illustrates the range of heat load that can be provided using electrically heated heat exchangers

Tab. 6.6: Kilowatt heat per square meter [40].

Heater	Watt density		Power range (kW)		Heat transfer area (m^2)		kW/m^2	
	W/in.^{2}	kW/m^2	Low	High	Low	High	Low	High
Frontier	26	40.3	3	36	0.07	0.89	42.86	40.45

Figure 6.2 is an illustration of using heat exchangers (plate and frame or electrically heated tubular heaters) as reactors. Where and how the reactants get added simplify

the reaction times and the chemistry. This is difficult to achieve in large reactors. Smaller reactors and in-line additions provide the flexibility to capitalize on reaction mechanisms by allowing the sequential addition of reactants to maximize yield.

Electric-powered tubular heat exchangers, compared to jacketed reactors and auxiliary-powered (steam or heated liquid) heat exchangers, provide the opportunity as discussed above, along with more heat per unit volume of the reaction space. With a high concentration of reactants and higher heat flux, reaction rates can be accelerated due to the typical Arrhenius behavior [44] as well as a possible reduction of kinetic reaction order, from higher (≥ 1) to, at best, zero-order. This should lead to improving/lowering reaction times, improving conversions, and reducing waste.

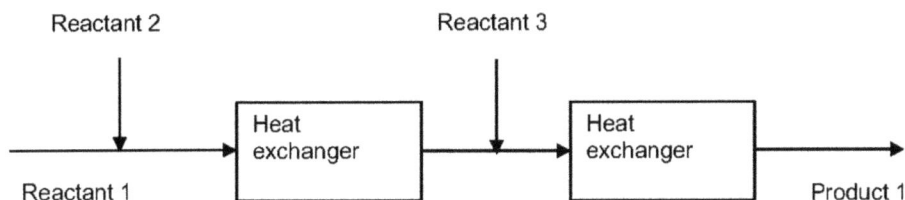

Fig. 6.2: Plate and frame or electrically heated tubular exchanger configuration as reactor.

-   **Screw conveyor/feeder:** Screw conveyors can be used to meter a solid in a reaction system. There are two ways these conveyors can be used. One method would be to use an electrically heated conveyor to melt the solid raw material in the reaction system, as illustrated in Fig. 6.3. Meter-feeding the reaction system will control the liquid. Used this way minimizes the needed solvent, which would be generally necessary in a conventional reaction system. An alternate way can be the total melting of the solid raw material that gets mixed with the reactant or solvent. Molten material can be metered in the process.

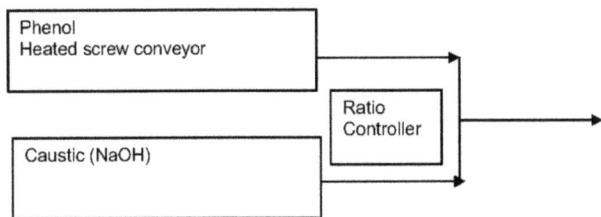

Fig. 6.3: Electrically heated screw conveyor with ratio control solvent (caustic).

-   **Use of Pumps as mixer/agitator and a temperature controller:** Reactors, generally, have an agitator to assure uniform mixing. Most of the time they work as expected. However, in many instances, feeding a reactant can have a localized exotherm, resulting in overheating and producing an undesirable product that can be a contaminant. This becomes an impurity, which cannot be removed. In such cases, pilot plant-sized reactors (60–100 liter) provide an easy and excellent method to eliminate the formation of such impurities. The suggested addition method (Fig. 6.4) also promotes immediate dispersion of the reactant, and a uniform and efficient reaction.

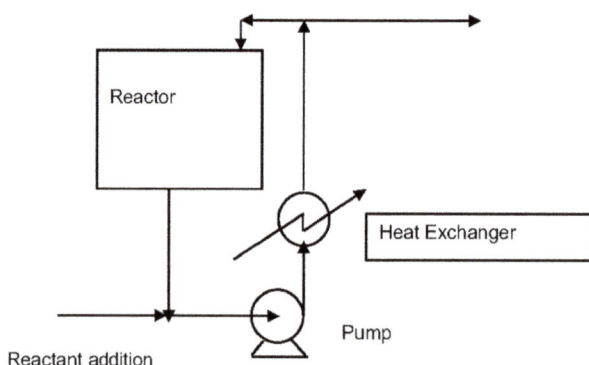

Fig. 6.4: Use of pump and heat exchanger for process control.

Such reactor systems can be skid-mounted and can be used where needed.

–   **Use of pump and screw feeder combination as a disperser of solid feed:** Adding solid feed to a reactor/tank that has solvent/liquid (cold or hot) or other reactant can be a challenge, especially when the solvent temperature is at near operating conditions. Even if it is not, the solid charging time extends the batch cycle. The solid charging process can be simplified by using an eductor as a feeder. Liquids, especially viscous, also can be charged in the venture, to maximize mixing. Solid dispersers [45] can be used to grind the solid in the solvent and facilitate reactions.

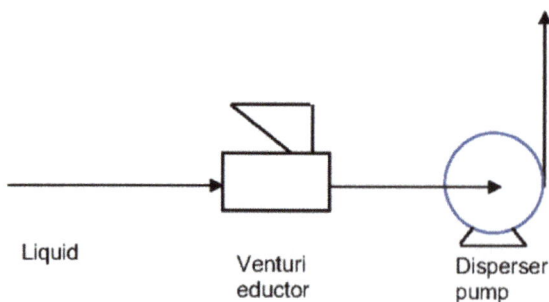

Fig. 6.5: Use of venturi eductor to change solids/liquids to a reaction system.

- **Decanter** [46]: There are two ways to separate water and organic liquids: azeotropic distillation, followed by gravity decanter, or just a gravity decanter.
  These are two common and the simplest ways. Each has its value. Azeotropic
  distillation is used to take out the water and solvent out of the reaction mass.
  Gravity decantation separates the liquids using differences in solubility and
  density. This is one of the simplest ways to separate the liquids. Properly designed decanters work brilliantly and are one of the simplest rendition of mutual insolubility and density differences to separate two liquids.

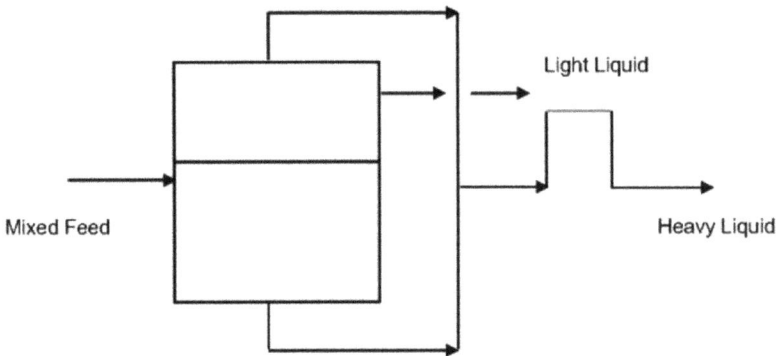

Fig. 6.6: Gravity separator (decanter).

The heavy-phase seal leg can be easily adjusted if the decanter shown in Fig. 6.6 is
used for the processing of different products. Engineers have incorporated process
controls for level and flow control. Density differences and settling time are excellent process controls and, generally, they minimize cost.

Solubility of the reaction products, along with decanters, can be exploited to
simplify processes. A hypothetical case would be the extraction of a desired product
from a process stream. Mixed feed, in Fig. 6.6, has the desired product in water
phase. In the subsequent reaction step, water is an undesirable phase. An organic
solvent can be added to the decanter feed and used to extract the desired product,
and be separated for further processing. This is a simple exploitation of the physical
properties of components of the reaction mass.

6.6 API examples

Using examples of few APIs, alternate operating philosophies are suggested. Some
of the equipment options discussed above are included. They can be modified to
suit an individual company's needs. For each process reviewed, it is necessary that
smaller pilot plant reactors, modular reactors, or heat exchangers that can be used

as reactors, be part of the design consideration. It is expected that these suggestions could significantly lower pharma's "E Factor" [5, 6] and improve asset utilization [30]. Both product quality and profitability, will improve through economies of scale.

6.6.1 Metoprolol

Of the many patents for metoprolol, two patents, USP 6252113 [47] and US 2009/ 0247642 A1 [48], were randomly selected and reviewed. Using the chemistry and processes, a potential simplified process is presented.

6.6.1.1 USP 6252113 [47] example

The process for preparation of 1-(2,3-epoxypropoxy)-4-(2-methoxyethyl) benzene (CAS 37350-58-6: metoprolol) is described.

p-(2-Methoxyethyl)phenol (~6.6 mol), epichlorohydrin (1.45 equiv.) and water (~2 kg.) were combined and the mixture heated to ~ 50 °C. Sodium hydroxide solution (50%; 1.4 equiv.) was added over 3 h and the temperature was elevated to reach approximately 60 °C during the addition. Formation of the title compound occurred during this period.

The batch was stirred for another 1 h at approximately 60 °C, cooled to approximately 50 °C, the phases separated, and the product was washed with water.

The residue was distilled at ≤190 °C at a pressure of ≤20 mm Hg and the distillate was collected. The yield of the title compound was 80% of the theoretical calculation. and the purity was 98%, according to GC analysis.

Metoprolol base:

1-(2,3-Epoxypropoxy)-4-(2-methoxyethyl)benzene (1 kg, 4.8 mol), isopropyl alcohol (~0.9 kg) and isopropylamine (0.8–1.7 kg, 3–6 equiv.) were mixed and reacted for 2–5 h at reflux. Formation of metoprolol base occurred during this period.

The reaction mixture was then concentrated at atmospheric pressure until the inner temperature reached ~100 °C. Water was added to the batch and then distilled off in vacuo until the inner temperature reached ~100 °C, to form a concentrate.

The resulting concentrate was diluted with isobutyl methyl ketone (~0.6 kg) and water (~2.2 kg), and concentrated sulfuric acid was added to adjust the pH to 4–6.

After separation, isobutyl methyl ketone (~1 kg) was added to the water layer and concentrated sodium hydroxide solution was added to adjust the pH to 13.

The organic layer was concentrated in vacuo at ≤ 80 °C until the distillation ceased, and the concentrated batch was redissolved in acetone (~1.6 kg) and filtered, to yield metoprolol base solution. The assay of metoprolol base in the solution was determined by titration. Yield: ~ 1.2 kg metoprolol base (100%) ~ 95% of theory. The purity of the metoprolol base was 96%.

The overall yield of metoprolol (USP 6252113) is ~ 76%. Laboratory process uses three different solvents (isopropyl alcohol, acetone, and isobutyl ketone). By using three different solvents, besides adding complexity to the process, their value or function is not clear. In addition, their recovery and recycling is difficult. The process described can be simplified. Alternate single solvent could be used.

6.6.1.2 US 2009/0247642 A1 [48] example

This patent describes the two following steps.

Step A: Preparation of 1,2-epoxy-3-(4-(2-methoxyethyl)phenoxy)propane

To a 400 L reactor containing 49.6 kg of deionized water, potassium hydroxide pellets (88.25%) of 7.93 kg (0.125 kmol) was added, while maintaining the temperature below 30 °C. The mixture was stirred until dissolution, and then 20 kg (0.131 kmol) of 4-(2-methoxyethyl)phenol was added. The reactor was then closed, inertized, and the mixture was stirred for 20 min, during which time, an opaline solution was obtained.

To the above mixture, 12.54 kg (0.135 kmol) of R,S-epichlorohydrin was added for over 30 min. The reaction mixture, which had two layers, was then heated to 35 ± 2 °C and kept at this temperature for 6 ±1 h. Thereafter, 0.41 kg (0.0064 kmol) of potassium hydroxide pellets (88.25%) and 0.38 kg of deionized water were added. The reaction mixture was then maintained at 35 ± 2 °C for 15 ±1 h. Thereafter, the water-epichlorohydrin mixture was distilled under vacuum until 160 L were collected and an orange-colored liquid residue was obtained.

Step B: Preparation of ± 1-(isopropylamino)-3-[p-(2-methoxyethyl)phenoxy]-2-propanol

The distillation residue obtained in Step A above was next cooled to between 0–5 °C and 62.4 kg (1.056 kmol) of isopropylamine was added for over 2–3 h, while maintaining the temperature below 15 °C, to yield an orange-colored suspension. The reaction mixture was then heated to a reflux temperature (50–55 °C) for over 40 min, and maintained at the reflux for 3 h. The reaction mixture was then cooled to room temperature (20–25 °C), and the reaction evolution was monitored by HPLC. Next, the reaction mixture was reheated and the excess isopropylamine was removed by distillation at atmospheric pressure, while not exceeding 70 ± 3 °C.

The obtained residue was next cooled to 20–30 °C; a vacuum pump was connected to the system and the reactor was heated again to continue the distillation (under vacuum), until reaching 70 ± 3 °C. The reactor was then maintained at this temperature for 30–40 min and an orange-colored oil was produced. Deionized water (7 kg) was then added to the obtained residue; the vacuum distillation was continued until a temperature of 70 ± 3 °C was reached, and the reactor was maintained at this temperature for 30–40 min.

Deionized water (30 kg) and toluene (38.2 kg) were then added to the resulting residue. The temperature of the mixture was then adjusted to room temperature

(20–25 °C), and the mixture was stirred for 30 min. The layers were allowed to decant for 30 min and were then separated. The toluenic layer was first washed with 15 kg of deionized water, followed by a second washing with 5 kg of deionized water and 0.22 kg of hydrochloric acid 35%, and was finally washed twice with 15 kg of deionized water. The resulting toluenic metoprolol base solution was weighed, and an aliquot was assayed for metoprolol base content.

6.6.1.3 Simplified metoprolol process

Chemistry of metoprolol synthesis is illustrated in Figs. 6.7 and 6.8. Table 6.7 illustrates the theoretical stoichiometry of the reaction. This can be used as a guide to economize the process. Excessive amount of reactants and solvents are being used, than necessary. Chemists and chemical engineers well versed in process development can create an excellent process.

Tab. 6.7: Theoretical stoichiometry of metoprolol synthesis.

Chemical	Theoretical
p-(2-Methoxyethyl) phenol	1.0
Epichlorohydrin	1.0
NaOH	1.0
Isopropyl amine	1.0

Process engineers and chemists familiar with the design of such processes will totally understand the simplicity of the synthesis. Processes outlined in USP 6252113 [47] and US 2009/0247642 A1 [48], as suggested, are complex and can be simplified.

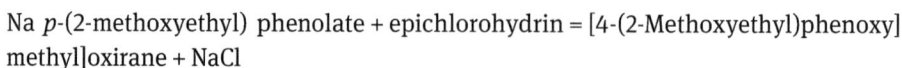

$$p\text{-(2-Methoxyethyl) phenol} + NaOH = Na\ p\text{-(2-methoxyethyl) phenolate} + H_2O$$

$$C_9H_{12}O_2 + NaOH = C_9H_{11}O_2Na + H_2O$$

Na p-(2-methoxyethyl) phenolate + epichlorohydrin = [4-(2-Methoxyethyl)phenoxy] methyl]oxirane + NaCl

$$C_9H_{11}O_2Na + C_3H_5ClO = C_{12}H_{16}O_3 + NaCl$$

[4-(2-Methoxyethyl)phenoxy]methyl]oxirane + isopropyl amine = metoprolol

$$C_{12}H_{16}O_3 + C_3H_9N = C_{15}H_{25}NO_3$$

Fig. 6.7: Metoprolol reaction chemistry.

Fig. 6.7: Metoprolol reaction chemistry [49] Using physical properties (Tab. 6.8) [50] of the chemicals used and produced, an excellent process can be developed and designed. Due to cost, sodium hydroxide would be preferred over potassium hydroxide.

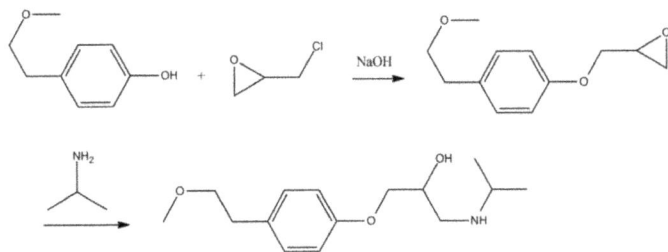

Fig. 6.8: Metoprolol synthesis [49].

Tab. 6.8: Physical and chemical properties of metoprolol chemicals [50].

Chemical	CAS no.	Chemical formula	Mol. wt	MP (°C)	BP (°C)
p-(2-Methoxyethyl) phenol	56718-71-9	$C_9H_{12}O_2$	152	42	125@ 3 mm Hg
Epichlorohydrin	106-89-8	C_3H_5ClO	92.5	−25.6	118
Caustic, 50%	1310-73-2	NaOH	40	12	140
[4-(2-Methoxyethyl)phenoxy] methyl]oxirane	56718-70-8	$C_{12}H_{16}O_3$	208	67	306
Isopropyl amine	75-31-0	C_3H_9N	59	−101	33
Metoprolol [1-[4-(2-Methoxyethyl)-phenoxy] -3-(isopropylamino)-2-propanol]	51384-51-1	$C_{15}H_{25}NO_3$	267	120	398

In metoprolol synthesis, sodium salt of p-(2-methoxyethyl) phenol is produced by reacting p-(2-methoxyethyl) phenol with commercially available caustic soda. It is subsequently reacted with epichlorohydrin and isopropyl amine to produce metoprolol, which is converted to its succinic salt, since toluene/water azeotrope, toluene would be an ideal solvent in the synthesis. Production of metoprolol succinate is a single-step synthesis, followed by crystallization. If tartrate is to be produced, it can be done similarly. Based on global demand [51], continuous crystallization process can be developed and commercialized. Technologies exist and have to be selected. Multiple formulation lines would be needed to fill the various dosage needs.

With the estimated global demand of ~ 1.4 million kg/year [51], about 51 plants [52] currently producing the API. The most likely batch production is the current preferred method. Each batch plant would be producing about 28,000 kg metoprolol per year.

Due to lack of economies of scale, most of these plants, compared to high yearly production rate plants (they could be two or three continuous process plants), will be inefficient and have their own waste.

For the synthesis of metoprolol, due to the simplicity of the chemistry and so much idle equipment at most sites, it should not be difficult for a company to repurpose the available equipment to create and commercialize a simple continuous manufacturing process. Thus, investment for a dedicated new equipment might not be needed. Such repurposing opportunities are reviewed.

Simplicity of chemistry and the estimated demand suggests that between one or two continuous plants can easily fulfill the global demand at 100–200 kg/h rate and operating about 7,500 h/year. Such plant/s, due to economies of scale, would have an efficient process and, most likely, will have about 30% of the waste generated by the 51 plants. The continuous process factory manufacturing cost will be significantly lower than the batch process. This would come through from economies of scale, purchasing power of the raw materials, inventory management (cash flow), and efficient manufacturing plant/s. In addition to fulfilling the global demand, the production rate can be modulated in a continuous plant to meet the changing API demand.

Figure 6.9 is a schematic of the metoprolol reaction and reviews the process equipment options that are a combination of the available conventional equipment and the equipment that is conventionally not used in the chemical industry. Due to the smaller size of the pilot plant equipment, the concept of solvent reduction can also be tested.

In Fig. 6.9, Rx1 and Rx2 are electrically heated tube heat exchangers [40] that can be used for the reaction process. These would be used as modular reactors and can be used for the synthesis of other products also. If electrically heated tube heat exchangers are not available, use of smaller, between 75 and 150 L or smaller jacketed reactors, could be evaluated. The idea is to minimize the solvent use by minimizing the reaction mass fluid. In addition, the heat flux to the liquid has to be maximized. This can hasten reaction times and reduce the batch cycle times, if batch process is selected for manufacturing.

The following are some of the design considerations for the modular metoprolol process. This process, if properly designed, can significantly reduce solvent use. It is highly possible that a properly designed process could be operated as a continuous process to fulfill the global metoprolol API demand. A solvent would be needed and it can be selected by experimentation. It would be necessary to generate solubility data for the selected solvent.

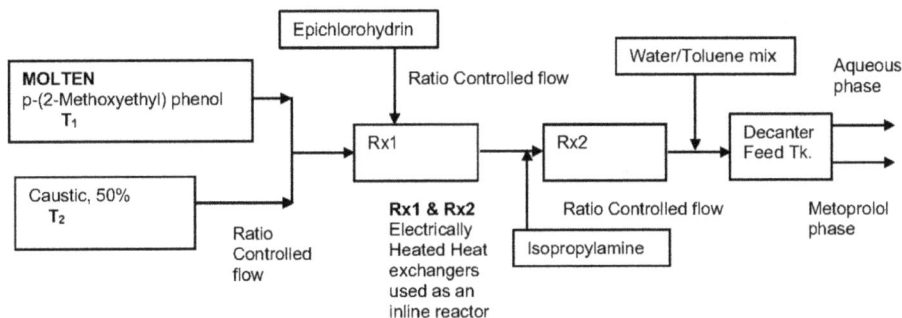

Fig. 6.9: Metoprolol simplified manufacturing process schematic.

1. Due to the low melting point of *p*-(2-methoxyethyel) phenol, it can be used as a melt; thereby reducing/minimizing the need for a solvent. Liquid phenol will react with caustic to produce the corresponding sodium phenolate, which can be further reacted with epichlorohydrin in an electrically temperature-controlled tubular heat exchanger to act as a reactor. The resulting product can similarly be reacted with isopropylamine in an electrically heat-controlled tube heat exchanger, acting as a reactor. Since the intermediates are low melting point materials, an opportunity to minimize the solvent use exists and needs to be explored. Temperature/solubility relationship of each step's reaction product will have to be generated. Similarly, reaction and the solubility parameters for the epichlorohydrin derivative will also have to be generated, and the properties used in the reaction process.

2. Since solvent use can be minimized, the reaction can be carried out at high temperatures. This would result in a zero-order reaction, thereby minimizing the reactor volume needed. Safety cannot be compromised. Instead of using a conventional reactor, an electrical heat-controlled tubular heat exchanger [40] can act as the reaction space. Use of an electric tubular heat exchanger, as a reactor, minimizes the large reactor space that, in conventional reactors, has to be filled with a solvent for proper mixing and heat transfer. If such heat exchangers are not available, small jacketed reactors need to be evaluated. Reaction residence time will determine heat exchanger configuration.

3. Isopropyl amine has a low boiling point and it can be used to improve mixing, thereby improving the reaction rate.

4. A water/toluene mix and a subsequent decanter [46] use will simplify the process further.

5. In the patents reviewed, intermediates are purified. That may be an acceptable practice for the laboratory process development, as their property information is an indicator of the in-process quality. This can be used to optimize the process and is necessary for an economic and viable process. Such processes, if commercialized, would have a low "E-factor."

Depending on the production capacity, *p*-(2-methoxyethyl) phenol for the metoprolol process has three feed options.

1. *p*-(2-Methoxyethyl) phenol could be received as liquid in a tank truck.
2. *p*-(2-Methoxyethyl) phenol, received as solid, could be fed using options sug-
    gested in Fig. 6.3 or 6.5.

6.6.2 Hydrochlorothiazide (HCTZ)

Synthesis of hydrochlorothiazide, Fig. 6.9, is another classic text book chemistry.
With its estimated global need of about 0.7 million kg/year [53], its production at
about 50 plants [54] suggests that the processes have been fitted in the existing plants
[16, 17]. This, most likely, has happened not due to the capabilities of the chemists
and chemical engineers but due to business model of the different companies and
the lack of processing options available.

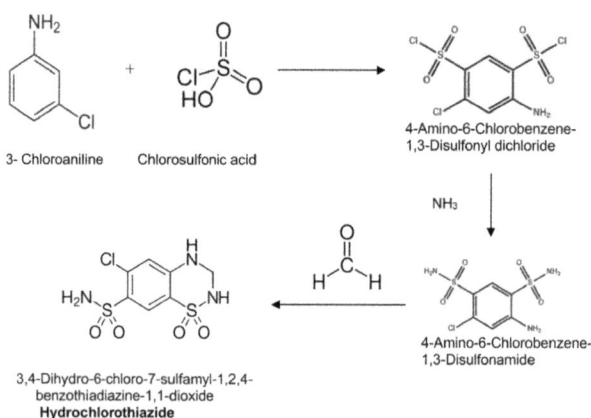

Fig. 6.10: Synthesis of hydrochlorothiazide [55].

Following are the steps outlined in the USP 2965675 [56].

Step 1: *m*-Chloroaniline (64 g, 0.5) is added dropwise, with stirring, to 375 mL
chlorosulfonic acid in a 3 L, round bottom, three-necked flask, cooled in ice bath.
Sodium chloride (350 g) is added, portion–wise, over a period of 1–2 h and the mixture is then heated gradually in an oil batch to 150 °C. After 3 h, at 150–160 °C, the
flask is cooled thoroughly in an ice bath and the contents treated with a liter of cold
water. The product is extracted with ether, and the extract washed with water and
dried over sodium sulfate. After removal of ether on the steam bath, the residual
5-chloroaniline-2,4-sulfonyl chloride is crystallized from benzene–hexane, MP
130–132 °C.

Step 2: The product from step 1 is cooled in an ice bath, treated with 150 mL of 28% ammonium hydroxide solution, and the mixture is heated to produce 4-amino -6-chlorobenzene-1,3-disulfonamide. It is crystallized.

Step 3: 88 g of 4-amino-6-chlorobenzene-1,3-disulfonamide is treated with 1.1 liter of formic acid, under reflux, to produce 6-chloro-3,4-dihydro-2H-1,2,4-benzothiadiazine -7-sulfonamide 1,1-dioxide (HCTZ). HCTZ would then be crystallized.

In other patents [57, 58], formaldehyde is added, twice, to complete the reaction. These syntheses are developed in the laboratory to show feasibility. Such intermittent additions are disruptive to the actual plant manufacturing process, if commercialized. They also prolong the batch cycle time, and also lead to frequent sampling of in-process material to check intermediate product quality and the reaction progress. These have to be minimized and avoided.

HCTZ chemistry needs to be streamlined and revised for a viable and economic commercial process. Information from the patents [55–58] and Tab. 6.9 [59] can be used to develop and define the operating conditions. The following need to be considered.

1. Operating conditions and methods mentioned in the patent show the feasibility of the chemistry, but they may not be the optimum processing conditions for the equipment to be used. They will have to be validated and refined for the operating conditions.
2. Laboratory operating conditions and process would have to be tested in the equipment to be used, to assure that the process uses minimum and safe solvent, which can be easily recycled. Solubility data would have to be generated for the selected solvent.
3. How can the solvent use be minimized and kept to a single solvent?
4. Stoichiometry suggested in the USP 2965675 [56] needs to be reviewed, as it suggests excessive amount of chlorosulfonic acid. Excess will have to be neutralized and disposed as waste.
5. Depending on the volume of the product to be produced at a site, equipment selection becomes critical.
6. Purification process would need to be developed and tested.

Physical properties Tab. 6.7 give clues about the process simplification.

Figure 6.12 is a suggested simple process schematic for the production of HCTZ. Reaction products after each step are solids at room temperature. Solvent selection and reaction conditions need to be defined.

It may be possible to carry out the first reaction step as a melt and maintain the temperature at about 135 °C to keep the reaction mass liquid. Use of sodium chloride, USP 2965675 [56], suggests that the inventors were trying to use the melting point of the disulfonyl dichloride as the reaction media. Preferably, a single solvent of high boiling point, instead, could be used as a solvent for the whole reaction. It would be added to the liquid from the first reaction step to assure a fluid mass

Tab. 6.9: Physical properties of chemicals for hydrochlorothiazide [59].

Chemical	CAS	Chemical formula	Mol. wt	MP (°C)	BP (°C)
3-Chloroaniline	108-42-9	C_6H_6ClN	127.5	−10.4	95–96
Chloro sulfonic acid	7790-94-5	$ClHO_3S$	116.5	−80	151–152
4-Amino-6-chlorobenzene-1,3 disulfonyl dichloride	671-89-6	$C_6H_4Cl_3NO_4S_2$	324.5	130–132	474
Ammonia	7664-41-7	NH_3	17	−78	−33
4-Amino-6-chlorobenzene-1,3-disulfonamide	121-30-2	$C_6H_8ClN_3O_4S_2$	285.5	257–261	
p-Formaldehyde	30525-89-4	$OH(CH_2O)_nH$ ($n = 8-100$)	90	120	
Hydrochlorothiazide	58-93-5	$C_7H_8ClN_3O_4S_2$	297.5	273	

when liquid ammonia is added to the process, to produce 4-amino-6-chlorobenzene-1,3-disulfonamide, which is a high melting point solid.

For amination, the use of liquid ammonia will significantly improve the reaction rate. It will also free up the reactor volume that would have been occupied by water if ammonium hydroxide solution is used. Ammonia addition rate can be used to control the heat of exotherm. Sulfonamide, from step two, reacts with formaldehyde to produce HCTZ, a solid at room temperature. Formaldehyde reaction can also use an electrical heat exchanger, reviewed earlier [40], with proper residence time as a reactor.

An experienced chemist and chemical engineer will totally understand the reaction mechanics and would do everything possible to have a simple process, similar to the one illustrated in Fig. 6.12. Purification process for HCTZ would have to be developed and tested.

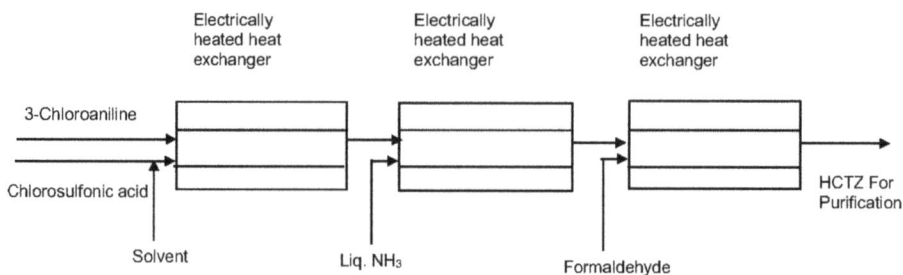

Fig. 6.11: Process for HCTZ manufacture.

Laboratory equipment can be used to test each individual step of the process concept outlined above. It will also allow the development of processing conditions that would need to be tested and improved.

Use of pilot plant would be necessary to authenticate the above outlined concept. Some of the equipment that have been discussed earlier should be considered to reduce the solvent use. Kinetics of each reaction step can be managed using electrically heated tubular heat exchanger or small jacketed reactors. These could significantly reduce the total volume of the reaction space if the product can be campaigned or produced continuously using modular equipment. Reduced reaction space would translate to reduction in solvent use.

Modular equipment should be considered for commercial production. If such equipment is not available, consideration should be given to using pilot plant equipment that is much smaller compared to traditional reactors, as modular reactors/equipment. Its use would reduce the solvent needed for the process. It might lead to a continuous process to produce the global demand at a single site. Multiple formulation lines would be needed to produce different dose tablets.

6.6.3 Dimethyl fumarate

Commercial production of dimethyl fumarate, like other fine/specialty chemicals, presents an opportunity to improve and simplify its manufacturing process. The processing parameters that are explored in the laboratory show the pathway, but need to be reviewed for a commercial operation. Figure 6.12 is the chemical synthesis for the reaction.

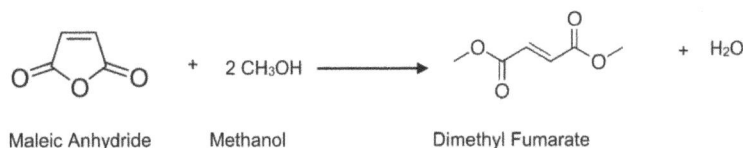

| Maleic Anhydride | Methanol | Dimethyl Fumarate |

Fig. 6.12: Maleic anhydride route.

Using maleic anhydride, various syntheses processes have been presented [60–63]. In these, anhydride is reacted with methanol using different catalysts (e.g., H_2SO_4, thiourea, $AlCl_3$, acetyl chloride, etc.) to produce mono methyl fumarate, which is then converted to dimethyl fumarate.

Dimethyl fumarate can also be produced from fumaric acid. However, this route, as explained in Chapter 5, based on the price difference of maleic anhydride versus fumaric acid, is not only more expensive but also produces additional water as a by-product, which increases the waste handling load.

Processing conditions for the maleic anhydride route, which includes the catalyst, need to be defined. References [60–63] indicate the conversion from maleic anhydride to be a two-step process route, suggested in Fig. 6.13, which is very viable. Smaller pilot plant reactors or modular reactors or electrically heated heat exchangers [35, 40] can be used for the process. Processing conditions for either of the process routes will need to be defined and optimized. Since the process can use alternate smaller-sized equipment, the solvent used in the process would be considerably reduced.

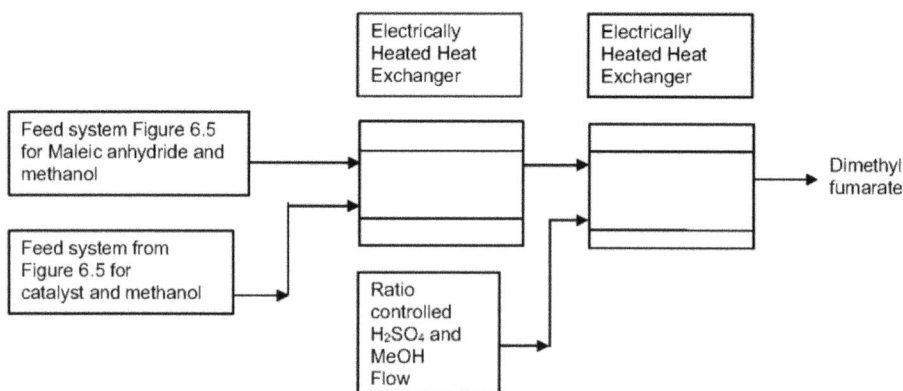

Fig. 6.13: Dimethyl fumarate process flow diagram.

6.6.4 Omeprazole

Omeprazole is a large-volume drug [64] used for acid reflux treatment. Its levo-enantiomer, sold under the Nexium name, is also a high-volume drug. Omeprazole, due to its high volume, most likely, would be produced in conventional-sized reactors that are used in the fine/specialty chemical industry. This is due to the similarity of manufacturing methods. As we all know, small-molecule active pharmaceutical ingredients are fine/specialty chemicals that have healthcare value. Otherwise, the chemistry and manufacturing methods are the same and similar.

A simplified block process diagram or similar diagram, illustrated in Fig. 5.23 [65–67], would, most likely, be used by most companies. However, the chemistry and the manufacturing process can be further improved if an attempt is made to minimize the solvent in the chemistry that would be commercialized using conventional processing equipment.

A review of the chemistry and various patents suggests the use of toluene as a solvent. Developers have done an excellent job of using the recovered wet toluene. It improves the yield and drives the process to lowering its "E-factor" [6]. Since solvents form a large volume of the reaction mass, their use can be reduced and the chemistry can be practiced using the pilot plant size or modular reactors as schematically shown in Fig. 6.14. The process can be operated in batch or continuous mode. It is very likely that the manufacturing company will adopt the continuous operation mode, once they have realized the value of benefits of capitalizing on the physical properties of the reactants.

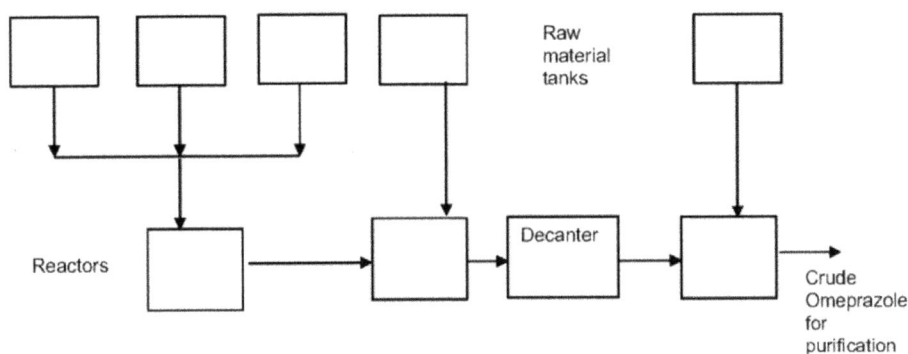

Fig. 6.14: Batch/continuous process for omeprazole.

For the continuous process, instead of using conventional large reactors, it is possible to use small reactors (75–150 L) or back mix plug flow reactors [68] that have been used in the fine/specialty chemical manufacturing. The produced product can be crystallized [69] continuously using the established criterion. Chemical engineers, who are familiar with process designs, would not have any problem commercializing such a process.

6.6.5 Modafinil

Synthesis of modafinil is another example of classical chemistry that can let an imaginative and creative chemist/chemical engineer, through exploitation of the physical and chemical properties and by selection of proper equipment, to commercialize a process that can have a significantly lower "E Factor" than the current practices.

Using the modafinil chemistry [70] (Fig. 6.15), a simplified process and flow diagram is explored in Figs. 6.16 and 6.17 [71].

Conventional reactors and equipment would be used for production. Consequently, they will have a high "E-factor." Based on modafinil's global demand, all of the needed API, using a continuous process, can be produced at a single plant [72]. However, multiple formulation plants would be required.

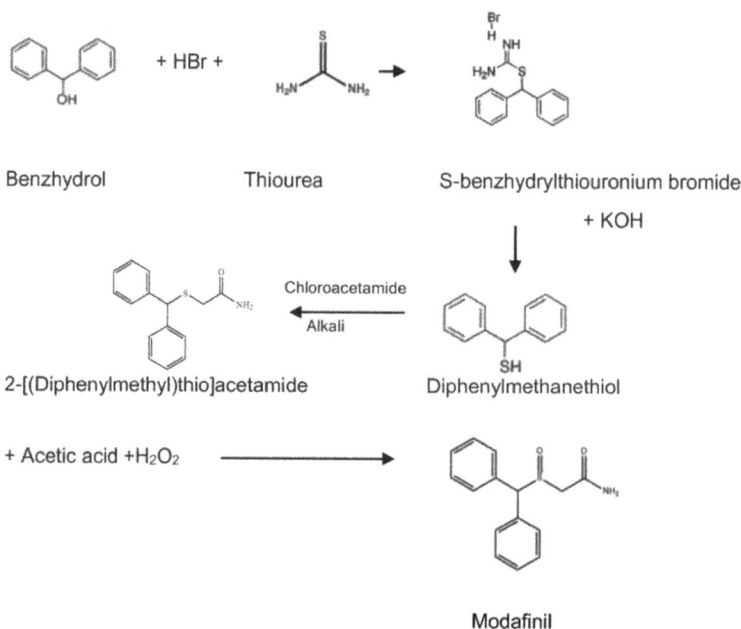

Benzhydrol   Thiourea   S-benzhydrylthiouronium bromide

+ KOH

Chloroacetamide
Alkali

2-[(Diphenylmethyl)thio]acetamide   Diphenylmethanethiol

+ Acetic acid +H₂O₂ ⟶

Modafinil

Fig. 6.15: Modafinil chemistry [70].

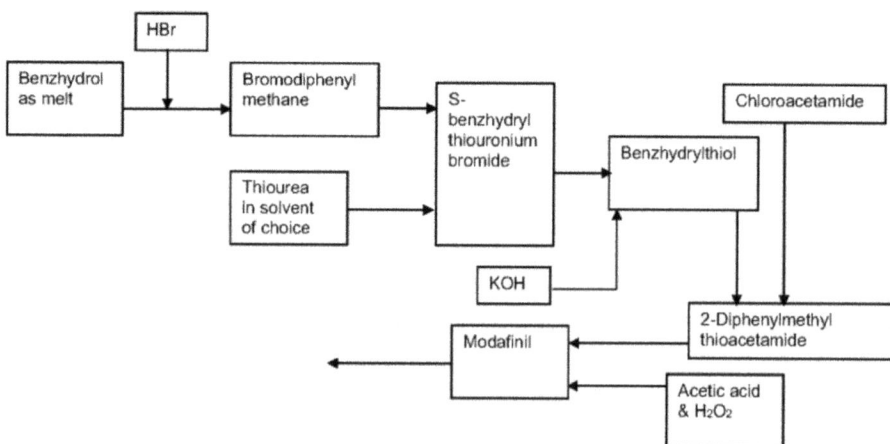

Fig. 6.16: Version ONE simplified modafinil process flow diagram [71].

Equipment and concepts, reviewed earlier in this chapter, in combination with modular reactors and smaller reactors versus conventional large reactors can be creatively used to significantly lower the solvent use.

As said earlier, each of the potential processing methods and equipment used has to be tested in laboratories and/or pilot plants. Chemical engineers and chemists,

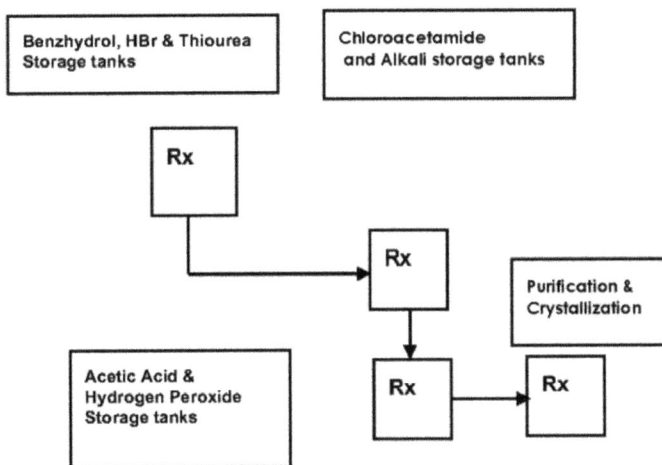

Fig. 6.17: Version TWO simplified modafinil process flow diagram [71].

who are well versed, can emulate, create, and test. They have to incorporate and exploit the physical properties to their advantage. It is necessary to emphasize that the solvent use has to be limited to one additional solvent, besides water, and volume use has to be minimized. Simple and efficient processes that will produce the highest quality product with best economics can be commercialized.

6.7 Pharma's continuous improvement process and why it has lagged

Having discussed the alternate equipment and methods that can be used to exploit manufacturing, an opportunity exists to lower the "E-factor" for the pharmaceutical industry Tab. 6.1. By definition, "continuous improvement" requires companies to put in an ongoing effort to improve their manufacturing technologies and processes. Unless the regulatory agencies facilitate such a change [5] and companies are held accountable for their product's performance, such an effort has not and will not be forthcoming.

The laws imposed by regulatory agencies, such as the FDA in the US, present substantial barriers to "continuous improvement" in the pharmaceutical industry. It is not possible for a manufacturer to implement any process change without a regulatory agency review. Regulatory agency review costs time and money; hence, there is no or little motivation for a manufacturer to engage in "continuous improvement." It is possible that more monies may be spent in a re-approval than the monies spent in improving the process.

Companies do make an effort. Sildenafil and Lyrica [74, 75], ibuprofen [75, 76] are excellent examples. Their "E-factor" numbers were lowered only after their patents expired. Even after "E-factor" reduction for these products, still better processes can be developed and commercialized. Review of other drugs present "continuous improvement" opportunities, but having too many producers [20] minimze any motivation for change.

The process of continuous improvement has to start as soon as the companies realize that they can lower the "E-factor," and not wait until the patents have expired and the drugs have become generic. As reviewed earlier, a village [18] has to get involved as soon as a process is developed. Control and lowering of each process's "E-factor" has to start from the inception of process development. Table 6.10 [73] is an illustration of a hypothetical lowering of "E-factor" numbers when applied to three drugs reviewed in this chapter.

Actual emissions for these drugs may be different, but the industry average is used to illustrate their impact. Bold numbers are the emission numbers if they were reduced to the lowest number for the fine chemicals, but unless pointed out, not many will understand the magnitude of emissions.

Tab. 6.10: Waste generated, for example, APIs [73].

E-factor	Omeprazole	Metoprolol	Modafinil	Total waste (kg/year)
kg waste per kg product	Waste for each drug at different levels (kg/year)			
100	218,400,000	142,350,000	34,164,000	394,914,000
50	109,200,000	71,175,000	17,082,000	197,457,000
25	54,600,000	35,587,500	8,541,000	98,728,500
E-factor numbers if emissions are reduced				
5	**10,920,000**	**7,117,500**	**1,708,200**	**19,475,700**

Methodologies (engineering, chemistry), discussed earlier, creativity, and imagination have to be employed to lower the "E-factor" numbers. Considerable effort is and would be needed to lower waste and emissions. Since the products are produced all over the world, establishment of global emission standards [77] could be a way to lower waste. However, there will be a considerable resistance to such standards.

Everyone recognizes the magnitude of emissions and waste generated from the pharmaceutical manufacturing, but no one has stepped up to do much about it, from the onset. There have to be rational reasons for not doing anything. It could be too much effort is needed per site as too many companies [20] are producing the same API and formulating them, and there is not sufficient return for the effort [19].

In the manufacture of fossil fuel products, bulk chemicals and fine/specialty chemicals, the process of continuous improvement is an ongoing effort. This is due to competitive pressures. Ongoing improvement process generally involves improvements in manufacturing process or technology. However, in brand pharma, processes for the brand drugs are locked in before Phase III clinical trial. Any change after that can invalidate the trial [5]. If any process change alters the product efficacy and performance, any improvements in brand API manufacturing, or their formulation processes are out-of-compliance as far as the regulators and the company is concerned. For this reason, no changes are made while the drug is under patent protection. The company's mission is to get their product to the market. All the costs related to low process yield and high solvent waste (high E-factor) are passed on to customer.

Generic manufacturers do improve the processes, but due to their occasional production of most APIs, they don't see value or return [19] in continuous improvement effort. Due to the short production runs/campaigns, continuous process optimization is avoided. In addition, the paperwork needed to inform the regulatory bodies of any improvements is considered an unnecessary burden.

Unlike brand drugs, generic producers can optimize their processes till they file ANDA (abbreviated new drug application), but they have to assure bioequivalence with the brand drug.

As has been discussed in the earlier chapters, generic companies do have the time to commercialize processes that have lower "E-factor," but the following interfere.
1. Too many producers, i.e., no economies of scale to benefit from any better manufacturing technologies [20]
2. Fitting the process in ill-suited existing equipment [16, 17]
3. Acceptable profits can be made using existing methods
4. Regulatory burden(time and money)

In most of the recent published articles (too many to cite), solvent reduction is the most discussed topic for any waste-related discussion. There are other ways waste can be reduced and they are:
1. Improved overall process yield
2. Use of the right equipment for the process
3. Using alternate process

These have been discussed earlier and in various chapters of this book. It is up to the chemists and chemical engineers to adopt and include them in their process development, scale-up, and commercialization. Again the village [18] has to get involved from the onset in the process development.

As said earlier, each of the potential processing methods and the equipment used have to be tested in laboratories and/or pilot plants. Chemical engineers and chemists, who are well versed, can emulate, create, and test. Besides the equipment and methods suggested, other existing equipment that are available and have not been used or suggested in the manufacture of API can be explored, tested, and used. They have to incorporate and exploit the physical properties to their advantage. Solvent use in reactions is necessary as they facilitate chemical reactions [78]. It is necessary to emphasize that the solvent use should be limited to one additional solvent, besides water, and their volume has to be minimized, as solvents have to be recycled. Simple and efficient processes that will produce the highest quality product with best economics can be commercialized.

Abbreviations

PAT   Process analytical technology
API   Active pharmaceutical ingredient
kg   Kilogram
KL   Kiloliter
sq. m. Square meter
mm Hg   Millimeter mercury
HCTZ   Hydrochlorothiazide
MP   Melting point
BP   Boiling point

References

[1] Small Molecule https://en.wikipedia.org/wiki/Small_molecule, Accessed January 29, 2021.
[2] Shreve, R. N. Unit Processes in Chemical Engineering, Industrial and Engineering Chemistry, 1954, 46, 4, pg., 672, Accessed June 22, 2020.
[3] Unit Operations, Wikipedia, https://en.wikipedia.org/wiki/Unit_operation Accessed March 11, 2021.
[4] Current Good Manufacturing Practice (cGMP) Regulations https://www.fda.gov/files/drugs/published/Q7-Good-Manufacturing-Practice-Guidance-for-Active-Pharmaceutical-Ingredients-Guidance-for-Industry.pdf September 2016 Accessed March 10, 2021.
[5] Roschangar et al Overcoming Barriers to Green Chemistry in the Pharmaceutical Industry - the Green Aspiration Level™, Green Chemistry, 2014, pgs. 1–17, Accessed December 10, 2020.
[6] Sheldon, R. A. The E Factor 25 Years On: The Rise of Green Chemistry and Sustainability, Green Chemistry https://pubs.rsc.org/en/content/articlelanding/2017/gc/c6gc02157c#!divAbstract, 2017, 19, 18–43, Accessed February 17, 2021.
[7] Burke, J. What does net zero mean? https://www.greenbiz.com/article/what-does-net-zero-mean, May 2, 2019 Accessed April 27, 2021.
[8] Larsson, D. G. J. et al Effluent from Drug Manufactures Contains Extremely High Levels of Pharmaceuticals, Journal of Hazardous Materials, 148, 3, 30 September 2007, Pages. 751–755..
[9] Malhotra, G. Pharmaceuticals, Their Manufacturing Methods, Ecotoxicology, and Human Life Relationship, Pharmaceutical Processing, November 2007, pgs. 24–26. Accessed August 10, 2009.
[10] Anastas, P. et. al. Green Chemistry: Principles and Practice, Chemical Society Reviews, https://doi.org/10.1039/1460-4744/1972,, 2010, 39, 301–312, Accessed April 21, 2021.
[11] Scott, A.: Good Chemistry" Chemical Week March 15, 2010 Accessed April 2, 2010.
[12] Pharmaceuticals Trouble in the making Economist, August 31, 2002, Accessed February 18, 2021.
[13] Malhotra, G.: API Manufacture-Simplification and PAT: Pharmaceutical Processing, November 2005, Pages 24–27; [Innovation and Continuous Improvement in Pharmaceutical

Manufacturing. The PAT Team and Manufacturing Science Working Group Report. http://www.fda.gov/cder/gmp/gmp2004/manufSciWP.pdf Accessed September 1, 2005.

[14] US FDA White Paper: Innovation and continuous improvement in pharmaceutical manufacturing: Pharmaceutical cGMP for the twenty-first Century https://wayback.archive-it.org/7993/20170405121836/https:/www.fda.gov/ohrms/dockets/ac/04/briefing/2004-4080b1_01_manufSciWP.pdf, Accessed February 18, 2021.

[15] Fortunak, J. M. Current and Future Impact of Green Chemistry on the Pharmaceutical Industry, https://www.future-science.com/doi/pdf/10.4155/fmc.09.60, Future Medicinal Chemistry, 2009, 1, 4, 571–575. 14 Aug 2009, Accessed Feb10, 2021.

[16] Malhotra, G.: Square Plug in A Round Hole: Does This Scenario Exist in Pharmaceuticals?, Profitability through Simplicity, https://pharmachemicalscoatings.blogspot.com/2010/08/square-peg-in-round-hole-does-this.html August 17, 2010.

[17] Malhotra, G.: Why Fitting a Square Plug in a Round hole is Profitable for Pharma and Most Likely Will Stay? https://pharmachemicalscoatings.blogspot.com/2014/08/why-fitting-square-plug-in-round-hole.html August 1, 2014.

[18] Malhotra, G.: The Good, the Bad, the Ugly (1)complexities of Pharmaceutical Manufacturing, Profitability through Simplicity, https://pharmachemicalscoatings.blogspot.com/2018/04/the-good-bad-ugly-1-complexities-of.html, April 9, 2018 Accessed November 1, 2020.

[19] Chapter Four: Table 4.6 Months for Return and Revenue.

[20] Malhotra, G.: Pharmaceutical Quality: Concepts, Misconceptions, Realities and Remedies, https://pharmachemicalscoatings.blogspot.com/2019/11/pharmaceutical-quality-concepts.html Accessed February 10, 2020.

[21] Chapter Five Figure 5.1 Steps involved in API Commercialization.

[22] Chapter Four Table: 4.3 Yield and number of reaction steps correlation.

[23] Chapter Five, Table 5.3 Cost Analysis of Dimethyl Fumarate.

[24] Dimethyl fumarate API Sites https://www.pharmacompass.com/active-pharmaceutical-ingredients/dimethyl-fumarate Accessed November 30, 2020.

[25] Hatch-Waxman Letters, https://www.fda.gov/drugs/abbreviated-new-drug-application-anda/hatch-waxman-letters, July 19, 2018 Accessed March1, 2021.

[26] WTO TRIPS agreements https://www.wto.org/english/tratop_e/trips_e/intel2_e.htm January 1, 1995 Accessed March 1, 2019.

[27] Xing, Y. et al. Do Lax Environmental Regulations Attract Foreign Investment?, https://citeseerx.ist.psu.edu/viewdoc/download?doi=10.1.1.571.7589&rep=rep1&type=pdf, Revised August 2000, Accessed March 10, 2021.

[28] Developing Countries, Wkipedia.org, https://en.wikipedia.org/wiki/Developing_country, Accessed February 10, 2021.

[29] Malhotra, G.: Why Have the Fine and Specialty Chemical Sectors Been Moving from the Developed Countries? https://pharmachemicalscoatings.blogspot.com/2009/02/why-are-fine-and-specialty-chemical.html February 9, 2009 Accessed January 12, 2021.

[30] Benchmarking Shows Need to Improve Uptime, Capacity Utilization, http://www.pharmamanufacturing.com/articles/2007/144/ accessed March 8, 2016.

[31] Levenspiel, O. Chemical Reaction Engineering, John Wiley & Sons Inc, 1972, Second Edition, Accessed December 20, 2020.

[32] Malhotra, G.: Process Centricity Is the Key to Quality by Design, Profitability through Simplicity, http://pharmachemicalscoatings.blogspot.com/2010/04/process-centricity-is-key-to-quality-by.html April 6, 2010 Accessed February 21, 2021.

[33] Malhotra, G.: A Radical Approach to Fine/Specialty API Manufacturing https://pharmachemicalscoatings.blogspot.com/2010/01/radical-approach-to-finespecialty-api.html, January 20, 2010 Accessed March 17, 2021.

[34] Malhotra, G.: Considerations to Simplify Organic Molecule (API) Manufacturing Processes: My perspective, https://pharmachemicalscoatings.blogspot.com/2019/04/considerations-to-simplify-organic_77.html, April 20, 2019 Accessed March 17, 2021.

[35] Shanley, A. Flexible Pharma: Puzzling Out the Plant of the Future, Pharmaceutical Manufacturing, Nov/Dec 2009, pg. 14–21. Accessed January 10, 2021.

[36] Erector Set, https://en.wikipedia.org/wiki/Erector_Set Accessed March 17, 2021.

[37] Less Is More in API Process Development, Pharmaceutical Manufacturing, https://www.pharmamanufacturing.com/articles/2005/294/, July/August 2005 pgs. 50–51 Accessed November 10, 2020.

[38] Malhotra, G. Focus on Physical Properties to Improve Processes, Chemical Engineering, 119, 4, April 2012 pgs. 63–66. Accessed Feb 10, 2020.

[39] Pfaudler: https://www.pfaudler.com/uploads/files/pfaudler-din-be-reactors-1.pdf Accessed March 20, 2021.

[40] Process Technology https://www.processtechnology.com/ accessed May 23, 2017.

[41] Alfa Laval Plate and Frame Heat exchangers, https://www.alfalaval.us/products/heat-transfer/plate-heat-exchangers/gasketed-plate-and-frame-heat-exchangers/ Accessed March 17, 2021.

[42] Tranter Plate and Frame Heat exchangers https://www.tranter.com/heat-exchangers/super changer-plate-frame-heat-exchanger/ Accessed March 17, 2021.

[43] Diabon graphite plate heat exchanger https://www.sglcarbon.com/pdf/SGL-Information-Sheet-PT-DIABON-Graphite-PHX-EN.pdf Accessed March 12, 2021.

[44] Arrhenius Equation, https://en.wikipedia.org/wiki/Arrhenius_equation Accessed March 21, 2021.

[45] High-capacity rotor/stator mixers https://www.mixers.com/insights/mti_76.pdf, accessed March 25, 2021.

[46] Perry, J. H. et al., Chemical Engineers' Handbook, Fourth Edition, McGraw-Hill Book Company, Decanter Pg., 21–18, Accessed June 10, 2020.

[47] USP 6252113.

[48] US 2009/0247642 A1.

[49] Chapter Five Fig.: 5.14 Metoprolol Synthesis.

[50] Chapter Five: Tab.: 5.15 Physical and Chemical Properties of Metoprolol chemicals.

[51] Chapter Five Table 5.14 Global Metoprolol API Demand.

[52] Number of Metoprolol plants https://www.pharmacompass.com/active-pharmaceutical-ingredients/metoprolol accessed March 14, 2021.

[53] Chapter V, Table Tab.5.6: Yearly HCTZ (API) Need.

[54] Number of plants producing Hydrochlorothiazide, https://www.pharmacompass.com/active-pharmaceutical-ingredients/hydrochlorothiazide Accessed December 17, 2020.

[55] Nasim, A.: http://medicinal-chemistry-notes.blogspot.com/2015/12/synthesis-of-hydrochlorthiazide.html Accessed July 4, 2017.

[56] Novello, F. C., USP 2965675 Aniline Compounds, Merck & Co. Dec. 20, 1960 Accessed Feb. 10, 2019.

[57] Ramakrishnan et. al. WO 2007/026376 A2, A Novel Process for the Preparation of Highly Pure Crystalline Hydrochlorothiazide March 7, 2007 Accessed November 11, 2010.

[58] Deo et. al. WO 2009/150497 A1 A Process for Preparation of Highly Pure Hydrochlorothiazide, December 17, 2009 Accessed November 11, 2010.

[59] Chapter Tab: 5.5 Physical Properties of Chemicals for Hydrochlorothiazide.

[60] Pullaguria et al. Biophore India Pharmaceuticals Inc. WO 2015/140811 A2, US 2017/009638 A1, USP 10636076 B2, An improved process for the synthesis of dimethyl fumarate April 20, 2020 accessed June 20 2020.

[61] Raillard et al. XenoPort Inc. USP 9302977, Method of making monomethylfumarate, April 5, 2016 accessed June 10, 2017.

[62] de Souza, R. O. M. A. et al Continuous-flow Synthesis of Dimethyl Fumarate: A Powerful Small Molecule for the Treatment of Psoriasis and Multiple Sclerosis, RSC Advances, 2020, 10, 2490. Accessed January 16, 2020.

[63] CN 86106666A One one-step preparation method of dimethyl fumarate, Accessed June 20, 2020.

[64] Chapter Five Tab. 5.13: Global Omeprazole API Demand.

[65] Chapter Five Figure 5.23 Simplified Flow for Omeprazole.

[66] Malhotra, G.: Alphabet Shuffle: Moving From QbA to QbD – An Example of Continuous Processing, https://www.pharmaceuticalprocessingworld.com/alphabet-shuffle-moving-from-qba-to-qbd/ February 23, 2009 Accessed March 31, 2020.

[67] Malhotra, G.: Analysis of API (Omeprazole): My Perspective, Poster Session: Pharmaceutical Engineering, 2009 AIChE Annual Meeting, November 11, 2009, Nashville, TN.

[68] Levenspiel, O. et. al. Backmixing in the Design of Chemical Reactors, Industrial & Engineering Chemistry, 1959, 51, 12, 1431–1434.

[69] Anousis, N. et. al., Merck & Co. Inc. WO 00/09497 Improved Omeprazole Process and Compositions Thereof, February 24 2000 Accessed April 22, 2021.

[70] Chapter Five Fig. 5.17 Modafinil Chemistry.

[71] Malhotra, G.: Review of Continuous Process for Modafinil, Continuous Processing in the Chemical and Pharmaceutical Industry II, 2009 AIChE Annual Meeting, November 10, 2009, Nashville, TN Accessed October 11, 2020.

[72] Chapter Five Tab. 5.17: Global Modafinil API Need, Kg./ Yr.

[73] Malhotra, G.: Active Pharmaceutical Ingredient Manufacturing (API) and Formulation Drive to NET ZERO (Carbon Neutral)? Profitability through Simplicity, https://pharmachemicalscoatings.blogspot.com/2021/04/active-pharmaceutical-ingredient.html, April 29, 2021 Accessed April 30, 2021.

[74] Sanderson, K. It's Not Easy Being Green, Nature, January 2011, 469, pg. 18–20. Accessed May 8, 2021.

[75] Dunn et. al. Green Chemistry in the Pharmaceutical Industry, Wiley-VCH Verlag GmbH & Co. KG.aA, Weinheim 2010.

[76] Jolliffe et. al Process Modelling and Simulation for Continuous Pharmaceutical Manufacturing of Ibuprofen, Chemical Engineering Research & Design, 97, 2015, pg. 175–191, Accessed May 8, 2021.

[77] Malhotra, G.: Can Uniform Safety, Health and Effluent and Manufacturing Standards Create Process Technology Innovation and Competition in Pharmaceuticals? Profitability through Simplicity https://pharmachemicalscoatings.blogspot.com/2017/01/can-uniform-safety-health-and-effluent.html January 10, 2017 Accessed January 12, 2021.

[78] Li et.al. Green Chemistry for Chemical Synthesis, PNAS September 9, 2008 105 (36) 13197–13202; https://doi.org/10.1073/pnas.0804348105 Accessed September 7, 2020.

Chapter 7
Process documentation and operating strategies

Each pharmaceutical operating company has its own business and operating strategy for its operations, and that includes products and manufacturing technologies. These strategies influence each step – starting with their product, process development, manufacturing, and investment. Regardless of their operating strategies, they have to assure that each produced API is of established repeatable quality.

Instead of going into a detailed discussion and review of how to file an NDA and/or ANDA application, the discussion, here, is about the process, its design basis, equipment, and process variables. All of this information can be very useful in filing these applications, can reduce the commercialization time and internal training, and can also reduce the time if manufacturing is to be outsourced. It is compilation of the knowledge of how the variability in stoichiometry and process operating conditions can change process yield and product quality. This information can be used to correct any process deviation and can also be used to resolve any problem or troubleshoot any issue that arises during the manufacture of the API. Creation of such information and capitalizing on it to resolve any processing and quality issue has to be the culture of each operating company. Safety and safe practices cannot be overlooked. Documentation of information is valuable and critical.

The generated information can also be used for continuous process improvement and expansion, such as transition from a batch process [1, 2] to a continuous process [3, 4]. Briefly, both operating scenarios have been discussed earlier but are also reviewed as part of the process documentation.

USFDA cGMP for API are excellent [5] guidelines and are necessary as they eliminate/minimize deviation from "good process design practice" that are imperative for quality products. Every company has to document the necessary information. Compiled information not only assists in following the good manufacturing practices but is also useful for many other segments of each company's businesses. They are reviewed here.

7.1 Operating strategies and process documentation

7.1.1 Operating strategies

Since diseases are a part of human life, it is necessary to cure them and to extend life. Initially their cures were found in different botanical plants. These medicines were extracted from plants and their dosages dispensed. To assure consistent availability of

https://doi.org/10.1515/9783110702842-007

quality products, plant-derived, disease-curing, active ingredients were synthesized in the laboratories. As the demand grew, they were commercialized for larger availability. Consistent and repeatable quality became a necessary requirement. Since most of the synthesis chemistries and methods were similar to that of fine/specialty chemicals, available equipment could be easily used to produce the needed APIs.

Paclitaxel (Taxol) derived from yellow yew tree for treatment of breast, lung, and ovarian cancer [6] is a recent success story of synthesizing a naturally occurring product for cancer treatment.

Invention and creation of new disease treatments has been in the genes of the pharmaceutical industry. However, development of new manufacturing equipment technologies for the API manufacturing and their formulations has never been a priority. There is good reason for it. Pharma development chemists and engineers have relied on the equipment that has been used for the development of fine/specialty chemicals. This works very well for API development and commercialization. Since the lab and the existing fine/specialty chemical processes can be easily modified and used for commercial scale production with or without optimization, the need to develop any special equipment or process for API manufacturing has not been considered a priority. As a result and as discussed in earlier chapters, the majority of API processes generate large quantities of waste [7].

Lack of better manufacturing technology and adoption has extended to formulations also. Mortar and pestle methods have prevailed for the formulations even when better technologies have existed. Since formulations are not part of this book, these technologies are not discussed any further.

Since the pharma operating companies can fit them in the existing fine/specialty chemical equipment [8, 9] by modifying the processes, they have avoided the investment that might be needed for individual products. This has resulted in pharma manufacturing not needing innovative equipment or processes for API manufacturing. These decisions are based on economics, product volumes, and manufacturing traditions and have been discussed in different chapters.

Since the processes are not designed for specific products, repeated "in-process" testing of the intermediates and the final products, even with science and engineering being applied for API manufacturing, has assured that product quality is met. Thus, the need for product-dedicated equipment is not considered important. If the product cannot be reworked to meet specs, it is discarded as waste. Companies make sure that only the products meeting product specifications get to the market.

If pharma has to stop or alter continued practice of "in-process" testing, a process of continuous improvement and re-evaluation has to be applied to each company's business model [10]. Ongoing evaluation would and should lead to manufacturing technology innovation.

Yearly product demand per site determines if a batch or a continuous process will be used. Each has an established definition.

Batch processes [11] are used for the products that do not have a large yearly demand, and their chemistry can be executed in existing equipment. Their production during the year is intermittent. Since the equipment can also be used to produce other products whose chemistry can be fitted in the equipment, it is necessary that it is thoroughly cleaned between each process. This is necessary to avoid contamination. Lack of continued use of the equipment at a site also results in low asset utilization [12–16], a well-recognized fact in pharma API manufacturing. This has been discussed in earlier chapters.

Continuous process [3, 4], unlike the batch process [1, 2] except for preventive maintenance and unexpected downtime, is operated throughout the year. At 100% on-stream time, such processes operate about 8,400 (=24 × 7 × 50) h/year, producing a single product. Equipment for such processes is designed for the single product chemistry. Other chemistries are difficult to fit in such process designs.

One easy differentiation between a batch and a continuous process is that in the batch process, in-processes material can be held for the upcoming processing step. Unlike the batch process, no material is held anywhere in a continuous process for any time.

Table 7.1 illustrates the estimated global need for selected APIs. Processes for a few APIs are reviewed. This analysis can be extended and used for other molecules. The first and foremost consideration for production of a product after its chemistry and the manufacturing process has been identified is the annual capacity. Annual capacity dictates the case for the type of process that will be used for production. Would it be a batch process or a continuous process? The following products are selected for the discussion [17].

Tab. 7.1: Annual estimated global need for selected APIs [17].

API	Global annual need (kg/year)	Production rate (kg/h)
Omeprazole	2,184,000	260
Dimethyl fumarate	1,349,040	~161
Metoprolol	1,423,500	~170
Hydrochlorothiazide	711,750	84.7
Modafinil	365,000	43.5
Levothyroxine	41,000	~4.8
Metformin	63,875,000	Multiple plants

Hours per year = 24 × 7 × 50 = 8,400 h @ 100% OST = on-stream time (hours available)

About 8,400 h/year represents 100% manufacturing operation, with no downtime. Generally, 15% downtime is allocated for planned maintenance and unexpected downtime. Downtime allocation is each company's choice and is used in standard product cost calculation.

Even as demand for omeprazole, dimethyl fumarate and metoprolol are high and could be well suited for continuous production, based on the current operating philosophies, they are still produced using batch processes [8, 9] in multiple plants. Each such process will have high "E-factor" emissions [7].

7.1.2 Process documentation

Since processes for many API are fitted in existing equipment [8, 9], pharma manufacturing companies need to have excellent documentation of the process and its operating rationale, chemistry, process equipment design and selection basis, physical and chemical properties of the reactants and intermediates, and their mutual behavior.

It is necessary and even critical to know the impact of every process operating condition and method of addition of reactants, as each deviation from a prescribed process could produce a product that necessarily does not meet the desired quality from the onset. Impact of each process deviation on the product quality can be significant. Thus, it is necessary to document impact of each deviation and its remedy.

This is not only necessary for every regulatory filing but can also significantly reduce commercialization time. This information can be extremely useful in transition, if the product manufacturing is going to be outsourced.

A road map of what can result in quality products without repeated analysis is reviewed. Readers can tailor the knowledge base for their products. Not only such information that facilitates operations but also information for continuous improvement and to expand, if need be, are given. For lack of a better term, this document "Process Design and Technical Manual" can be modified to suit needs. Such a document would include all of the process design rationale, justification, and calculations. This manual is an excellent tool to understand the design rationale for the equipment, process controls, and the whole operating philosophy for the process. Process design considerations and parameters assist in troubleshooting and replacing the equipment.

Documentation is also an excellent tool for internal training. It could be used for regulatory filings. Since the same equipment is used for multiple products, it is necessary to emphasize that a detailed documentation of the equipment related to a given product is essential.

If modular processing equipment or other configurations, discussed in earlier chapters, are used in the manufacture of APIs, such documentation will be extremely valuable, as it will detail the rationale and logic related to every product and process. High product quality will be sustained, and there will be no errors.

Table 7.2 lists the "table of contents" that should be part of documentation for each product and its process. This information is each designer's thinking, justification, and rationale for the process. It is also each company's documentation of the process. This is an excellent tool for speedier commercialization and technology transfer, if the product is to be outsourced. It can also be used for most regulatory filings, trouble-shooting the process, and expansion. This document has to be a live document and needs to be updated continuously.

Each company can edit/tailor "table of contents" to suit its needs. Some camps within each company might consider such a document difficult to produce, cumbersome, or having all of the confidential information in one place as a security risk. This compilation also may not be considered necessary or critical by many. However, having it provides an extraordinary value to the companies. In the long run, its value is incalculable. All of the rationale and the information that is needed for the process, troubleshooting, and de-bottlenecking are easily accessible, as it is created by each company. It just has to be compiled. In real business situations, even the early naysayers, who have seen its usefulness, have been convinced of its value.

Examples for each section of "table of contents" are reviewed to explain their value. As indicated earlier, practitioners can add/expand to include content to suit their manufacturing, safety, and regulatory needs. To illustrate usefulness, information discussed is for different chemistries and processes.

The focus, here, is to share the value of the information and let the chemists and chemical engineers document their rationale and creativity, so that it can be used to exploit and resolve any adverse situation and "continuously improve" the manufacturing process. It is hoped that all involved will see the value of improvement to lower the product's "E-factor" [7] and improve quality. It is also expected that information compiled will be helpful in process development and commercialization of the company's additional products.

Tab. 7.2: Table of contents: process design and technical manual.

Section	Table of contents
1	Process and its description
2	Process charge and recovery
3	Raw material specifications
4	Process chemistry Heat and mass balance

Tab. 7.2 (continued)

Section	Table of contents
5	Process equipment, description, and their design criterion and details This section should include the necessary design calculations for each process equipment and process controller. Necessary process control parameters should also be included.
6	Process conditions and effect of variables
7	Suggested operating instructions
8	Laboratory synthesis procedure
9	Analytical methods
10	Thermodynamic and physical properties of raw materials, intermediates, and final products
11	Final product specifications
12	cGMP and cleaning practices
13	Safety Material safety data sheets Methods and instructions of safe handling
14	US Pharmacopeia Standards

7.2 Section 1: process and its description

Omeprazole and metformin hydrochloride chemistries and processes [17] are used as illustrations for this section. Theoretical stoichiometry is the mass balance at 100% yield. Its inclusion is helpful, as it identifies the by-products that are produced in the reaction.

7.2.1 Omeprazole

Omeprazole is produced by reacting pyrmethyl alcohol with thionyl chloride to produce pyrmethyl chloride. Wet toluene is used as a solvent. Wet toluene not only improves the yield, but also simplifies the process by not having to dry it [18]. Pyrmethyl chloride is reacted with metmercazole and caustic to produce pyrmetazole, which is oxidized using m-chloroperoxybenzoic acid. The resulting omeprazole is processed and used for further processing using traditional methods.

Figure 7.1 presents the theoretical stoichiometry.

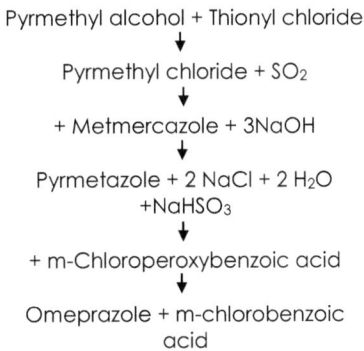

Pyrmethyl alcohol + Thionyl chloride
↓
Pyrmethyl chloride + SO_2
↓
+ Metmercazole + 3NaOH
↓
Pyrmetazole + 2 NaCl + 2 H_2O
+NaHSO$_3$
↓
+ m-Chloroperoxybenzoic acid
↓
Omeprazole + m-chlorobenzoic
acid

Fig. 7.1: Theoretical omeprazole chemistry.

As reviewed in earlier chapters, using the best yearly product demand and the company's business model, the manufacturing process (batch or continuous) gets defined. Process designers have to know and understand the process parameters, operating conditions, and their impact on the product quality, so that the produced product will meet the established quality specifications.

Batch and continuous processes will be different, as illustrated in Figs. 7.2 and 7.3. Based on process (batch or continuous) selected, raw material addition and handling will influence the economics.

Volume of solvent and the sequence of reactant addition for each chemical have to be defined for each processing step. Solubility will influence how the process is designed. The amount of toluene used as a solvent in the process is not identified. It has to be the lowest volume to ensure that the reaction mass can be pumped. To this end, knowledge of solubilities of various reactants and reaction products in the reaction is necessary. Companies have to create their solubility data. The value of solvent minimization has been explained in Chapter 6.

7.2.1.1 Batch process

In the batch process, each reaction step would be carried out in individual reactors, and after the reaction is complete, the mass will be transferred to the next reactor for further processing. Each processing company could select and carry out their batch reactions based on its preference.

Omeprazole's batch process, if conducted in reactors that range from 1,500 liters and up, would be similar to the laboratory process, as the process would not have dedicated equipment. This would also mean extra solvent use and cleaning time. Batch process could also be done in the same reactor but would be a complex process.

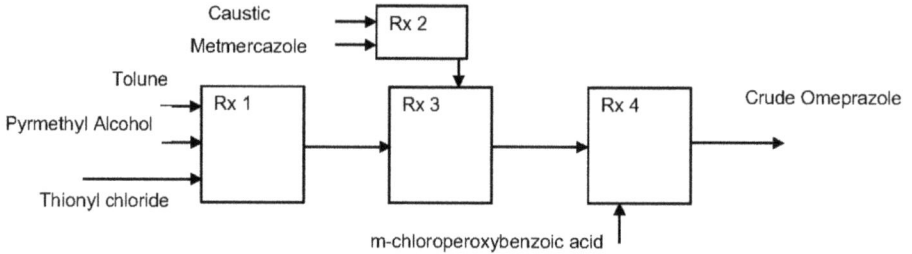

Fig. 7.2: Batch omeprazole process.

7.2.1.2 Continuous process

Pyrmethyl alcohol is a liquid at room temperature. Toluene, alcohol, and thionyl chloride can be metered to react to produce pyrmethyl chloride slurry in toluene. Metmercazole is a solid at room temperature and would have to be slurried in caustic solution. Toluene is used in the process and, thus, could be used as the solvent.

Reactions can be done in a circulating plug flow or similar reactor with proper mechanism to control the heat of reaction. Reaction temperatures can be higher than the temperatures suggested in the patent. This can be advantageous, as the higher reaction temperature accelerates the reaction rate and minimal impurities would be formed. No intermediate sampling would be necessary. Pyrmetazole and omeprazole of the subsequent steps can be similarly produced, using circulating plug flow or similar reactors. Reaction rates can be precisely controlled.

If omeprazole is produced using a continuous process, equipment size, compared to the conventional equipment, would be considerably reduced. Companies would require investment for a dedicated process. Since continuous process equipment is smaller in size compared to conventional process equipment, companies can use pilot plant and associated equipment as an alternate, or other modular equipment, as discussed in earlier chapters, as a distinct possibility.

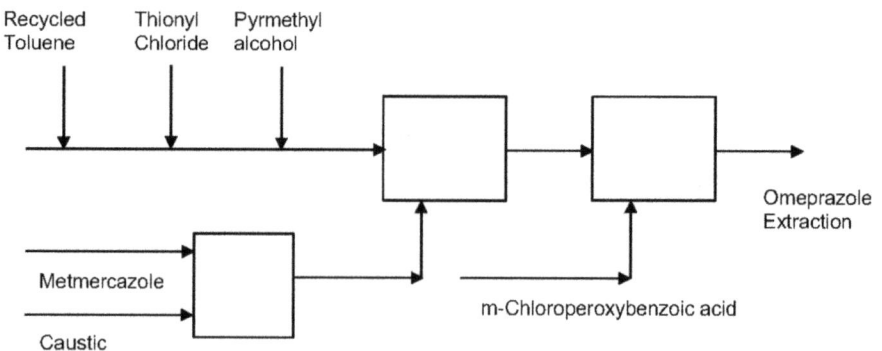

Fig. 7.3: Continuous omeprazole process.

7.2.2 Metformin hydrochloride

Theoretical metformin hydrochloride chemistry is outlined in Fig. 7.4.

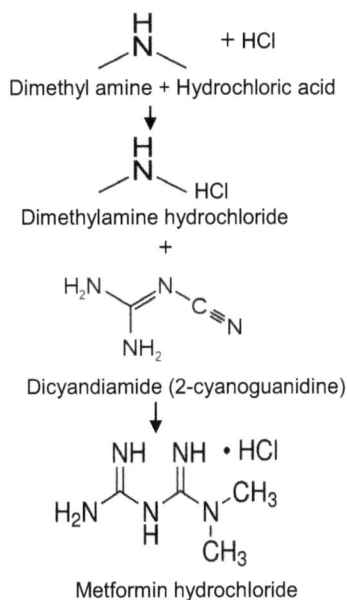

Metformin hydrochloride   **Fig. 7.4:** Theoretical metformin hydrochloride chemistry.

With its global demand (Table 7.1), one would expect it would be manufactured using a continuous process. However, there are about 81 API supplier companies [19]. Most likely, the majority of their production uses the batch process. Fundamental chemistry is per Fig. 7.4, but how it is executed (reaction, purification, and drying) to produce the dry API can be different, and that makes it a batch or a continuous process.

7.2.2.1 Batch/continuous process

Metformin chemistry can be practiced in three different ways. Each is described. Companies selecting either route have to review the economics of each route and select the one that suits their needs.

Process #1

Use molten dimethyl amine hydrochloride and dicyandiamide (2-cyanoguanidine) in stoichiometric quantities to produce metformin hydrochloride, which will be purified and dried. Schematics of an all-melt process could be similar to Fig. 7.5.

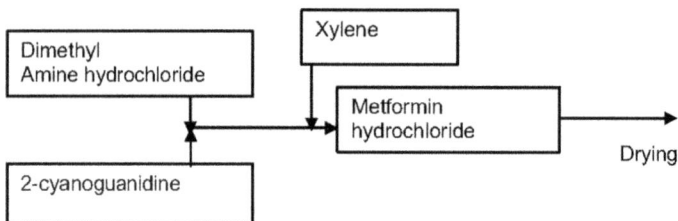

Fig. 7.5: Metformin all-melt process.

An all-melt process necessitates that each raw material is available as a solid or can be melted. If molten raw materials are available, each producer will have to acquire necessary melt equipment. Most likely, this process would only work for a continuous process. Economic justification for investment in the melt equipment would be difficult for a batch process.

Process #2

Figure 7.6 is a batch metformin process that uses dimethyl amine (DMA), liquefied gas, and commercially available hydrochloric acid. Hydrochloride is centrifuged and further reacted with dicyandiamide to produce metformin hydrochloride.

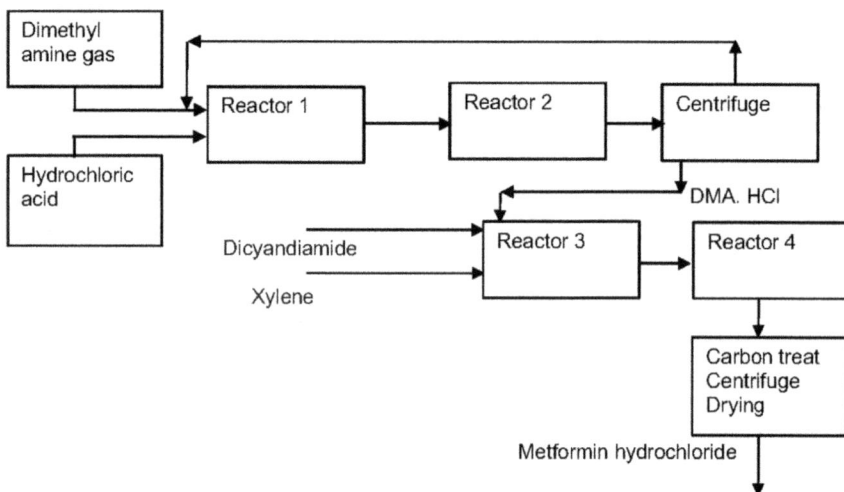

Fig. 7.6: Metformin batch process.

Commercial bulk availability of DMA would dictate this process route. Normal processes to decolorize and drying follow. Most of the chemists and chemical engineers are familiar with such processes, and they are not reviewed. This process is being commercially followed [20]. Process charge for this process is reviewed later in this chapter.

Process #3
Process described in Fig. 7.6 can be a continuous process and is schematically described in Fig. 7.7. Every chemist and/or chemical engineer understands that continuous processes are more economical and produce less waste, compared to batch processes [21].

In this process, DMA can be used as liquid or as a 40% solution. All of the physical and chemical properties of each raw material and reaction product can be exploited to design and commercialize this process. Heat and mass balance would assist in the process design. Heat of exotherm of DMA and hydrochloric acid can be used to raise the reaction at the water/xylene azeotropic temperature. If any additional heat is needed, it can be provided after DMA addition to drive the reaction to completion. Xylene would be separated using a decanter and recycled. Water from the decanter can be treated.

Any additionally needed xylene would be used to slurry dicyandiamide. This process uses 40% commercially available dimethyl amine hydrochloride. If DMA is to be used as gas, water mass balance would have to be carefully reviewed. Again the economics of the process would have to be justified.

7.3 Section 2: process charge and recovery

Process charge and recovery for the metformin process schematically shown in Fig. 7.6 is included. The referenced company [20] chose to explain different stages of the process as illustrated in Figs. 7.8–7.11. Tab. 7.3, Tab. 7.4, Tab. 7.5, Tab 7.6 Tab. 7.7 are the mass balances of processes illustrated in Fig. 7.8, Fig. 7.9, Fig. 7.10 and Fig. 7.11.

These figures are also the process and instrumentation diagrams of various steps of metformin hydrochloride process [20].

Metformin hydrochloride 1,200 tons per month

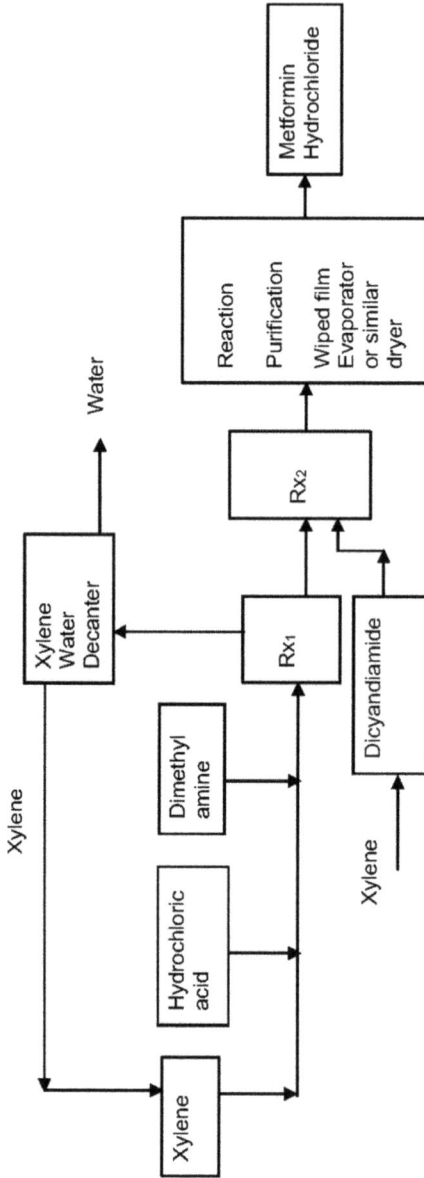

Fig. 7.7: Metformin continuous process.

Tab. 7.3: Stage 1 dimethyl hydrochloride [20].

Stage 1 DMA HCl		No. of batches: 50 per day	
Input (kg)		Output (kg)	
DMA	320	DMA HCl	580
HCl 30%	866	Loss on drying	35
		Evaporation loss	571
Total	1,186		1,186

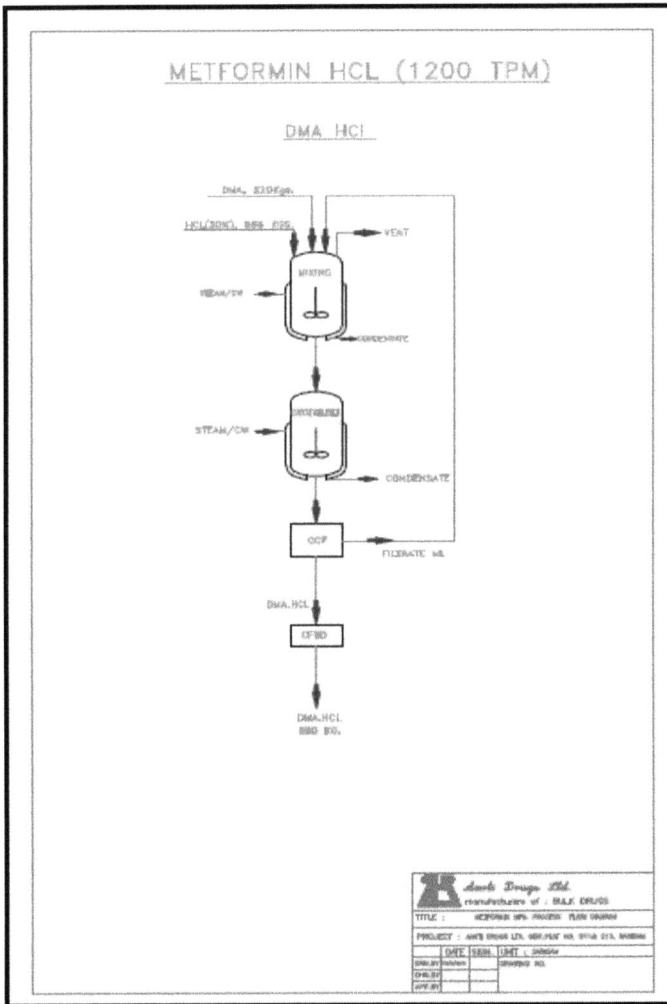

Fig. 7.8: Production of DMA hydrochloride for metformin process [20].

Tab. 7.4: Stage 1 crop of metformin process [20].

Stage 1: first crop		No. of batches: 50 per day	
Input (kg)		Output (kg)	
DMA HCl	580	Metformin hydrochloride	1,000
Dicyanodiamide	500	Loss on drying	56
Xylene	423		Recovery
Charcoal	4	Xylene	425
Water (recycle)	1,220.5	Xylene loss	7
		Recycle water	1,063.5
		Purge ML for second crop	179
		Residue	
		Solid waste (charcoal)	6
	2,736.5	Total output	2,736.5

Tab. 7.5: Second crop of metformin process [20].

Stage II: second crop		Batches per month 48	
Input (kg)		Output (kg)	
Purge mother liquor	4,475	Product, wet second crop	900
Methanol	640	Recycle water	2,950
		Mother liquor for MeOH recycle and third crop	1,265
Total input	5,115	Total output	5,115

Tab. 7.6: Third crop of metformin process [20].

Stage II: Third crop		Batches per month 24	
Input (kg)		Output (kg)	
Purge mother liquor	2,530	Product, wet third crop	300
		Recycle water	450
		Recovered methanol	1690
		Residue solid waste	90
Total input	2,530	Total output	2,530

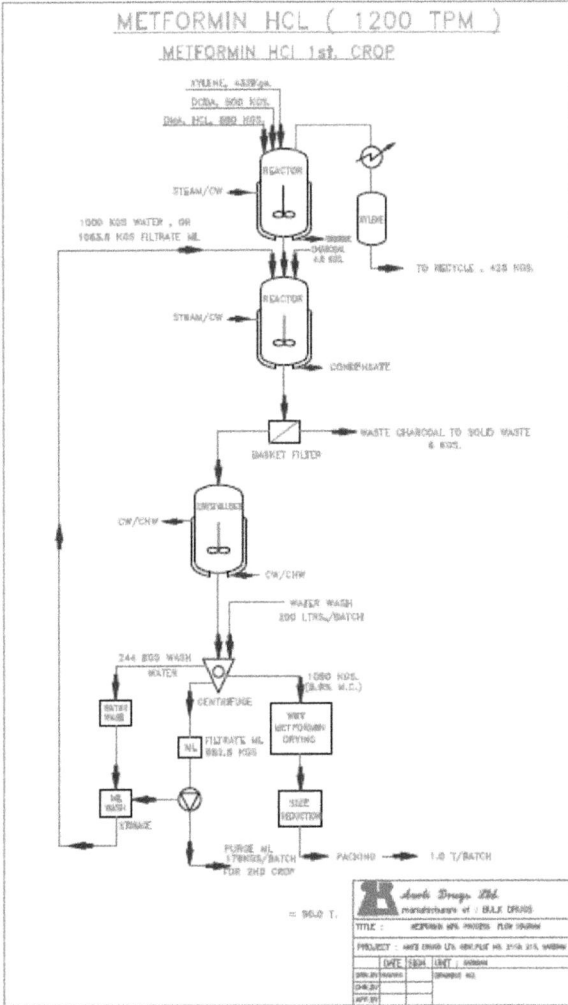

Fig. 7.9: Stage 1 crop of metformin process [20].

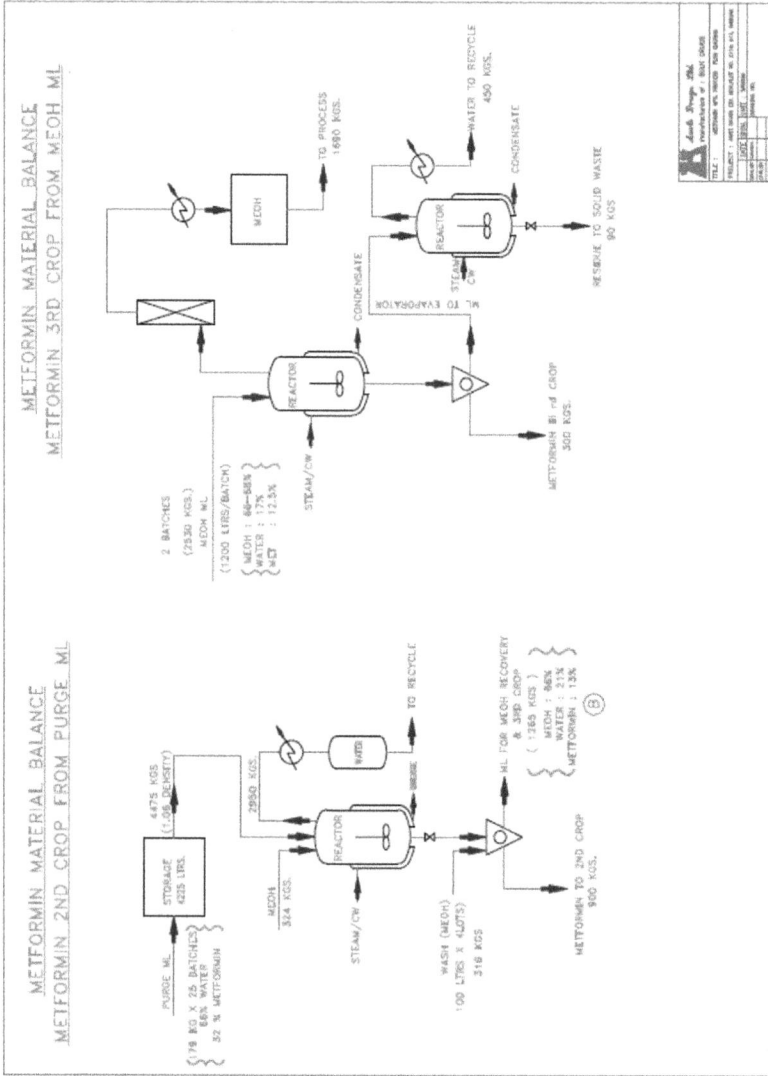

Fig. 7.10: Second and third crop of metformin process [20].

Tab. 7.7: Purification of second and third metformin crop [20].

Purification of second and third crop		Batches per month: 44	
Input (kg)		**Output (kg)**	
		Product	
Metformin second and third crop	1,200	Metformin HCl	700
Methanol	632	Loss on drying	48.5
Freshwater	2,000	Recycle water	1,881
Charcoal	5	Recovery	
		Methanol mother liquor for second crop	1,200
		Solid waste (charcoal)	7.5
Total input	3,837	Total output	3,837

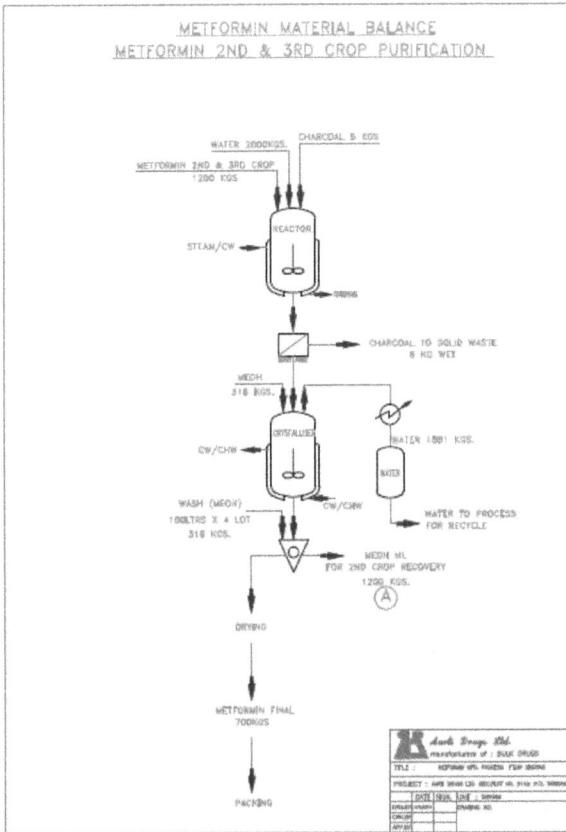

Fig. 7.11: Purification of second and third metformin crop [20].

7.4 Section 3: raw material specifications/properties

Generally, at the start of detailed process development, as reviewed in earlier chapters, the chemistry of the reaction is detailed and the process is tested for feasibility and economics. Once the feasibility is proven, the real process development begins, in earnest. It is necessary that the API laboratory process development and manufacturing be based on commercially available raw materials. This will assist in developing the best process. In addition, this facilitates the API manufacturers to develop relationships with suppliers. For commercial production, unless the specialty supply houses/companies are as competitive as the commercial suppliers, it is important not to source raw materials from them as, most likely, they have a limited supply chain. Their materials are expensive and do not represent the reality, as they have very little semblance to commercially available materials.

Sourcing of commercially available raw materials can be a challenge. This can be due to multiple reasons. To have raw material specifications as part of the Technical Manual spells out every rationale and reason for the selection. The following are some examples. These are accessed from different suppliers online technical data sheets.

Compiling physical properties, for example, viscosities, heat of formation, solubilities, heat of reaction, etc. of raw materials and intermediates can be a significant challenge. In the middle of the twentieth century, vendors shared many of the properties of the key starting raw materials.

In theory, the internet should be able to provide rudimentary physical and chemical properties [22] of products supplied by vendors. Most vendors have engaged a data mining strategy on information gathered from inquiries for their own marketing benefit. Often, vendors ask questions that either cannot be answered at the moment or are considered proprietary to the manufacturer. This approach can be an impediment to process innovation and can significantly extend process development time.

Some of the starting raw materials for many of the APIs are fine/specialty chemicals that are strictly used for pharma. Since their sales volumes are not large, producing companies do not spend time on developing and studying every property or their solubilities in different solvents. With the challenges of accessing many of the physical properties, user companies might have to generate their properties and solubility data. They may have to develop temperature/property relationships. Correlations for solubility predictions have been published. They will have to be verified and improved in the laboratory and used in process design [23–25]. Published solubility data for different chemical compounds is available and is useful [26, 27]. Some of the properties for different chemicals, that are available, are shared. Tables 7.8, 7.9, 7.10, 7.11, 7.12, 7.13, 7.14, 7.15, 7.16, 7.17, 7.18 and 7.19 are examples of raw material specifications of some of the chemicals used in different processes. Physical properties of these chemicals used in each reaction need to be acquired so that they can facilitate process design.

7.4.1 4-(2-Methoxyethyl)phenol

Tab. 7.8: Physical properties of p-(2-methoxyethyl) phenol [28].

4-(2-Methoxyethyl)phenol $C_9H_{12}O_2$	
CAS	56718-71-9
Mol. wt	152.19
B.P. (°C)	41–45
Solubility	Methanol

7.4.2 Epichlorohydrin

Tab. 7.9: Physical properties of epichlorohydrin [29].

Epichlorohydrin C_3H_5ClO	
CAS	106-89-8

Epichlorohydrin is a colorless, volatile, and flammable liquid with an irritating, chloroform-like odor, which emits toxic fumes when heated to decomposition. It is only slightly soluble in water but is soluble in many organic solvents.

Mol. wt	92.53
Sp. gr. kg/m^3@ 20 °C	1.18
Boiling point (°C)	117.9
Melting point (°C)	−48

7.4.3 Sodium hydroxide

Tab. 7.10: Physical properties of sodium hydroxide [30, 31].

Chemical formula	NaOH
CAS	1310-73-2
Molar mass	39.997
Density (kg/m^3)	1.52
Melting point (°C)	−12
Boiling point (°C)	135

7.4.4 Ammonia

Tab. 7.11: Physical properties of ammonia [32].

Ammonia	NH_3
CAS	7664-41-7
Molecular weight	17.031
Density (kg/m^3)	0.73
Boiling point (°C)	−33.34
Melting point (°C)	−77.73

7.4.5 Methanol

Tab. 7.12: Physical properties of methyl alcohol [33].

Methanol (methyl alcohol) CH_3OH	
CAS	67-56-1
Mol. wt	32
Sp. gr. kg/m^3@ 20 °C	791
Boiling point (°C)	65
Melting point (°C)	−97.7

7.4.6 Maleic anhydride

Tab. 7.13: Physical properties of maleic anhydride [34].

Maleic anhydride $C_4H_2O_3$	
CAS	108-31-6
Mol. wt	98.06
Sp. gr. kg/m^3	1430.0
Boiling point (°C)	201

Tab. 7.13 (continued)

Maleic anhydride $C_4H_2O_3$	

Melting point (°C)	53–54

Appearance: Maleic anhydride, 98+%; maleic anhydride, 98+%; colorless needles, white lumps, or pellets with an irritating, choking odor; colorless or white solid with an acrid odor.

7.4.7 Thiourea

Tab. 7.14: Physical properties of thiourea [26, 35, 36].

Thiourea	

Chemical formula	CH_4N_2S
CAS	62-56-6
Molar mass	76.12 g/mol
Appearance	White crystalline solid
Density	1,405 kg/m³
Melting point	182 °C
Boiling point	Decomposes
Solubility in water	13.7 g/100 ml (25 °C)14.2 g/100 ml (25 °C)
Solubility	Reacts with acids soluble in ethanol and methanol.Almost insoluble in diethyl ether and hexane
Solubility in ethanol	36 kg/m³ (20 °C)47 kg/m³ (31.9 °C)63 kg/m³ (45 °C)85 kg/m³ (58 °C)98 kg/m³ (64.7 °C)
Solubility in methanol	119 kg/m³ (25 °C)164 kg/m³ (40.7 °C)220 kg/m³ (53.7 °C)246 kg/m³ (61.9 °C)

7.4.8 Toluene

Tab. 7.15: Physical properties of toluene [37].

Synonym	Methylbenzene
CAS number	108-88-3
Molecular formula	$C_6H_5CH_3$
Molecular weight	92.14
Density	865
Boiling point (kg/m^3)	110–111 °C
Melting point	–93 °C
Flash point	4.4 °C
Viscosity (kg/m s)	0.00056
Toluene and water form a minimum boiling azeotrope (20.2% water @ 85 °C) [39]	

Azeotropic properties can be selectively used to improve manufacturing processes. Use of toluene azeotrope has been reviewed in this chapter and earlier chapters. Toluene azeotropes with water and has about 0.06% moisture. Use of wet toluene improves the first-step conversion in the manufacture of Omeprazole [18]. Properties of other chemicals can be explored and exploited to simplify processes.

7.4.9 Xylene

Tab. 7.16: Physical properties of xylene [39].

Molecular formula C_8H_{10}	
CAS no.	108-38-3
Molecular weight	106.16 g/mol
Density (kg/m^3)	864

Tab. 7.16 (continued)

Molecular formula C_8H_{10}	
Boiling point (°C)	138.5
Melting point (°C)	−47.4 (meta)
Water/xylene azeotrope Temp. (°C) [38]	94.5

Similarly, azeotropic property of xylene/water can be used for the production of metformin hydrochloride. In this process, dimethyl amine is reacted with commercially available hydrochloric acid, with xylene as the solvent. This has been reviewed in this Chapter 5.

7.4.10 Dimethyl amine

Tab. 7.17: Physical properties of dimethyl amine [40, 41].

Molecular formula C_2H_7N	
CAS no.	124-40-3
Molecular weight	45.08
Density (kg/m³)	680
Boiling point (°C)	6.8
Melting point (°C)	−92.2

Dimethyl amine, being a gas at room temperature, is commercially available as a 40% solution and 100% liquid. Both forms can be used in a reaction, and their use has been reviewed earlier in this chapter. Economics and availability will help make a decision.

Dimethylamine hydrochloride is miscible in water, and this property can be of value in the production of metformin hydrochloride.

7.4.11 Ammonia

Tab. 7.18: Physical properties of ammonia [42].

Ammonia	NH_3
Boiling point (°C)	−33.33
Density, liquid at −28 °F/−33.35 °C (1 atm., kg/m^3)	696
Density, liquid at 70 °F/21.1 °C (kg/m^3)	617.5
Heat (enthalpy) of evaporation, ΔH_v, at boiling point (kJ/kg)	1372.0
Heat(enthalpy) of formation, ΔH_f (gas) (kJ kg)	−2,695
Heat (enthalpy) of fusion/melting, ΔH (m, kJ/kg)	332.3
Melting (freezing) point (°C)	−77.73
Molecular weight	17
Specific heat, C_p (gas) (kJ/kg K)	2.175
Specific heat, C_p (liquid) (kJ/kg K)	4.744

7.4.12 Hydrochloric acid

Tab. 7.19: Physical properties of hydrochloric acid [43, 44].

Hydrochloric acid	HCl, 20 Be
CAS	3747-01-0
Mol. wt	36.5
Melting point (°C)	−35
Boiling point (°C)	57

7.5 Section 4: process chemistry and heat balance

7.5.1 Process chemistry

In this section, the chemistry of the process is described. Metformin hydrochloride chemistry is used as an example. It is necessary to outline the chemistry and its rationale, as it is used for the process design. Of the three potential processes schematically shown earlier, a batch process (Fig. 7.6) [20] is used.

Hydrochloric acid is charged to the designated reactor and then, the desired amount of DMA gas is bubbled into the hydrochloric acid. Reaction temperature is controlled. After the desired residence time, the batch is transferred to the next reactor, which is used as a centrifuge feed tank. The filtrate is recycled, if necessary, to ensure that there is no unreacted dimethyl amine. Wet dimethyl amine hydrochloride is dried.

In Reactor 3, the desired amount of xylene is charged, and desired amounts of dimethylamine hydrochloride and dicyandiamide are charged. The reaction mass is heated, and the desired temperature is maintained. Evaporated xylene is condensed and recycled. The reaction mass is transferred to Reactor 4, where carbon is added to decolorize the metformin hydrochloride solution. The reaction mass temperature is maintained to ensure that metformin hydrochloride is dissolved in xylene. After a designated time to ensure that the solution has met the color specifications, the reaction mass is transferred to a reactor where the reaction mass is cooled to precipitate all of the metformin hydrochloride, which is centrifuged and dried.

7.5.2 Heat and mass balance

Table 7.20 is a typical heat and mass balance sheet for a batch process. It is extremely important that the physical properties of each process stream are included, as they are needed to design every part of the process, e.g., piping, pumps, heat exchangers, and other material handling equipment. Details of properties and flow rate at the operating temperature are critical to design the process.

All the streams and those that include recovery and disposal have to be included. In this example, to protect the confidentiality of the process, components have not been defined. However, a chemist and/or a chemical engineer will fully understand the value of such mass balance.

Since mass balance and processing conditions for most batch processes are different, it is necessary to know that proper equipment is used for an efficient process. However, that is normally not the case. Mostly, the equipment that is available on site is used [8, 9]. They generally lead to prolonged batch cycle times, inefficient process, and lower-than-desired yields. This has been discussed in earlier chapters.

Fig. 7.12 is an illustration of a continuous process. Heat and mass balances for a continuous process are similar to batch process heat and mass balance. In this example, Table 7.21, to protect proprietary information, components have not been defined. However, chemists and chemical engineers will understand the value of such information. Additional properties and other relevant information that are necessary for equipment design should be included.

In every chemical process, design, reactors, heat exchangers and pumps are involved. Other specialized equipment might be needed. Using the process operating parameters, design rationale of each equipment becomes critical. This is important for every continuous process.

Tab. 7.20: Heat and mass balance of a batch process.

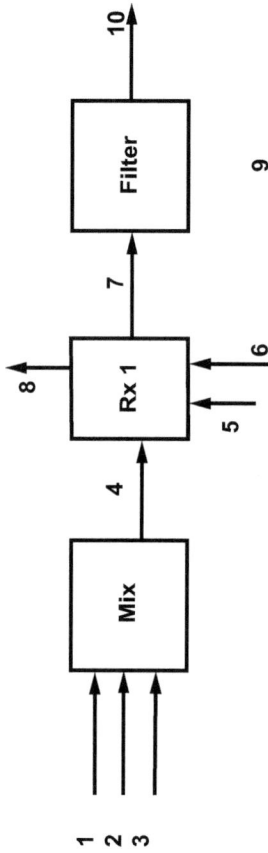

Stream Number	1	2	3	4	5	6	7	8	9	10
Component										
A	5000									
B		1248				592				
C		1248	16325			592				
D					5360					19754
E										
F										
Total	5000	2496	16325	23821	5360	1184		420	20	29925
Teemo °C	50		25	47				85	80	80
Pressure										
Specific gravity	1.232	1.525	1		1.334	1.525				
Viscosity, Pascal second					0.0017					
Sp. Heat, cal/kgC		0.783	1		0.38					
pH										

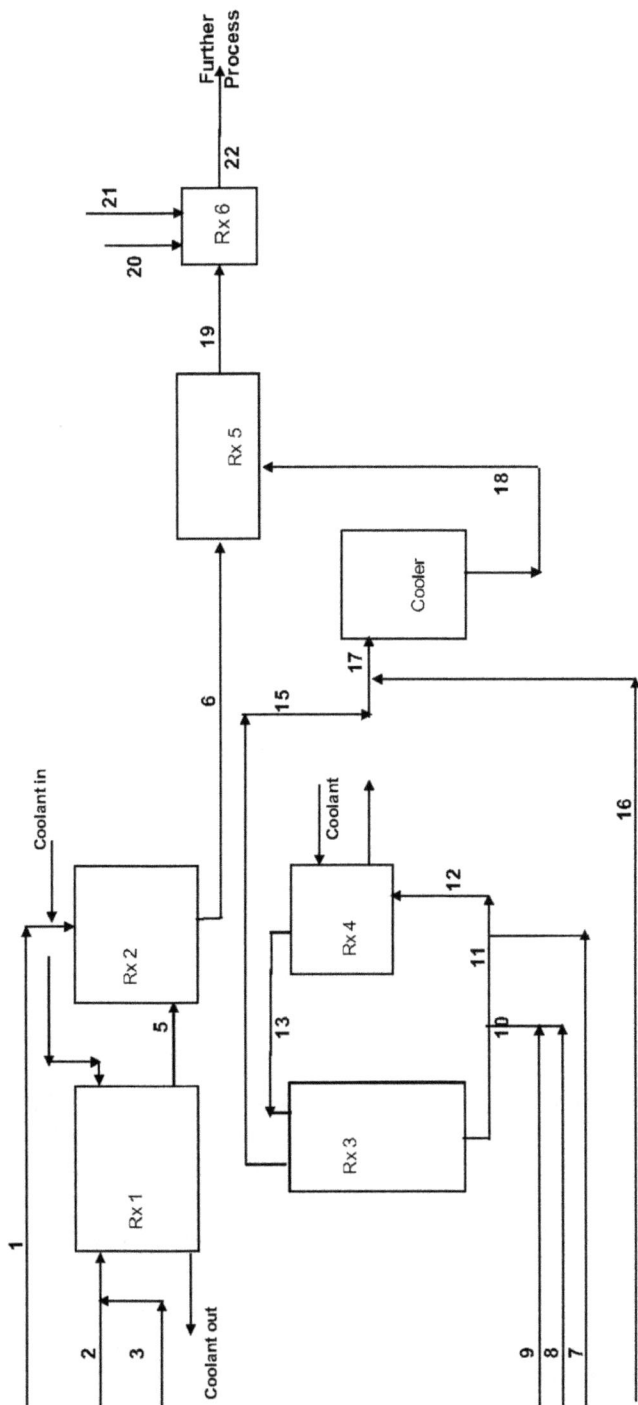

Fig. 7.12: Process flow diagram of a continuous process.

However for batch processes, the equipment is not necessarily designed for a specific product. Documentation of the process and other details mentioned in Tab. 7.2 can be a challenge. In these circumstances, extra care and scrutiny are needed at the start of each batch campaign. Product quality becomes paramount, in such cases.

7.6 Section 5: process equipment

For the manufacture of APIs as mentioned earlier, batch or continuous process are the two choices. However, based on the total global volume of each API, many plants produce the same API [45]. This suggests that each company could produce the API using their own process, chemistry, and equipment, which could be dissimilar to those of the next company or their own site. If the company outsources the API to a CMO (contract manufacturing organization), their equipment and methods could be different, too.

The manufacturer's approved API ANDA would be based on the equipment that is available at their site and is used for commercial production. It is expected that the process engineers would design a process that fits like a glove in new or existing equipment. Documentation of the process design and its rationale is critical for consistent product quality. It is important for each producing company to know the characteristics of each piece of equipment, so that the selected equipment and the process produce the desired quality product, from the onset.

If the same equipment is not available for routine product campaigns on an ongoing basis, it is very likely that the product profile could be different for each batch in each campaign. It is possible that the operating parameters and the operating procedures might have to be altered to assure quality. Thus, the documented information can be used to troubleshoot the process, if there is need. In addition, some of the "in-process" analysis can be minimized if every nuance of the reaction operating parameter and equipment performance is understood and is accounted for, in the operating procedure.

The basis of equipment selection and design criteria for the reactors, pumps, heat exchangers, centrifuge, and distillation unit have to be documented in the technical manual. Though the practice of documenting the design details might be considered obsolete, without the documentation of "why and what" of the process, no one will understand the rationale for the equipment design and its operations. Considering job reassignments, retirement, and job changes of operating production staff, such documentation is critical.

Process operating instructions are written on the basis of process chemistry and equipment design and authenticated. But if they are not practiced batch after batch or for the continuous process, it is very likely that the product will have batch-to-

Tab. 7.21: Heat and mass balance of a continuous process.

	1	2	3	4	5	6	7	8	9	10	11	12	13	14	15	16	17	18	19	20	21	22
Flow rate, Kg/hr	690	3,473	414	3,887	3,887	4,550	334	760	1,140	1,900	30,070	30,404	30,404	28,170	2,233	4,770	7,003	7,003	11,564	3,028	259	14,783
Temp. °C	250	15	30	23	4	31	30	30	15	48	40	45	40	40	15	s	23	8	33	15	30	50
Sp. Gr.																						
Composition																						
A	682																					
B			193	193	193	8		380		380	425	425	48	45	4		4	4	11			
C												334										
D	8	3,473	220	3,694	3,694	3,755	334	380	1,140	1,520	21,757	21,756	21,841	20,237	1,604	4,770	6,374	6,374	10,128	3,028		13,185
E											4,419	4,419	4,770	4,419	350		350	350				
F											3,470	3,470	3,745	3,470	276		275	275	276			549
G						751																
H						36																
I																						
J																					259	349
K																						700
L																			1,149			57

batch or lot-to-lot product quality variation. This cannot be tolerated, as it will result in lesser product quality and recalls. Thus, it is necessary that the equipment used is operated as close as possible to the process design criterion documented in the product's process operating manual. Process performance cannot deviate from the established process.

It is important that the functionality of the process equipment used for each processing step be described in the manual. If any equipment is specially designed and/or fabricated for use in the process, its description should be included. Agitated tanks that are used for holding materials need to be included. If a gravity-based decanter is used for phase separation, its design and process operating parameters need to be included. Basically, the rationale for use of each equipment design has to be described,

If a product requires use of a centrifuge or any filter or a crystallizer, their design basis and their suggested operating procedure need to be included.

Some examples of the equipment information, for example, heat exchangers (Fig. 7.13, 7.14), centrifugal pumps (Fig. 7.15, 7.16, 7.17, 7.18), and reactor (7.19) need to be included in the "Process Design and Technical Manual" are illustrated. All of the necessary design basis and calculations for each process equipment have to be included. This section would also include design basis information for the process controllers. Process control logic has to be included.

It is necessary that the process control design basis and the logic have to be included in the manual. The process control logic might have to be modified if different equipment than the original design is used in the manufacture of APIs.

In a batch or a continuous process, as has been discussed throughout the book, total command of the process is a must. This especially applies to the meters and valve that are used for the flow control of liquids and solids. They have to perform at their optimum continuously, when they are being used. The workmanship of the meters and control valves, over the years, have been perfected to precisely deliver the materials at the desired rates. However, they are machines that on occasions will not perform optimally. Thus, it is essential that they are calibrated on a planned schedule. Some might consider this an unnecessary perturbation, but experienced chemists and chemical engineers will understand the value of such an exercise. It will save the headaches of dealing with less-than-expected yield and off- specification product, which translate to financial loss.

Shell and Tube Heat Exchanger Specification Sheet

1	Company:						
2	Location:						
3	Service of Unit:		Our Reference:	Design by:			
4	Item No.: 1		Your Reference:				
5	Date:	Rev No.:	Job No.:				
6	Model	/ in	Type	Connected in		parallel	series
7	Surf/unit(eff.)	ft2	Shells/unit	Surf/shell (eff.)			ft2
8	**PERFORMANCE OF ONE UNIT**						
9	Fluid allocation			Shell Side		Tube Side	
10	Fluid name						
11	Fluid quantity, Total		lb/h				
12	Vapor (In/Out)		lb/h				
13	Liquid		lb/h				
14	Noncondensable		lb/h				
15							
16	Temperature (In/Out)		F				
17	Dew / Bubble point		F				
18	Density (Vap / Liq)		lb/ft3	/	/	/	/
19	Viscosity		cp	/	/	/	/
20	Molecular wt, Vap						
21	Molecular wt, NC						
22	Specific heat		BTU/(lb*F)	/	/	/	/
23	Thermal conductivity		BTU/(ft*h*F)	/	/	/	/
24	Latent heat		BTU/lb				
25	Pressure		psi				
26	Velocity		ft/s				
27	Pressure drop, allow./calc.		psi				
28	Fouling resist. (min)		ft2*h*F/BTU				
29	Heat exchanged		BTU/h		MTD corrected		F
30	Transfer rate, Service		Dirty		Clean		BTU/(h*ft2*F)
31	**CONSTRUCTION OF ONE SHELL**				Sketch		
32			Shell Side		Tube Side		
33	Design/Test pressure	psi	/ Code		/ Code		
34	Design temperature	F					
35	Number passes per shell						
36	Corrosion allowance	in					
37	Connections	In	/		/		
38	Size/rating	Out	/		/		
39	in	Intermediate	/		/		
40	Tube No.	OD	Tks- avg	in	Length	ft	Pitch	in
41	Tube type		Material			Tube pattern	
42	Shell	ID	OD	in	Shell cover		
43	Channel or bonnet				Channel cover		
44	Tubesheet-stationary				Tubesheet-floating		
45	Floating head cover	-			Impingement protection		
46	Baffle-crossing		Type		Cut(%d)	Spacing: c/c	in
47	Baffle-long		Seal type			Inlet	in
48	Supports-tube				Type		
49	Bypass seal			Tube-tubesheet joint			
50	Expansion joint			Type			
51	RhoV2-Inlet nozzle		Bundle entrance		Bundle exit		lb/(ft*s
52	Gaskets - Shell side		Gasket Grade	Tube Side	Gasket grade		
53	Floating head						
54	Code requirements			TEMA class			
55	Weight/Shell		Filled with water		Bundle		lb
56	Remarks						
57							
58							

Fig. 7.13: Shell and tube heat exchanger data sheet [46].

```
                Heat Exchanger Selection, version 8.5.2

   Job Name: 234094                        Engineer:
   Company:                                Representative:
   Customer: Power Plus                    Salesman:

                       ** INPUT PARAMETERS **

            TUBESIDE                                SHELLSIDE
   Fluid Type:            Water         Fluid Type:            Steam
   Flow Rate (gpm) :        110         Steam Press. (psig):      30
   Entering Temp (F):       145         Sat.Steam Temp(F):    274.03
   Leaving Temp (F):        170
   Fouling Factor:       0.0005
   Load (MBh):          1346.09

   Tube Material:   304 S.S. .049 in.
   Max Length (ft):          7
   LMTD:                116.08
   Lbs/hr. Steam        1449.2

                       ** SELECTION RESULTS **

        Dia. Num.  Length  Baff.   Tube     Tube   Shell    Shell
        (in) Passes  (ft)   Pitch  Vel(fps) Pd(ft) Vel(fps) Pd(ft)
          8    2       3             4.40    0.69
```

TUBESIDE	FLUID	FLOW (gpm)	TEMP. IN (F)	TEMP. OUT (F)	PRESS. DROP (ft)	VELOCITY (fps)
	Water	110	145	170	0.69	4.40

SHELLSIDE	STEAM	LBS. PER HOUR	SAT. STEAM TEMP.	STEAM PRESSURE		
		1449.2	274.03	30		

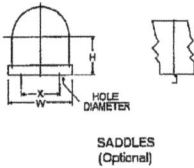

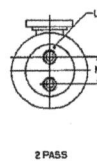

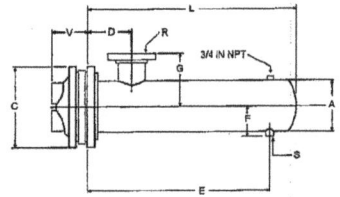

* Dimensions in inches, heating surface in sq.ft., weight in lbs.

M	U		V	A	C	D	E	F	G	L	R	S	Heating Surface	Shipping Weight
5	3T		3-3/8	8-5/8	13-1/2	8	30	6	6-7/8	36	2-1/2T	1T	25.1	234

SADDLE DIMENSIONS: H-7-5/16; W-11-1/4; X-9; Hole Diameter.-5/8

Fig. 7.14: Shell and tube heat exchanger data sheet [47].

ASSOC. PIP RESP73H/V	DATA SHEET	RESP73-DM
ASME CENTRIFUGAL PUMPS (SI UNITS)		PAGE 1 OF 3 MAY 2018

ISSUED FOR: ☐ PROPOSAL ☐ PURCHASE ☐ AS BUILT

FACILITY NAME/LOCATION: _____

ITEM NAME: _____	PURCHASER/LOCATION: _____
ITEM TAG NO.: _____	JOB NO: _____
SERVICE: _____	PURCHASER ORDER NO.: _____
UNIT: _____	SUPPLIER/LOCATION: _____
P&ID NO.: _____	SUPPLIER ORDER/SERIAL NOS.: _____ / _____

DATA PROVIDED BY: ● PURCHASER ■ SUPPLIER ◆ SUPPLIER IF NOT BY PURCHASER

● **GENERAL**

NO. PUMPS REQ.: _____ ◆ PUMP SIZE: _____ ◆ PUMP MODEL: _____ ◆ PUMP TYPE: _____

NUMBER MOTOR DRIVEN: _____	NUMBER TURBINE DRIVEN: _____	GEARBOX ITEM NUMBER: _____
MOTOR ITEM NUMBER: _____	TURBINE ITEM NUMBER: _____	GEARBOX PROVIDED BY: _____
MOTOR PROVIDED BY: _____	TURBINE PROVIDED BY: _____	GEARBOX MOUNTED BY: _____
MOTOR MOUNTED BY: _____	TURBINE MOUNTED BY: _____	

● **OPERATING CONDITIONS**

	RATED	MAX.	NORMAL	MIN.	
CAPACITY:					M^3/H
SUCTION PRESSURE:	KPA				
DISCHARGE PRESSURE:	KPA				
DIFFERENTIAL PRESSURE:	KPA				
DIFFERENTIAL HEAD:	M @ MINIMUM S.G.				
HYDRAULIC POWER:	KW				
OPERATING TIME:					HR/YR
NPSH AVAILABLE:					M
SUCTION SPECIFIC SPEED, MAXIMUM					

SYSTEM DESIGN:
☐ STAND ALONE OPERATION ☐ PARALLEL OPERATION
☐ SERIES OPERATION WITH ITEM NUMBER: _____
SUCTION PRESSURE: MIN./MAX.: _____ / _____ KPA

SERVICE:
☐ CONTINUOUS ☐ INTERMITTENT: _____ STARTS/DAY

SYSTEM CONTROL METHOD:
☐ SPEED ☐ FLOW ☐ LEVEL ☐ TEMPERATURE
☐ PRESSURE ☐ PIPE FRICTION RESISTANCE ONLY

■ **PERFORMANCE**

PERFORMANCE CURVE NO.: _____ ◆ SPEED: _____ RPM

	RATED	MAX.	NORMAL	MIN.	
MEASURED AT CAPY.:					
NPSH REQ'D.:					M

TOTAL DIFFERENTIAL HEAD @ RATED IMPELLER: _____ M
MAX. DIFFERENTIAL HEAD @ RATED IMPELLER: _____ M
MINIMUM CONTINUOUS FLOW:
THERMAL: _____ M^3/H STABLE: _____ M^3/H
ALLOWABLE OPERATING REGION: _____ TO: _____ M^3/H
BEST EFFICIENCY POINT FOR RATED IMPELLER: _____ M^3/H
SUCTION SPECIFIC SPEED: _____
IMPELLER DIA.: RATED: _____ MAX.: _____ MIN.: _____ MM
PUMP RATED POWER: _____ KW EFFICIENCY: _____ %
MAXIMUM POWER @ RATED IMPELLER: _____ KW

CASE PRESSURE RATING:
☐ MAX. ALLOWABLE WORKING PRES.: _____ KPA @ _____ °C
☐ HYDROSTATIC TEST PRESSURE: _____ KPA

● **PUMPED FLUID**

PUMPED FLUID: _____

	RATED	MAX.	NORMAL	MIN.	
PUMPING TEMP.:					°C
AT DESIGNATED TEMP.:	RATED	MAX.	NORMAL	MIN.	
SPECIFIC GRAVITY:					
VAPOR PRESSURE:					KPA ABS
VISCOSITY:					CP
SPECIFIC HEAT:					KJ/KG°C

INITIAL BOILING POINT: _____ °C @ _____ KPA ABS
LIQUID: ☐ HAZARDOUS ☐ FLAMMABLE
☐ OTHER: _____
_____
_____
CORROSION / EROSION CAUSED BY: _____
% SOLID: _____ MAX. PARTICLE SIZE: _____ MM

● **SITE CONDITIONS**

LOCATION:
☐ INDOOR ☐ OUTDOOR
ALTITUDE: _____ M
RANGE OF AMBIENT TEMPS.: MIN./MAX.: _____ / _____ °C
ELECTRICAL CLASSIFICATION:
CL.: _____ GR.: _____ DIV.: _____
☐ NON HAZARDOUS

● **GENERAL REMARKS**
_____
_____
_____
_____
_____
_____
_____
_____
_____

NO.	DATE	REVISION DESCRIPTION	BY	APVD.

Fig. 7.15: Centrifugal pump data sheet [48].

PiP	ASSOC. PIP RESP73H/V	DATA SHEET	RESP73-DM
	ASME CENTRIFUGAL PUMPS (SI UNITS)		PAGE 2 OF 3 MAY 2018

MECHANICAL DATA

◆ IMPELLER TYPE:
☐ CLOSED ☐ OPEN ☐ SEMI-OPEN
☐ OTHER: _____

◆ CASING MOUNTING:
☐ FOOT ☐ CENTERLINE
☐ VERTICAL IN-LINE ☐ CLOSE-COUPLED

BEARINGS:
◆ BEARING MANUFACTURER: _____
■ RADIAL BEARING TYPE: _____ NO. _____
■ THRUST BEARING TYPE: _____ NO. _____
■ BEARING ISOLATORS:
■ SHAFT STIFFNESS RATIO (L^3/D^4): _____

● LUBRICATION:
☐ FLOOD ☐ PURE MIST ☐ SHIELDED (GREASE)
☐ GREASE ☐ PURGE MIST ☐ SEALED (GREASE)
☐ CONSTANT LEVEL OILER REQUIRED
■ ☐ HOUSING VENT REQUIRED
☐ MAGNETIC DRAIN PLUG IN HOUSING REQUIRED
◆ ☐ OIL COOLER REQUIRED
☐ SEAL SPRAY GUARD REQUIRED
◆ ☐ OIL VISCOSITY: ISO GRADE: _____ OTHER: _____

NOZZLE CONNECTIONS:
■ SIZE ◆ RATING ◆ FACING
SUCTION:
DISCHARGE:

● AUX. CASE CONNECTION: ☐ DRAIN REQUIRED
◆ SIZE: _____ MM
☐ THREADED ☐ WELDED & FLANGED

◆ MATERIALS

MATERIAL CLASS CODE: _____
CASING: _____
IMPELLER: _____
COVER: _____
SHAFT: _____
SHAFT SLEEVE: _____
BASEPLATE: _____
CASING GASKET: _____
IMPELLER GASKET: _____
CASING FASTENERS IF NOT PER 4.8.1: _____
BEARING HOUSING: _____
BEARING HOUSING ADAPTER: _____
BEARING HOUSING END SEALS: _____
COUPLING GUARD: _____
MECH. SEAL GLAND: _____
MECH. SEAL GLAND FASTENERS: _____
O-RING MATERIAL (4.6.1.3): _____

◆ DRIVER

HORSEPOWER RATING: _____ KW SPEED: _____ RPM
DRIVE HP SELECTED FOR MAX. S.G.: _____ & MAX. VISC.: _____ CP
REMARKS: _____
_____

◆ COUPLING BETWEEN PUMP AND DRIVER

MANUFACTURER: _____
TYPE: _____
SIZE: _____
MODEL: _____
SPACER LENGTH: _____ MM
COUPLING GUARD TYPE:
☐ MANUFACTURER'S STANDARD
☐ BASEPLATE MOUNTED
☐ NON-SPARK COUPLING GUARD REQUIRED
REMARKS: _____

● BASEPLATE

TYPE: ☐ GROUTED ☐ FREE STANDING
☐ PRECAST POLYMERS
◆ ☐ CENTERLINE OF PUMP TO STILT BOTTOM: _____ MM
☐ VERTICAL PUMP CASE SUPPORT BRACKET
DESIGN: ☐ PIP STANDARD RESP002 (DATA SHEET ATTACHED)
☐ MANUFACTURER'S STANDARD
REMARKS: _____

● PAINT AND SHIPMENT PREPARATION

PUMP: BASEPLATE:
☐ MANUFACTURER'S STD. ☐ MANUFACTURER'S STD.
☐ OTHER: ☐ OTHER:
SHIPMENT:
☐ DOMESTIC ☐ EXPORT ☐ EXPORT BOXING REQ'D.
NUMBER OF MONTHS OF STORAGE: _____
■ TOTAL WEIGHT: _____ KG

■ INSPECTION AND TESTING

☐ FINAL INSPECT. REQD.: DAYS NOTIFICATION REQD.: _____

TEST:	NON-WITNESSED	WITNESSED	CERTIFICATE
HYDROSTATIC:	☐	☐	☐
PERFORMANCE:	☐	☐	☐
NPSHR:	☐	☐	☐
VIBRATION:	☐	☐	☐
OTHER:	☐	☐	☐

☐ DISMANTLE AND INSPECT AFTER TEST
☐ CASTING REPAIR PROCEDURE APPROVAL REQUIRED
MATERIAL CERTIFICATION REQUIRED:
☐ CASING ☐ COVER ☐ IMPELLER ☐ SHAFT
☐ OTHER:
INSPECTION REQUIRED FOR CONNECTION WELDS:
☐ MANUFACTURER'S STANDARD ☐ VISUAL INSPECTION
INSPECTION REQUIRED FOR CASTINGS:
☐ MANUFACTURER'S STANDARD ☐ VISUAL INSPECTION
☐ OTHER:

● MANUFACTURER DOCUMENTATION REQUIREMENTS

FOR SUPPLIER DATA REQUIREMENTS REFER TO: _____
REMARKS: _____

Fig. 7.16: Centrifugal pump data sheet [48].

PiP	ASSOC. PIP RESP73H/V	DATA SHEET	RESP73-DM
	ASME CENTRIFUGAL PUMPS (SI UNITS)		PAGE 3 OF 3 MAY 2018

MECHANICAL SEAL

- FURNISHED BY: ☐ SUPPLIER ☐ PURCHASER
- MOUNTED BY: ☐ SUPPLIER ☐ PURCHASER
- ◆ SEAL CLASSIFICATION CODE: _____
- ◆ MANUFACTURER: _____
- ◆ MODEL: _____
- ◆ MANUFACTURER CODE: _____
- SEAL TYPE: ☐ CARTRIDGE ☐ COMPONENT
- SEAL DESIGN: ☐ SINGLE ☐ DUAL ☐ DRY GAS
  - ☐ PRESSURIZED ☐ UNPRESSURIZED
- SEAL CHAMBER: ☐ TAPER BORE ☐ CYLINDRICAL BORE
- SEAL CHAMBER SIZE: ☐ OVERSIZED ☐ STANDARD
- ◆ SLEEVE MATERIAL: _____
- ◆☐ PUMPING RING REQUIRED: _____
- ◆☐ THROAT BUSHING REQUIRED: _____
  - MATERIALS: _____
- ◆ REMARKS: _____
  - _____
  - _____

SEAL GLAND

- TAPS REQUIRED: ☐ QUENCH ☐ FLUSH ☐ DRAIN
  - ☐ OTHER: _____
- ◆☐ THROTTLE BUSHING:
  - MATERIALS: _____

SEAL FLUSHING

- PRIMARY FLUSH PLAN NO.: _____
- EXTERNAL FLUSH LIQUID: _____
- SUPPLY TEMP. MIN./MAX.: _____ / _____ °C
- SPECIFIC GRAVITY: _____
- SPECIFIC HEAT: _____ KJ/KG°C
- VAPOR PRESSURE: _____ KPA ABS @ _____ °C
- ■ FLOW RATE REQD. MIN./MAX.: _____ / _____ M³/H
- ■ PRESSURE REQD. MIN./MAX.: _____ / _____ KPA
- ■ TEMP. REQD. MIN./MAX.: _____ / _____ °C
- PRIMARY SEAL FLUSH PIPING:
  - ☐ TUBING ☐ PIPE ☐ OTHER: _____
- MATERIAL:
  - ☐ 316 SS ☐ CARBON STEEL ☐ OTHER: _____
- PIPING/TUBING SIZE: _____ MM
- PIPING ASSEMBLY:
  - ☐ THREADED ☐ UNIONS ☐ FLANGED
  - ☐ TUBE TYPE FITTINGS ☐ SOCKET WELDED
- ◆ REMARKS: _____
  - _____
  - _____
  - _____
  - _____
  - _____

BARRIER/BUFFER FLUSH SYSTEM

- BARRIER FLUSH PLAN NO.: _____
- BARRIER FLUSH LIQUID: _____
- TEMPERATURE MIN./MAX.: _____ / _____ °C
- SPECIFIC GRAVITY: _____
- SPECIFIC HEAT: _____ KJ/KG°C
- VAPOR PRESSURE: _____ KPA ABS @ _____ °C
- ◆ PRESSURE REQD. MIN./MAX.: _____ / _____ KPA
- ■ MAWP OF SECONDARY SEAL SYSTEM: _____ KPA
- TEMP. REQD. MIN./MAX.: _____ / _____ °C
- SECONDARY SEAL FLUSH PIPING:
  - ☐ TUBING ☐ PIPE ☐ OTHER: _____
- MATERIAL:
  - ☐ 316 SS ☐ CARBON STEEL ☐ OTHER: _____
- PIPING/TUBING SIZE: _____ MM
- PIPING ASSEMBLY:
  - ☐ THREADED ☐ UNIONS ☐ FLANGED
  - ☐ TUBE TYPE FITTINGS ☐ SOCKET WELDED
- SEAL POT IF NOT PER 4.6.7.1: ☐
- ◆ REMARKS: _____

SEAL SYSTEM INSTRUMENTATION

PRIMARY SEALS:	GAUGES	SWITCHES	TRANSMITTERS
FLOW:	☐	☐	☐
TEMPERATURE:	☐	☐	☐
PRESSURE:	☐	☐	☐

SECONDARY SEALS:			
FLOW:	☐		☐
PRESSURE:			
LEVEL:			

REMARKS: _____
_____

COOLING OR HEATING PIPING PLANS

- PIPING PLAN NO.: _____
- NAME OF FLUID: _____
- INLET TEMPERATURE: _____ °C
- OUTLET TEMPERATURE: _____ °C
- ■ RATED FLOW: _____ M³/H
- ■ SUPPLY PRESSURE: _____ KPA
- ■ MAX. ALLOWABLE ΔP: _____ KPA
- ☐ GALVANIZED PIPE ☐ SS TUBING
- ☐ SIGHT FLOW INDICATOR
- ☐ OUTLET SHUT-OFF VALVE
- ◆ REMARKS: _____
  - _____
  - _____
  - _____
  - _____

Fig. 7.17: Centrifugal pump data sheet [48].

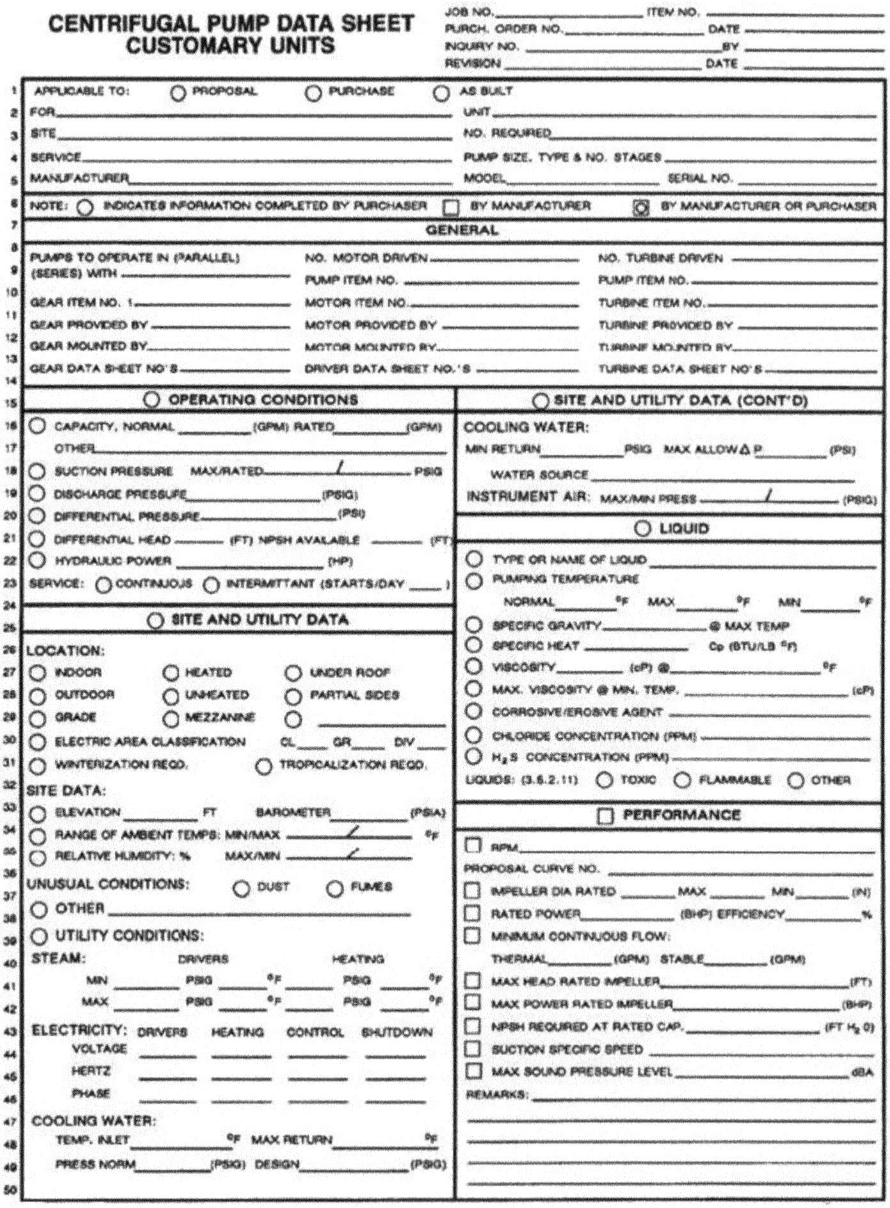

CENTRIFUGAL PUMP DATA SHEET
CUSTOMARY UNITS

JOB NO. _____ ITEM NO. _____
PURCH. ORDER NO. _____ DATE _____
INQUIRY NO. _____ BY _____
REVISION _____ DATE _____

1 APPLICABLE TO: ◯ PROPOSAL ◯ PURCHASE ◯ AS BUILT		
2 FOR _____	UNIT _____	
3 SITE _____	NO. REQUIRED _____	
4 SERVICE _____	PUMP SIZE, TYPE & NO. STAGES _____	
5 MANUFACTURER _____	MODEL _____ SERIAL NO. _____	
6 NOTE: ◯ INDICATES INFORMATION COMPLETED BY PURCHASER ☐ BY MANUFACTURER ◉ BY MANUFACTURER OR PURCHASER		

GENERAL

8 PUMPS TO OPERATE IN (PARALLEL)	NO. MOTOR DRIVEN _____ NO. TURBINE DRIVEN _____
9 (SERIES) WITH _____	PUMP ITEM NO. _____ PUMP ITEM NO. _____
10 GEAR ITEM NO. 1 _____	MOTOR ITEM NO. _____ TURBINE ITEM NO. _____
11 GEAR PROVIDED BY _____	MOTOR PROVIDED BY _____ TURBINE PROVIDED BY _____
12 GEAR MOUNTED BY _____	MOTOR MOUNTED BY _____ TURBINE MOUNTED BY _____
13 GEAR DATA SHEET NO'S _____	DRIVER DATA SHEET NO.'S _____ TURBINE DATA SHEET NO'S _____

◯ OPERATING CONDITIONS | **◯ SITE AND UTILITY DATA (CONT'D)**

15 |
16 ◯ CAPACITY, NORMAL _____ (GPM) RATED _____ (GPM) | COOLING WATER:
17 ◯ OTHER _____ | MIN RETURN _____ PSIG MAX ALLOW Δ P _____ (PSI)
18 ◯ SUCTION PRESSURE MAX/RATED _____/_____ PSIG | WATER SOURCE _____
19 ◯ DISCHARGE PRESSURE _____ (PSIG) | INSTRUMENT AIR: MAX/MIN PRESS _____/_____ (PSIG)
20 ◯ DIFFERENTIAL PRESSURE _____ (PSI) |
21 ◯ DIFFERENTIAL HEAD _____ (FT) NPSH AVAILABLE _____ (FT) | **◯ LIQUID**
22 ◯ HYDRAULIC POWER _____ (HP) | ◯ TYPE OR NAME OF LIQUID _____
23 SERVICE: ◯ CONTINUOUS ◯ INTERMITTANT (STARTS/DAY ___) | ◯ PUMPING TEMPERATURE
24 | NORMAL _____ °F MAX _____ °F MIN _____ °F
25 **◯ SITE AND UTILITY DATA** | ◯ SPECIFIC GRAVITY _____ @ MAX TEMP
26 LOCATION: | ◯ SPECIFIC HEAT _____ Cp (BTU/LB °F)
27 ◯ INDOOR ◯ HEATED ◯ UNDER ROOF | ◯ VISCOSITY _____ (cP) @ _____ °F
28 ◯ OUTDOOR ◯ UNHEATED ◯ PARTIAL SIDES | ◯ MAX. VISCOSITY @ MIN. TEMP. _____ (cP)
29 ◯ GRADE ◯ MEZZANINE ◯ _____ | ◯ CORROSIVE/EROSIVE AGENT _____
30 ◯ ELECTRIC AREA CLASSIFICATION CL___ GR___ DIV___ | ◯ CHLORIDE CONCENTRATION (PPM) _____
31 ◯ WINTERIZATION REQD. ◯ TROPICALIZATION REQD. | ◯ H_2S CONCENTRATION (PPM) _____
32 SITE DATA: | LIQUIDS: (3.6.2.11) ◯ TOXIC ◯ FLAMMABLE ◯ OTHER
33 ◯ ELEVATION _____ FT BAROMETER _____ (PSIA) |
34 ◯ RANGE OF AMBIENT TEMPS: MIN/MAX _____/_____ °F | **☐ PERFORMANCE**
35 ◯ RELATIVE HUMIDITY: % MAX/MIN _____/_____ |
36 UNUSUAL CONDITIONS: ◯ DUST ◯ FUMES | ☐ RPM _____
37 | PROPOSAL CURVE NO. _____
38 ◯ OTHER _____ | ☐ IMPELLER DIA RATED _____ MAX _____ MIN _____ (IN)
39 ◯ UTILITY CONDITIONS: | ☐ RATED POWER _____ (BHP) EFFICIENCY _____ %
40 STEAM: DRIVERS HEATING | ☐ MINIMUM CONTINUOUS FLOW:
41 MIN _____ PSIG _____ °F _____ PSIG _____ °F | THERMAL _____ (GPM) STABLE _____ (GPM)
42 MAX _____ PSIG _____ °F _____ PSIG _____ °F | ☐ MAX HEAD RATED IMPELLER _____ (FT)
43 ELECTRICITY: DRIVERS HEATING CONTROL SHUTDOWN | ☐ MAX POWER RATED IMPELLER _____ (BHP)
44 VOLTAGE _____ _____ _____ _____ | ☐ NPSH REQUIRED AT RATED CAP. _____ (FT H_2O)
45 HERTZ _____ _____ _____ _____ | ☐ SUCTION SPECIFIC SPEED _____
46 PHASE _____ _____ _____ _____ | ☐ MAX SOUND PRESSURE LEVEL _____ dBA
47 COOLING WATER: | REMARKS: _____
48 TEMP. INLET _____ °F MAX RETURN _____ °F |
49 PRESS NORM _____ (PSIG) DESIGN _____ (PSIG) |
50 |

API-610 Pump data sheet (customary units)

Fig. 7.18: Centrifugal pump data sheet (API-610) [49].

DOCUMENT NO.	AGITATED VESSEL PROCESS DATA SHEET			SHT : 1
AB2201-R1101	PROJECT : XXXXXXXX		LOCATION : XXXXXXX	

Extreme Engineering Technology Consultant LLP.

AGITATED REACTOR R-1101					
	01-02-2021	For Inq.	0	RG	DA	KM
	DATE	ISSUE	REVISION	PREP	CHKD	APPD

EQUIPMENT NAME :	Reactor		QTY	1
SUPPLIER	M/S	TYPE: FLANGED / MONOBLOCK		

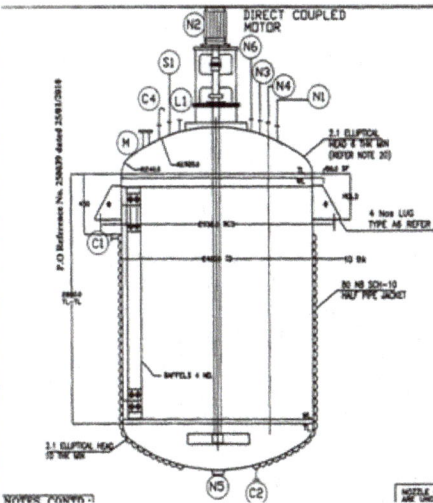

OPERATING / MECHANICAL DATA

	Vessel	Jacket/Limpet
FLUID	Organic	Thermic fluid
Operating Temperature (°C)	180	230
Operating Pressure (kg/cm²g)	1.5 / Full Vacuum	2
Design Temperature (°C)	225	290
Design Pressure (kg/cm²g)	3	3
Hydrotest Pressure (kg/cm²g)	6	6
Construction Code	ASME Sec. VIII, Div. 1	ASME Sec. VIII, Div. 1
Stress Relieving	Acc. to code	Acc. to code
Joint Efficiency	85%	70%
Radiography	Spot	Nil
Total Volume		
Working Volume	3 kL	-
Shell ID	-	mm
Length. TL-TL	-	mm
Jacket D (mm)	-	mm
Jacket Height from Bottom TL	-	mm
Height of Support from Bottom TL	-	mm
Corrosion Allowance (Vessel/Jkt)		mm
Insulation Thikness		mm
Shell Thickness	6	mm
Dish Thickness (Top / Bottom)	8 / 8	mm
Jacket Shell Thk	-	mm
Empty Wt.	*	kgs
Water Filled Wt.	*	kgs
Number of baffles	4	nos
Baffle thickness	6	mm
Baffle (width x length)	100 x 1500	mm

Type of support	Bracket
Number of supports	4

Remarks:

1> Mechanical design gurantee of equipment will be in vendor's scope.

2> All spare nozzles, handholes & manholes will be provided with blind flanges.

3> Insulation cleats will be provided where ever needed.

4> Spare gaskets & fasterners will be supplied with the equipment.

5> CS equipments / parts will be painted with primer.

MATERIAL OF CONSTRUCTION

Main Shell	SA 240 Gr.SS316
Dish Ends	SA 240 Gr.SS316
Nozzle Flanges	SA 240 Gr.SS316
Nozzle Pipes	SA 240 Gr.SS316
Limpet Coil	CS
Baffle	SA 240 Gr.SS316
Support	IS 2062
Nut & Bolts	MS
Gasket	CAF

NOZZLE SCHEDULE (ANSI B16.5, SCH 40,150#)

NOZZLE	SIZE (NB)	SERVICE	QTY
N1	80	Inlet	1
N2	80		1
N3	50		1
N4	40	Thermowell	1
N5	100	Bottom Outlet	1
N6			1
L1			1
C2	80	Utility outlet	1
C1	80	Utility inlet	1
M		Manhole / Hand hole	1

Fig. 7.19: Agitated vessel process data sheet [50].

7.7 Section 6: process conditions and effect of variables

It is important to document the cause and effect of every deviation of the process condition. This is critical as the majority of API manufacturing processes are batch operations. In addition, if different equipment, on the same or other site is used to produce the API, significant process variability can be introduced in the API manufacturing process, as the behavior of every process equipment can be different, even when the design criterion and fabrication are the same.

Reaction mechanisms, generally, give excellent clues about the sequence of addition of chemicals for most reactions. One of the simplest examples is the formation of a diazo salt of an aromatic amine. If hydrochloric acid is to be used in the reaction, addition of acid to amine (it could be dissolved in water or appropriate solvent) will produce the amine hydrochloride, which, in turn, reacts with sodium nitrite to form the diazo salt that is used in subsequent reaction. However, if the hydrochloric acid is added to amine and sodium nitrite mix, NO_x will be formed, which is not desirable. In addition, impurity formation can lower the yield.

Process stoichiometry has to be precisely controlled. It is necessary to understand and document the influence of stoichiometry variation on the process yield and the impurity profile. The same applies to temperature operating parameters, feed rates and raw material addition methods, and the points of addition. Any and every deviation will be manifested by change in overall reaction profile, and that can change the process yield and impurity profile. Mole ratios of process streams cannot deviate outside the established control limits. "What if" analysis and understanding of each processing step and stoichiometry are important and have to be documented.

Chemists and chemical engineers can use some of the following to create their own guidelines for each process. It is critical that the changes to a commercial process are not made without authorization. If changes are to be made, they need to be tested and authenticated.

As has been suggested in earlier chapters, changes to the process and equipment can influence process dynamics and costs.

7.7.1 Liquid/solid feeds

It is extremely critical that the liquid and solid feed systems are the same for the designated API manufacture. Not only the feed mechanism, but also the location and method of feed in the reaction scheme can influence the process. The rationale for this is that the raw materials react with each other to produce a different product. Variations between equipment on the same site or other sites can influence the process stoichiometry and the outcome. Metering systems for the solids and liquids have to be calibrated respectively on a routine basis for the liquid and the solid that will be used in the process. Optimum performance is critical.

7.7.2 Batch process

In majority of batch processes, a solvent is used to facilitate the reaction. The solvent is generally heated to a prescribed temperature. If the amount of solvent use is changed without testing and authorization, process results can be different from the approved process. Reactants are added in the predetermined order to complete the reaction. It is necessary that the temperature profiles (heating and cooling), agitation, and feed methods are the same for each batch of the API. This is critical. Not having the same heating and cooling profile and the residence time can greatly influence the process yield and the impurity profile. Since the equipment size change is not translational, processing conditions change and care is needed to understand the influence on operating conditions.

Batch stoichiometry has to be controlled well within the design parameters. If the product can be or has to be reworked to an acceptable and saleable product, it can cost as much as 25% of the standard factory API cost [51].

In many of the API processes, gaseous raw materials are used. They are used as a liquid because the liquid feed rate can be precisely controlled. Operating conditions have to be precisely defined. Since the gas will evaporate and need heat for evaporation, any reaction exotherm can be easily absorbed. Calibration of gaseous raw materials can be a challenge but can be done.

7.7.3 Continuous process

Continuous processes are very different from batch processes; so are their operating philosophies. Per established definitions [3, 4], they are supposed to operate $24 \times 7 \times 50 = 8,400$ h/year. Each operating company, based on their established practices, allocates a certain time for maintenance and unexpected shutdowns. Unlike batch processes, the liquid and solid intermediate materials are not held in any storage facility.

Since the liquids are added in an established sequence to capitalize on their mutual properties and reactivity, it is necessary that the same sequence and stoichiometry are also followed in a commercial operation. Temperature and pressure profiles have to be precisely followed.

In continuous processes, stoichiometry has to be controlled extremely well. Deviations outside the established parameters will result in off-spec product and lower-than-expected yield, resulting in significant financial loss through nonsalable product, cleanup, and disposal costs. Rework of the reactive in-process materials is difficult, if not impossible. If the cause of deviation is not remedied quickly, equipment will be idle until the cause of error is figured and remedied. All this can happen as the flow of the materials is time-independent.

Most companies understand the impact of batch vs. continuous processes and have to make sure every cause and effect of process deviation from the set parameters is understood well. Technology and methodology of precise process control has been, and, is being, practiced every day, in the fine/specialty chemical industry, the older cousin of the pharmaceutical industry, which includes API manufacturing and their formulations. Thus, precise controls are not a new art or technology.

In each type of manufacturing process manual, the chemist or the chemical engineer should include the influence of stoichiometry, temperature, and feed rate variation in the product yield and quality. Documentation of impact of these variables is extremely useful in troubleshooting the product quality and yield variation.

7.8 Section 7: suggested operating conditions

Operating conditions for a batch or a continuous process have to be defined precisely and documented. They need to be strictly followed and cannot be changed without testing and authorization.

For a batch process, the raw materials have to be charged in the predetermined sequence, at the defined weights, and heated/reacted at the operating temperatures.

For a continuous process, process operating manual has to have a startup and a shutdown procedure. Once the process is operating, the feed rate of each reacting process raw material has to be at the established process feed controllers/meters settings. Heating and cooling parameters have to be monitored and controlled. Cleaning procedures and their sequence have to be defined well.

7.9 Section 8: laboratory synthesis procedure

Process development starts with writing the chemistry on paper and progressing to visualize its commercialization. It may sound or look to be simple and is a simple exercise. However, on the contrary, it is the most difficult task that needs to be perfected, as it is the first brick of the process foundation.

Different process schemes and methods are imagined and tested in the laboratory. The process selected is the most viable, safe, economical, and has to be easily executable on commercial scale. If we review the processes that are outlined in the patents, they are based on the traits of the chemicals used and can produce the desired product. However, generally, the methodology outlined in the patent is for batch processes, is based on laboratory equipment, and may not be the one that is used in the commercial process.

The following example is for the diazonium salt formation of methyl anthranilate. The synthesis for a batch process is outlined in USP 4464537 [52]:

$$RNH_2 + 2HCl + NaNO_2 = RN_2Cl + 2H_2O + NaCl \quad \text{Eq. 7.1}$$

About 76 g (0.5 mol) of methyl anthranilate and 135 g of an aqueous sodium nitrite solution (containing 0.5 mol of $NaNO_2$) were added gradually in the course of 30 min to 196 gm. of 30% strength by weight aqueous hydrochloric acid (=1.61 mol of HCl) at from – 5.0 to + 10 °C. The diazonium salt solution is processed further.

The laboratory continuous process is as follows. Table 7.22 is the stoichiometry for the laboratory process.

Tab. 7.22: Laboratory synthesis of diazonium salt methyl anthranilate.

Raw material	Mol. wt	Mole ratio	g/g of MA-T	Unit charge (g/test)
Methyl anthranilate	151.1	1.00	1.00	37.8
Water	18	8.89	1.058	40.0
HCl (100%) use 37%	36.5	2.41	0.583	22.02 (50 mL)
Sodium nitrite	69	1.10	0.503	19.0

Synthesis for the same synthesis emulating a continuous process is as follows:

Place 50 mL of HCl (37%) in 250 mL beaker equipped with stirrer and a thermometer. Add 5 mL of water and cool to 0–5 °C using suitable ice bath. Dissolve sodium nitrite in 30 mL of water and place in an addition funnel with a long tube. Charge about 4 mL of methyl anthranilate with a dropper into the stirred acid followed by dropwise addition of sodium nitrite, below the acid surface, Continue alternate addition of the amine and the nitrite. The appearance of brown fume (NO_x) should be avoided by controlling the addition of nitrite. With all of the amine in, the remaining nitrite solution should be added to starch-iodide end point. The color of anthranilate diazo is golden yellow, and it is a clear solution.

Process design for the batch process will be different from that for the continuous process.

7.10 Section 9: analytical methods

It is necessary to document each analytical method used to measure and correct every stoichiometric and process deviation. In the batch processes, any process deviation due to stoichiometry imbalance or temperature deviation within limits can be corrected and the manufacturing process controlled to produce a product that meets its specifications. However, in a continuous process, deviation from established stoichiometry and temperature parameters can result in significant financial loss, because the incorrect chemistry or process operating parameters can produce significant quantities of off-spec material that might have to be disposed.

Analytical methods for every process step that is considered necessary for the manufacture of the product should be detailed. An HPLC (high-performance liquid) analysis method for dimethyl fumarate [53] is detailed below.

HPLC analysis of dimethyl fumarate [53]
Reagents, solvents, and standards:

Water (Milli Q or equivalent)
Perchloric acid 70% (AR grade)
Acetonitrile (HPLC grade)

Chromatographic conditions:

Apparatus: A high-performance liquid chromatograph equipped with quaternary gradient pumps, variable wavelength UV detector attached with data recorder, and integrator software.

Column: GL Science, Inertsil ODS 3 V, 250 × 4.6 mm, 5 μ

Column temperature: 30 °C
Sample cooler temperature: 15 °C
Mobile phase:
Mobile phase A = buffer
Buffer: 0.5 mL of perchloric acid dissolve in 1,000 mL of water (0.05% perchloric acid in water)
Mobile phase B = acetonitrile

Time (min)	% mobile phase A	% mobile phase B
0.01	90	10
3	90	10
30	50	50
40	90	50
43	90	10
50	90	10

Diluent: acetonitrile:water (20:80, v/v)
Flow rate: 1.0 mL/min
Detection: UV 210 nm
Injection volume: 20 μL

The retention time of dimethyl fumarate is about 20.0 min under these conditions. Relative retention time for fumaric acid is about 0.23 and for monomethyl fumarate is about 0.57, with respect to dimethyl fumarate under these conditions.

The present invention provides dimethyl fumarate, obtained by the above process, wherein the amount of an anion of a mineral acid is determined by analytical techniques; for example, chloride ion content can be monitored by ion chromatography, potentiometric titration with silver nitrate, turbidimetric method; sulfate ion content can monitored by ion chromatography, turbidimetric method.

The present invention provides dimethyl fumarate, obtained by the above process, wherein the chloride ion content was determined by USP method as described below:

Preparation of standard solution:

Take 0.2 mL of 0.02 N hydrochloride acid solution in 30–40 mL of water and mix. Filter this solution through 0.45 μ Nylon filter. Add 1 mL each of nitric acid and 0.1 N silver nitrate TS and sufficient water to make 50 mL. Mix and allow to stand for 5 min protected from direct sunlight.

Preparation of test solution:

Take 0.35 g sample in suitable test tube. Add 30–40 mL of water and sonicate for 3–4 min and filter through 0.45 μ Nylon filter. Add 1 mL each of nitric acid and 0.1 N silver nitrate TS and sufficient water to make 50 mL. Mix and allow to stand for 5 min protected from direct sunlight.

Then, compare the turbidity of test and standard solution.

The present invention provides dimethyl fumarate obtained by the above process, having dimethyl maleate content, as described, analyzed by using high performance liquid chromatography (HPLC), with the conditions described below:

Reagents, solvents, and standards:

Water (Milli Q or equivalent)
Potassium dihydrogen phosphate (AR grade)
Acetonitrile (HPLC grade)
o-Phosphoric acid (AR grade)

Chromatographic conditions:

Apparatus: A high-performance liquid chromatograph equipped with quaternary gradient pumps, variable wavelength UV detector attached with data recorder, and integrator software.

Column: GL-Science, Inertsil ODS 3 V, 250 × 4.6 mm, 5 mu
Column temperature: 30 °C
Sample cooler temperature: 15 °C
Mobile phase:
Mobile phase A = buffer
Buffer: 0.01 M potassium dihydrogen phosphate. Adjust pH 4.0 with o-phosphoric acid.
Mobile phase B = acetonitrile

Time (min)	% mobile phase A	% mobile phase B
0.01	80	20
03	80	20
30	50	50
32	80	20
40	80	20

Diluent: acetonitrile:water (40:60, v/v)
Flow rate: 1.0 mL/min
Detection: UV 210 nm
Injection volume: 50 μL

The retention time of dimethyl maleate is about 10.0 min under these conditions.

The present invention provides dimethyl fumarate, a compound of formula I, containing less than 400 ppm of an anion of a mineral acid.

In one embodiment, the present invention provides dimethyl fumarate, a compound of formula I, containing less than 400 ppm of chloride ion.

In one preferred embodiment, the present invention provides dimethyl fumarate, a compound of formula I, containing less than 100 ppm of chloride ion.

The present invention provides dimethyl fumarate, a compound of formula I, containing less than 400 ppm of an anion of a mineral acid and free of dimethyl sulfate.

In one embodiment, the present invention provides dimethyl fumarate, a compound of formula I, containing less than 400 ppm of chloride ion and free of dimethyl sulfate.

In one preferred embodiment, the present invention provides dimethyl fumarate, a compound of formula I, containing less than 100 ppm of chloride ion and free of dimethyl sulfate.

The present invention provides dimethyl fumarate obtained by the above process, having dimethyl sulfate content, as described, determined by residual solvent analysis using gas chromatography (GC), with the conditions described below:

Chromatographic parameters:

Instrument: Gas chromatograph equipped with FID detector and autosampler.
Column: DB-1, 30 m × 0.32 mm, 1.0 μm
Column temperature: 40 °C (hold for 2 min) to 240 °C
@20 °C/min, hold at 240 °C for 20 min

Injector/detector: 150 °C/270 °C
Carrier gas: nitrogen
Linear velocity: 30 cm/s
Split ratio: (2:1)
Injection volume: 5.0 μL
Diluent: dichloromethane (Merck, HPLC grade)

Similar analytical methods need to be included in the manual for the points in the process, so that quality product is produced. The methods documented and practiced cannot be changed without testing and authorization.

7.11 Section 10: thermodynamic and physical properties

It is useful to have thermodynamic and physical properties of the reactants and the reaction intermediates. These are useful in process design. The following are examples of the information for randomly selected chemicals. Names of some of the chemicals and reaction products are not shared to protect confidentiality.

7.11.1 Heat of reaction

Heat of reaction of each reaction product is documented. This, again, is of value in the design of equipment. Table 7.23 lists heat of reaction for different reaction products of a certain synthesis. In order to maintain confidentiality of the reaction and the chemicals involved, the products are not identified.

Tab. 7.23: Example of typical heat of reaction.

Product	Kilocalories/kilogram
Amine hydrochloride	27.8
Amine diazo	227.3
Product #1	361.3
Product #2	200
Product #3	305.7

Heat of reaction and process production rate are needed for temperature control and heat exchanger design.

7.11.2 Physical properties

Most of the APIs per site are not large volume products. Thus, it is possible that all of the necessary physical and thermodynamic property information might not be readily available from different databases. If that is the case, they might have to be generated internally and/or calculated using mathematical models. Mathematical data has to be confirmed.

Density, solubility, specific heat, heat of reaction, viscosity, etc. are a must for each process stream, as they are used and needed for sizing of the reactors, heat exchangers, and pumps. It is very likely that these are not readily available. Again, they might have to be generated in the laboratory.

Having most of these properties might not be considered necessary, but a process that is not designed on the basis of science and engineering fundamentals might not produce quality product from the onset. Having detailed knowledge of the design parameters is also useful for troubleshooting the process.

The following properties for chemical materials are generally available. For the process described in Tab. 7.20 dimethyl sulfate, sodium hydroxide and sulfuric acid are used. Their physical properties [Table. 7.24, Table. 7.25 and Table 7.26 respectively] are included. Their inclusion in every technical manual is a must. Tables 7.8–7.19 list physical properties of additional chemicals.

7.11.2.1 Dimethyl sulfate: $(CH_3)_2SO_4$

Tab. 7.24: Physical properties of dimethyl sulfate [54].

CAS number	77–78-1
Mol. wt	126.1
Melting point (°C)	−32
Boiling point (°C)	188
Solubility in water (g/mL @18 °C)	0.028
Heat of vaporization (kcal/g mol)	10.8
Sp. heat (kcal/kg °C)	0.38

Temp. (°C)	Specific gravity (kg/m³)	Viscosity (g/m-s)
15	1.334	0.0205
20	1.328	0.0185
25	1.322	0.0170
30	1.316	0.0155
35	1.310	0.0145
40	1.304	0.0135

Data here is compiled from different source

7.11.2.2 Sodium hydroxide: NaOH (solid)

Tab. 7.25: Physical properties of sodium hydroxide [55].

CAS no.	1310-73-2
Mol. wt	40
Melting point (°C)	318.4
Boiling point (°C)	1390
Solubility in water (g/mL @20 °C)	109
Sp. heat (kcal/kg °C)	0.782
Sp. gr. kg/m^3 (50%)	1,525.3

7.11.2.3 Sulfuric acid: H$_2$SO$_4$ (96%)

Tab. 7.26: Physical properties of sulfuric acid [56].

CAS no.	7664-93-9
Mol. wt	98
Melting point (°C)	10
Boiling point(°C)	340 decomposes
Solubility in water (kg/m^3 @20 °C)	In all proportions
Sp. heat (kcal/kg °C)	0.363
Sp. gr. kg/m^3 (96%)	1,840.0

Viscosity and density of the raw materials and intermediates can change with temperature. Appropriate charts should be included in the process manual. They are of value in process design and operations.

7.12 Section 11: final product specifications

Each produced product has its final specifications. Table 7.27 is one of the ways the final product specifications and its test methods should also be included in the Technical Manual.

Tab. 7.27: Product specification of a drug product.

Material code	20001659	Customer name	XYZ
Batch no	ABNH008478	Date of manufacture	06/2020
Batch quality	159.060 kg	Expiry date	05/2025
A.R. no.	1002FP20002521	Date of analysis	29-06-2020-14:00
Reference	USP/in-house	Specification number	S-08-MO-USP/03

Specifications

S. no.	Test	Specification	Result
1	Description	Slightly yellow to yellow	Slightly yellow
2	Solubility		
2.1	In 0.1 N sodium hydroxide	Soluble in 0.1 N sodium hydroxide	Soluble in 0.1 N sodium hydroxide
2.2	In water	Sparingly soluble in water	Sparingly soluble in water
2.3	Methanol	Sparingly soluble in methanol	Sparingly soluble in methanol
3	Identification		
3.1	By IR	To match with the working standard	Matches with the working standard
3.2	By HPLC	The retention time of the major peak in the chromatogram of the assay preparation corresponds to that in the chromatogram of the standard preparation, as obtained in the assay	The retention time of the major peak in the chromatogram of the assay preparation corresponds to that in the chromatogram of the standard preparation, as obtained in the assay
3.3	Chloride	Responds to the test for chloride	Responds to the test for chloride
4	Specific rotation of 20°	Between −125 °C and −138 °C	−133°
5	Hydrochloride content	Should be between 8.0% w//w and 9.0% w/w	8.3% w/w
6	pH (0.2 in 100)	Between 3.9 and 4.6 in a solution	4.4
7	Water	Not >1.0%	0.1%
8	Residue on ignition	Not >0.1%	0.0%

Tab. 7.27 (continued)

Material code 20001659		Customer name	XYZ
9	Sulfate	Not >0.4% w/w	<0.4% w/w
10	Related substance by HPLC		
10.1	Moxifloxacin-related compound A	Not >0.1%	Not detected
10.2	6,8-Dimethoxy (moxifloxacin-related compound B)	Not >0.1%	Not detected
10.3	8-Ethoxy (moxifloxacin-related compound C)	Not >0.1%	Not detected
10.4	6-Methoxy-8-fluoro (moxifloxacin-related compound d)	Not >0.1%	Not detected
10.5	8-Hydroxy (8-hydroxy moxifloxacin-related compound e)	Not >0.10%	Not detected
10.6	N-Methyl piperazine derivative of moxifloxacin	Not >0.10%	Not detected
10.7	8-Hydroxy quinolinic acid	Not >0.10%	Not detected
10.8	8-Methoxy of quinolinic acid	Not >0.10%	Not detected
10.9	Ethyl quinolinic ester	Not >0.10%	Not detected
10.10	Other invalid impurity	Not >0.10%	0.01%
10.11	Total impurity	Not >0.50%	0.01%
11	RR isomer content	Not >0.15%	Less than LOQ (LOQ: 0.03)
12	Assay by HPLC (on anhydrous basics)	Not <98.0% w/w and not >102.0% w/w	99.4% w/w
13	X-ray powder diffraction pattern	To match with moxifloxacin hydrochloride working standard	Matches with HCl working standard
14	Method I		
14.1	Pet ether	Not >100 ppm	Not detected
14.2	Ethyl acetate	Not >100 ppm	Not detected
14.3	Methanol	Not >100 ppm	Not detected
14.4	2-Propanol	Not >100 ppm	Not detected

Tab. 7.27 (continued)

Material code	20001659	Customer name	XYZ
14.5	Toluene	Not >100 ppm	Below the limit of detection (LOD:12)
14.6	Dimethyl formamide	Not >100 ppm	Not detected
15	Method II	Not >100 ppm	Not detected
15.1	Ethanol	Not >100 ppm	Not detected
15.2	Acetonitrile	Not >100 ppm	Not detected
15.3	Dichloromethane	Not >100 ppm	Not detected
15.4	Methyl isobutyl ketone	Not >3,000 ppm	1,454 ppm
16	Microbial count		
16.1	Total aerobic microbial count	Not >1,000 cfu/g	Nil cfu/g
16.2	Total yeast and mold count	Not >100 cfu/g	Nil cfu/g
17	Specified microorganisms		
17.1	Bile-tolerant Gram-negative bacteria	Should be absent	Absent
17.2	*Escherichia coli*	Should be absent	Absent
17.3	*Salmonella* species	Should be absent	Absent
17.4	*Staphylococcus aureus*	Should be absent	Absent
17.5	*Pseudomonas aeruginosa*	Should be absent	Absent
18	Endotoxins *S-08-MF-IH-05*	Not >0.4 EU/mg	<0.1 EU/mg

Additional comments, if any: Store conditions: preserve in tight containers and protect from moisture. Store at controlled room temperature 20–25 °C (excursions allowed between 15 and 30 °C). Always replenish with fresh silica gel whenever opened, purge with nitrogen, and seal the triple laminated bag. Packing condition: materials are packed in a clear polybag filled with nitrogen and tie with nylon strap. Keep this in triple laminated bag along with silica gel bag, insert the vacuum hose pipe into the triple laminated bag, apply vacuum to the bag, evacuate and fill with nitrogen, and seal it with heat sealer. Then keep this in HDPE container.

7.13 Section 12: cleaning and cGMP practices

7.13.1 Cleaning

API manufacturing has to adhere to very strict manufacturing and cleaning practices. Strictness is necessary because of the batch manufacturing practice in which different active pharmaceutical ingredients (API) can be produced in the same equipment. This is the case not only at the contract manufacturing organizations [57], but also true in the captive brand and/or generic API manufacturing firms as they do not have dedicated equipment for every product. If they did, it would add to the current excess of idle equipment [58].

Equipment cleaning between product changeovers is a must. Cleaning procedures have to be validated [59]. Established cleaning practices that have been approved cannot be changed without testing, validation, and authorization. It is necessary to understand these requirements, and if they are not followed, can result in financial losses.

Cleaning methodology of the continuous process equipment [60] will be different from the cleaning of the batch processes. Batch cleaning is intermittent and often between batches. However, continuous process [3, 4] not being stop-and-go may not require rigors of batch cleaning. Most likely, such process would require limited cleanings per year during their planned maintenance. They have to be thought through and validated.

Equipment cleaning rationale for continuous manufacturing operations, most likely, would be based on batch cleaning methodology but would be different and may not be as detailed as batch cleaning. This would be because the equipment is only exposed to chemicals that are part of the chemistry being executed in the dedicated equipment. Cleaning of continuous process equipment would have to be tested and validated.

Since APIs are manufactured using conventional batch processes, their cleaning solvent volume between products requires at least one turnover of the equipment volume. If the used solvent volume is large, even after recovery and recycling, it can be a significant volume. Multiple cleanings of different equipment used for batch processing per year will add to the operating company's environmental factor, "E-factor" [7]. This presents an opportunity for alternate manufacturing processes.

Modular plants, which are not part of the API manufacturing vocabulary, as has been reviewed in earlier chapters, if used, will change the operating landscape. With longer API runs, their in-between cleaning will be less frequent than batch process scenario cleanings. This, in turn, could mean significantly lower the "E-factor" [7] for each of the API operating sites.

7.13.2 cGMP

Chemists and chemical engineers are taught, from the onset, to create, design, and commercialize a process for a product that will produce repeatable quality product. Properly written "Process Technical Manual" covers the suggested cGMP [5] guidance that are needed for a batch [1, 2] or a continuous process [3, 4]. They are a live document for every process.

Regulators have proposed guidelines that they consider "good manufacturing practices," which should be commonly practiced for every manufacturing process. Actually, chemists and chemical engineers use them as the building blocks of every process. Thus, they should be easy to practice and should not be circumvented.

7.14 Section 13: safety, MSDS (now called SDS), and material handling

Safety is extremely important while handling chemicals. In the last one hundred years, there has been evolution of the safe handling of chemicals. Concept of material safety data sheets (MSDS) was introduced in the last century. They used to be two pages per chemical. Now, they are about seven pages each, suggesting that safety has become increasingly more important. Information contained in SDS (safety Data Sheets) is extremely valuable for safe handling of chemicals. However, still, it is up to the users of chemicals to take extra steps to respect and handle them carefully.

What was considered safe chemical yesterday could be considered as or is toxic today. Toxicology [61] is in its adolescence and will continue to evolve. In addition, development of better drugs over the years have led to use of exotic chemistries. Environmental impact of many APIs, their reaction intermediates, and by-products are not completely known and understood. Many of these are showing up in effluent streams and impacting soil, aquatic, bird, and animal life.

Studies [62–65] have been and are being done to understand the health effects of land, animal, and aquatic life after the ill-health effects were noticed. However, even with the knowledge of potential dangers and hazards of the chemicals, safety and safe handling, at times, slip through, resulting in significant and serious consequences. Thus, safe and proper handling of every chemical is extremely important.

Every "Process Technical Manual" should include material safety data sheets of raw materials, every process intermediate, and the finished product. This is critical especially for the batch processes, the preferred method of producing APIs, because the intermediates are generally isolated and stored for further processing. This information can be of value in reactor cleaning also.

Table 7.20 is the first page of safety data sheet for Pyrmetazole (omeprazole sulfide) [66], an intermediate in omeprazole synthesis. It needs to be noted that some information on the data sheets may not be complete as the information might not be available. Thus, it is necessary to take protective safety measures. Tab. 7.28 and Tab. 7.29 are safety data sheet [66] and product information sheets [67] for omeprazole sulfide.

Since the APIs can be used at their own site or could be transported out of manufacturer's site, they should have their respective safety data sheets e.g. metfromin hydrochloride Tab. 7.30 [68].

Tab. 7.28: Safety data sheet for pyrmetazole (omeprazole sulfide) [66].

SAFETY DATA SHEET
Omeprazole sulfide

Page: 1 of 6

Revision: 05/17/2016

according to Regulation (EC) No. 1907/2006 as amended by (EC) No. 1272/2008

Section 1. Identification of the Substance/Mixture and of the Company/Undertaking

1.1

Product Code: 18885

Product Name: Omeprazole sulfide

Synonyms: 6-methoxy-2-[[(4-methoxy-3,5-dimethyl-2-pyridinyl)methyl]thio]-1H-benzimidazole;

1.2 Relevant identified uses of the substance or mixture and uses advised against:

Relevant identified uses: For research use only, not for human or veterinary use.

1.3 Details of the Supplier of the Safety Data Sheet:

Company Name: Cayman Chemical Company
1180 E. Ellsworth Rd.
Ann Arbor, MI 48108

Web site address: www.caymanchem.com

Information: Cayman Chemical Company +1 (734)971-3335

1.4 Emergency telephone number:

Emergency Contact: CHEMTREC Within USA and Canada: +1 (800)424-9300
CHEMTREC Outside USA and Canada: +1 (703)527-3887

Section 2. Hazards Identification

2.1 Classification of the Substance or Mixture:

Skin Corrosion/Irritation, Category 2

Serious Eye Damage/Eye Irritation, Category 2

Skin Sensitization, Category 1

Specific Target Organ Toxicity (single exposure), Category 3

Aquatic Toxicity (Chronic), Category 1

2.2 Label Elements:

GHS Signal Word: Warning

GHS Hazard Phrases:

H315: Causes skin irritation.

H317: May cause an allergic skin reaction.

H319: Causes serious eye irritation.

H335: May cause respiratory irritation.

H410: Very toxic to aquatic life with long lasting effects.

GHS Precaution Phrases:

P261: Avoid breathing {dust/fume/gas/mist/vapors/spray}.

P264: Wash {hands} thoroughly after handling.

P272: Contaminated work clothing should not be allowed out of the workplace.

P273: Avoid release to the environment.

P280: Wear {protective gloves/protective clothing/eye protection/face protection}.

GHS Response Phrases:

P302+352: IF ON SKIN: Wash with plenty of soap and water.

P304+340: IF INHALED: Remove victim to fresh air and keep at rest in a position comfortable for breathing.

P305+351+338: IF IN EYES: Rinse cautiously with water for several minutes. Remove contact lenses, if present and easy

Multi-region format

Tab. 7.29: Product datasheet omeprazole sulfide [67].

PRODUCT INFORMATION

Cayman
CHEMICAL

Omeprazole sulfide
Item No. 18885

CAS Registry No.:	73590-85-9
Formal Name:	6-methoxy-2-[[(4-methoxy-3,5-dimethyl-2-pyridinyl)methyl]thio]-1H-benzimidazole
Synonyms:	Pyrmetazole, Ufiprazole
MF:	$C_{17}H_{19}N_3O_2S$
FW:	329.4
Purity:	≥98%
UV/Vis.:	λ_{max}: 300 nm
Supplied as:	A crystalline solid
Storage:	-20°C
Stability:	As supplied, 2 years from the QC date provided on the Certificate of Analysis, when stored properly

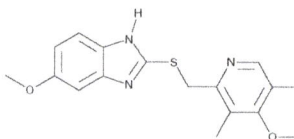

Laboratory Procedures

Omeprazole sulfide is supplied as a crystalline solid. A stock solution may be made by dissolving the omeprazole sulfide in the solvent of choice. Omeprazole sulfide is soluble in organic solvents such as ethanol, DMSO, and dimethyl formamide (DMF), which should be purged with an inert gas. The solubility of omeprazole sulfide in ethanol is approximately 25 mg/ml and approximately 30 mg/ml in DMSO and DMF.

Omeprazole sulfide is sparingly soluble in aqueous buffers. For maximum solubility in aqueous buffers, omeprazole sulfide should first be dissolved in DMF and then diluted with the aqueous buffer of choice. Omeprazole sulfide has a solubility of approximately 0.5 mg/ml in a 1:1 solution of DMF:PBS (pH 7.2) using this method. We do not recommend storing the aqueous solution for more than one day.

Description

Omeprazole sulfide is an intermediate used in the production of the gastric proton pump inhibitors, omeprazole (Item No. 14880) and esomeprazole (Item No. 17326).[1] As a degradation product, it is reported to be a direct-acting inhibitor of cytochrome P450 2C19 in pooled human liver microsomes (IC_{50} = 9.7 µM).[2]

References

1. Olbe, L., Carlsson, E., and Lindberg, P. A proton-pump inhibitor expedition: The case histories of omeprazole and esomeprazole. *Nat. Rev. Drug Discov.* **2(2)**, 132-139 (2003).
2. Ogilvie, B.W., Yerino, P., Kazmi, F., *et al.* The proton pump inhibitor, omeprazole, but not lansoprazole or pantoprazole, is a metabolism-dependent inhibitor of CYP2C19: Implications for coadministration with clopidogrel. *Drug Metab. Dispos.* **39(11)**, 2020-2033 (2011).

Tab. 7.30: Safety data sheet for metformin hydrochloride [68].

Safety Data Sheet
acc. to OSHA HCS

Printing date 01/25/2021   Revision date 01/25/2021

1 Identification

· **Product identifier**

· **Trade name: Metformin (hydrochloride)**
· **Synonym**
N,N-dimethyl-imidodicarbonimidic diamide, monohydrochloride
1,1-Dimethylbiguanide hydrochloride

· **Article number:** 13118
· **CAS Number:**
1115-70-4
· **EC number:**
214-230-6
· **Application of the substance / the mixture** For research use only, not for human or veterinary use.

· **Details of the supplier of the safety data sheet**
· **Manufacturer/Supplier:**
Cayman Chemical Co.
1180 E. Ellsworth Rd.
Ann Arbor, MI 48108
USA

· **Information department:** Product safety department
· **Emergency telephone number:**
During normal opening times: +1 (734) 971-3335
US/CANADA: 800-424-9300
Outside US/CANADA: 703-741-5970

2 Hazard(s) identification

· **Classification of the substance or mixture**

GHS07

Acute Tox. 4 H302 Harmful if swallowed.
Eye Irrit. 2A   H319   Causes serious eye irritation.

· **Label elements**
· **GHS label elements**
The substance is classified and labeled according to the Globally Harmonized System (GHS).
· **Hazard pictograms**

GHS07

· **Signal word** Warning
· **Hazard statements**
H302 Harmful if swallowed.
H319 Causes serious eye irritation.

(Contd. on page 2)
— US —

7.15 14: US Pharmacopeia standards

For every regulatory NDA [new drug application] and ANDA, [abbreviated new drug application] all of the above documented information, along with US Pharmacopeia standards, are needed. Attached are two typical standards. Table 7.31 and Table 7.32 are the US Pharmacopoeia standards for metformin hydrochloride and sodium levothyroxine.

Tab. 7.31: USP standard metformin hydrochloride.

USP standard: metformin hydrochloride
  Metformin hydrochloride [69]
$C_4H_{11}N_5 \cdot HCl$ 165.62

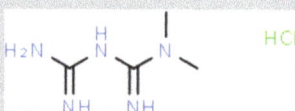

Imidodicarbonimidic diamide, N,N-dimethyl-, monohydrochloride.
1,1-Dimethylbiguanide monohydrochloride [1115-70-4].
» Metformin hydrochloride contains not less than 98.5% and not more than 101.0% of $C_4H_{11}N_5$. HCl, calculated on the dried basis.

Add the following:
Packaging and storage: Preserve in well-closed containers. Store it at room temperature. USP29
USP Reference standards 11: USP metformin hydrochloride RS. USP metformin-related compound A RS.

Identification:
A: Infrared absorption 197 K.
B: It meets the requirements of the tests for Chloride 191.
Loss on drying 731: Dry it at 105 for 5 h: it loses not more than 0.5% of its weight.
Residue on ignition 281: Not more than 0.1%.
Heavy metals, Method I 231: 0.001%.

Related compounds:
Mobile phase: Prepare a solution in water, containing 17 g of monobasic ammonium phosphate per L, adjust with phosphoric acid to a pH of 3.0, and mix.

Standard solution:
Prepare a solution of USP metformin-related compound A RS in water, having a known concentration of about 0.2 mg/mL. Transfer 1.0 mL of this solution to a 200-mL volumetric flask, dilute with mobile phase to volume, and mix (Note: Metformin-related compound A is 1-cyanoguanidine).
Test solution: Transfer about 500 mg of metformin hydrochloride, accurately weighed, to a 100-mL volumetric flask, dissolve in and dilute with mobile phase to volume, and mix.

Diluted test solution: Transfer 1.0 mL of the test solution to a 10-mL volumetric flask, dilute with mobile phase to volume, and mix. Transfer 1.0 mL of this solution to a 100-mL volumetric flask, dilute with mobile phase to volume, and mix.

Resolution solution: Transfer about 10 mg of melamine to a 100-mL volumetric flask, and dissolve in about 90 mL of water. Add 5.0 mL of the test solution, dilute with water to volume, and mix. Transfer 1.0 mL of this solution to a 50-mL volumetric flask, dilute with mobile phase to volume, and mix.

Chromatographic system (see Chromatography 621): The liquid chromatograph is equipped with a 218-nm detector and a 4.6-mm Å ~ 25-cm column containing packing L9. The flow rate is about 1.0–1.7 mL/min. Chromatograph the resolution solution, and record the peak responses as directed for procedure: the resolution, R, between melamine and metformin is not less than 10.

Procedure: Separately inject equal volumes (about 20 μL) of the test solution, the standard solution, and the diluted test solution into the chromatograph, record the chromatograms for not less than twice the retention time of metformin, and measure the peak areas. The area of a peak corresponding to metformin-related compound, A in the chromatogram of the test solution is not greater than the area of the corresponding peak in the chromatogram of the Standard solution: not more than 0.02% of metformin related compound A is found.

The area of any other secondary peak in the chromatogram of the test solution is not greater than the area of the major peak in the chromatogram of the diluted test solution; and, the sum of the areas of all secondary peaks in the chromatogram of the test solution is not greater than five times the area of the major peak in the chromatogram of the diluted test solution: not more than 0.1% of any other impurity is found; and not more than 0.5% of total impurities is found.

Residual solvents 467: Meets the requirements (Official January 1, 2007)

Assay: (Note: To avoid overheating of the reaction medium, mix thoroughly throughout the titration, and stop the titration immediately after the endpoint has been reached). Dissolve about 60 mg of metformin hydrochloride, accurately weighed, in 4 mL of anhydrous formic acid. Add 50 mL of acetic anhydride. Titrate with 0.1 N perchloric acid VS, determining the endpoint potentiometrically. Perform a blank determination, and make any necessary correction (see Titrimetry 541).

Each mL of 0.1 N perchloric acid is equivalent to 8.28 mg of $C_4H_{11}N_5$.HCl.

Auxiliary information: Staff liaison: Elena Gonikberg, Ph.D., scientist
Expert Committee: (MDGRE05) Monograph Development-Gastrointestinal Renal and Endocrine
USP29–NF24: Page 1364
Pharmacopeial forum: Volume no. 31[4], page 1092
Phone number: 1-301-816-8251

Tab. 7.32: USP standard sodium levothyroxine.

USP standard: levothyroxine sodium
Levothyroxine sodium [69]

$C_{15}H_{10}I_4NNaO_4$.xH_2O (anhydrous) 798.85

L-Tyrosine, *O*-(4-hydroxy-3,5-diiodophenyl)-3,5-diiodo-, monosodium salt, hydrate.

Monosodium L-thyroxine hydrate [25,416-65-3].

Anhydrous [55-03-8].

» Levothyroxine sodium is the sodium salt of the levo isomer of thyroxine, an active physiological principle obtained from the thyroid gland of domesticated animals used for food by man or prepared synthetically. It contains not less than 97.0% and not more than 103.0% of $C_{15}H_{10}I_4NNaO_4$, calculated on the anhydrous basis.

Packaging and storage: Preserve in tight containers, protected from light.
USP Reference standards 11: USP levothyroxine RS. USP liothyronine RS.

Identification:

A: Ignite about 50 mg in a platinum dish over a flame; it decomposes and liberates iodine vapors.
B: To about 0.5 mg add 7.5 mL of acid sodium chloride solution (prepared by mixing 300 mL of water, 250 mL of alcohol, 100 mL of 1 N sodium hydroxide, and 100 mL of hydrochloric acid) and 1 mL of sodium nitrite solution (1 in 100). Allow to stand in the dark for 20 min, and add 1.25 mL of ammonium hydroxide; a pink color is produced.

Specific rotation 781S: between 5° and 6°.

Test solution: an amount equivalent to 30 mg of anhydrous levothyroxine sodium per mL, in a mixture of alcohol and 1 N sodium hydroxide (2:1).

Water, Method III 921: Dry about 500 mg, accurately weighed, over phosphorus pentoxide at 60 and at a pressure not exceeding 10 mm Hg for 4 h; it loses not more than 11.0% of its weight.

Limit of inorganic iodides:

Extracting solution: Prepare a 1 in 100 solution of sulfuric acid in water.

Reference solution: Dissolve an accurately weighed quantity of potassium iodide in water to obtain a stock solution containing 0.131 mg equivalent to 0.100 mg of iodide per mL. Transfer 0.6 mL of this stock solution into a 1,000 mL volumetric flask, dilute with the extracting solution to volume, and mix. Each mL of the reference solution contains 0.06 μg of iodide (Note: Prepare this solution on the day of use).

Test solution: Transfer 7.5 mg of levothyroxine sodium to a beaker, add 100 mL of the extracting solution, and sonicate for 5 min.

Electrode system: Use an iodide-specific, ion-indicating electrode and a silver-silver chloride reference electrode connected to a pH meter capable of measuring potentials with a minimum reproducibility of ±1 mV (see pH 791).

Procedure: Transfer the reference solution to a beaker containing a magnetic stirring bar. Rinse and dry the electrodes, insert in the solution, stir for 5 min or until the reading stabilizes, and read the potential in mV. Repeat this process using the test solution. The requirements of the test are met if the test solution has a higher potential in mV than the reference solution: the limit is 0.08%.

Limit of liothyronine sodium:

Mobile phase and chromatographic system: Proceed as directed in the assay.
Standard solution: Prepare as directed for Standard preparation in the assay.
Test solution: Proceed as directed for the assay preparation.

Procedure: Proceed as directed in the assay. Calculate the quantity, in µg, of liothyronine sodium ($C_{15}H_{11}I_3NNaO_4$) in the sample taken by the formula: $(672.96/650.98)(10 \ C)(r_U/r_S)$, in which 672.96 and 650.98 are the molecular weights of liothyronine sodium and liothyronine, respectively; C is the concentration, in µg/mL, of USP liothyronine RS in the standard preparation; and r_U and r_S are the liothyronine peak responses obtained from the Test solution and the Standard solution, respectively: not more than 2.0% of liothyronine is found.

Residual solvents 467: meets the requirements. (Official January 1, 2007)

Assay:
Mobile phase: Prepare a degassed and filtered mixture of water and acetonitrile (60:40) that contains 0.5 mL of phosphoric acid in each 1,000 mL. Make adjustments, if necessary (see System Suitability under Chromatography 621).

0.01 M methanolic sodium hydroxide: Dissolve 400 mg of sodium hydroxide in 500 mL of water. Cool, add 500 mL of methanol, and mix.

Levothyroxine stock solution: Dissolve an accurately weighed quantity of USP levothyroxine RS in 0.01 M Methanolic sodium hydroxide to obtain a solution having a known concentration of about 0.4 mg of levothyroxine per mL. Liothyronine stock solution: Dissolve an accurately weighed quantity of USP liothyronine RS in 0.01 M methanolic sodium hydroxide to obtain a solution having a known concentration of about 0.4 mg of liothyronine per mL. Make a 1:100 dilution of this solution using mobile phase.

Standard preparation: Transfer appropriate volumes of Levothyroxine stock solution and Liothyronine stock solution to a suitable container, and dilute quantitatively and stepwise, if necessary, with Mobile phase to obtain a solution having known concentrations of about 10 µg of levothyroxine per mL and 0.2 µg of liothyronine per mL.

Assay preparation: Transfer an accurately weighed portion of about 100 µg of levothyroxine sodium into a centrifuge tube, add 2 glass beads, pipet 10 mL of mobile phase into the tube, and mix using a vortex mixer for 3 min. Centrifuge to obtain a clear supernatant, filtering if necessary.

Chromatographic system (see Chromatography 621): The liquid chromatograph is equipped with a 225-nm detector and a 4.6-mm Å ~ 25-cm column that contains packing L10. The flow rate is about 1.5 mL per minute. Chromatograph the Standard preparation, and record the peak responses as directed for Procedure: the resolution, R, between liothyronine and levothyroxine is not less than 5.0; and the relative standard deviation for replicate injections is not more than 2.0% for levothyroxine.

Procedure: Separately inject equal volumes (about 100 µL) of the standard preparation and the assay preparation into the chromatograph, record the chromatograms, and measure the responses for the major peaks. Calculate the quantity, in µg, of $C_{15}H_{10}I_4NNaO_4$ in the portion of Levothyroxine Sodium taken by the formula: $(798.85/776.87) (10 \ C) (r_U/r_S)$, in which 798.85 and 776.87 are the molecular weights of levothyroxine sodium and levothyroxine, respectively;

C is the concentration, in μg/mL, of USP levothyroxine RS in the standard preparation; and r_U and r_S are the levothyroxine peak responses obtained from the assay preparation and the standard preparation, respectively.

Auxiliary Information: Staff Liaison: Elena Gonikberg, Ph.D., scientist
Expert committee: (MDGRE05) Monograph Development-Gastrointestinal Renal and Endocrine
USP29–NF24, page 1248
Pharmacopeial forum: Volume no. 30[5], page 1630
Phone number: 1-301-816-8251

All of the information reviewed and discussed are necessary and need to be documented. Since the design considerations and the basis make the building block, they may be needed for troubleshooting and expansion of outsourcing of products. They have a long-term value.

Abbreviations

NDA	New drug application
ANDA	Abbreviated new drug application
API	Active pharmaceutical ingredient
USFDA	United States Food and Drug Administration
cGMP	Current good manufacturing practice
E-factor	Environmental factor
SDS	Safety data sheet
MSDS	Material safety data sheet
kg	Kilogram
CAS	Chemical Abstracts Service No.
Cp	Specific heat
NOx	Generic term for nitrogen oxides
IR	Infrared
HPLC	High-performance liquid chromatography
USP	United States Pharmacopeia

References

[1] Batch production, https://en.wikipedia.org/wiki/Batch_production, Accessed June 1, 2019.
[2] Batch Production Vs. Mass Production: https://simplicable.com/new/batch-production-vs-mass-production, February 1, 2017, Accessed July 3, 2020.
[3] Continuous Production, https://en.wikipedia.org/wiki/Continuous_production, Accessed June 1, 2019.
[4] The Pharmaceutical Manufacturing Process – Steps, Tools, and Considerations https://www.thomasnet.com/articles/chemicals/the-pharmaceutical-manufacturing-process-steps-tools-and-considerations/ Accessed July 5, 2020.

[5] Q7 Good Manufacturing Practice Guidance for Active Pharmaceutical Ingredients https://www.fda.gov/media/71518/download April 2018 Accessed May 6, 2021.

[6] Taxol® (NSC 125973) https://dtp.cancer.gov/timeline/flash/success_stories/s2_taxol.htm Accessed May 7, 2021.

[7] Malhotra, G.: Active Pharmaceutical Ingredient Manufacturing (API) and Formulation Drive to NET ZERO (Carbon Neutral)? Profitability through Simplicity, https://pharmachemicalscoatings.blogspot.com/2021/04/active-pharmaceutical-ingredient.html April 29, 2021 Accessed May 6, 2021.

[8] Malhotra, G.: Square Plug in A Round Hole: Does This Scenario Exist in Pharmaceuticals? Profitability through Simplicity, https://pharmachemicalscoatings.blogspot.com/2010/08/square-peg-in-round-hole-does-this.html, August 10, 2010, Accessed May 19, 2020.

[9] Malhotra, G.: Why Fitting a Square Plug in a Round Hole Is Profitable for Pharma and Most Likely Will Stay? https://pharmachemicalscoatings.blogspot.com/2014/08/why-fitting-square-plug-in-round-hole.html, August 1, 2014, Accessed April 5, 2020.

[10] Malhotra, G.: Strategies for Improving Batch or Creating Continuous Active Pharmaceutical Ingredients (API) manufacturing Processes, February, 2011 Informa UK, Revised March 2017, https://pharmachemicalscoatings.blogspot.com/2017/03/strategies-for-enhancing-active.html, Accessed January 14, 2021.

[11] Malhotra, G.: Batch, Continuous or "Fake/false" Continuous Processes in Pharmaceutical Manufacturing, Profitability through Simplicity, https://pharmachemicalscoatings.blogspot.com/2017/07/batch-continuous-or-fakefalse.html, July 20, 2017, Accessed May 10, 2021.

[12] Benchmarking Shows Need to Improve Uptime, Capacity Utilization, https://www.pharmamanufacturing.com/articles/2007/144/ Sep 20, 2007, Accessed May 19, 2020.

[13] Malhotra, G.: Continuous Pharmaceutical Processes and Their Demands, Contract Pharma, https://www.contractpharma.com/issues/2016-04-01/view_features/continuous-pharmaceutical-processes-and-their-demands/?userloggedin=true, April 5, 2016, Accessed May 15, 2020.

[14] Tyson, T.: Solving the Industry's Capacity Utilization Problem, https://www.pharmasalmanac.com/articles/solving-the-industrys-capacity-utilization-problem, October 28, 2019.

[15] Malhotra, G.: Opportunities for Generic Pharma to Clear the Quality Stigma, https://pharmachemicalscoatings.blogspot.com/2019/05/opportunities-for-generic-pharma-to.html, May 23, 2019, Accessed April 25, 2020.

[16] Couper, J. R. et al. Chemical Process Equipment, Selection and Design, Third, Elsevier, ISBN: 9780123969590, 2010.

[17] Chapter 5.

[18] Anders, G., AstraZeneca, A. B., USP 7227024, Method for the Preparation of A Benzimidazole Compound, June 5, 2007 Accessed June 11, 2008.

[19] https://www.pharmacompass.com/active-pharmaceutical-ingredients/metformin-hydrochloride Accessed June 14, 2021.

[20] M/s. Aarti Drugs Ltd. http://environmentclearance.nic.in/writereaddata/form-1A/2013_8_06_Aug_2013_1035311301214_13.pdf Accessed December 30, 2020.

[21] Malhotra, G.: Batch, Continuous or "Fake/false" Continuous Processes, American Pharmaceutical Review, https://www.americanpharmaceuticalreview.com/Featured-Articles/343596-Batch-Continuous-or-Fake-False-Continuous-Processes/ Sunday, October 15, 2017, Accessed June 14, 2021.

[22] Malhotra, G.: Information Challenges for Product, Process Development and Process Design: A Reality Check, https://pharmachemicalscoatings.blogspot.com/2011/04/information-challenges-for-product.html, April 10, 2011 Accessed May 10, 2021.

[23] Hansen Solubility Parameters https://hansen-solubility.com Accessed June 1, 2021.

[24] Hildebrand solubility parameter https://www.chemeurope.com/en/encyclopedia/Hilde brand_solubility_parameter.html Accessed June 1, 2021.

[25] Abraham et Al. Prediction of Solubility of Drugs and Other Compounds in Organic Solvents, Journal of Pharmaceutical Sciences, 2010, 99, 1500–1515, Accessed June 1, 2021.

[26] Shnidman, L. THE SOLUBILITY OF THIOUREA IN WATER, METHANOL, AND Ethanol, Journal of Physical Chemistry, 1933, 37, 6, 693–700, Accessed December 14, 2020.

[27] Liu et al. Liu et al. Measurement and correlation of the solubility of maleic anhydride in different organic solvents, Fluid Phase Equilibria, Vol. 367 2014, pgs, 1-6 Accessed November 15, 2020.

[28] Physical Properties p-(2-Methoxyethyl) Phenol https://www.chemblink.com/products/56718-71-9.htm Accessed May 20, 2021.

[29] Epichlorohydrin Properties http://www.inchem.org/documents/ehc/ehc/ehc33.htm#Section Number:2.1 Accessed May 20, 2021.

[30] Sodium Hydroxide Properties, https://byjus.com/jee/sodium-hydroxide/ Accessed May 30, 2021.

[31] Caustic Soda Handbook https://www.oxy.com/ourbusinesses/chemicals/products/docu ments/causticsoda/caustic.pdf Accessed May 30, 2021.

[32] Physical Properties Ammonia, https://www.engineeringtoolbox.com/ammonia-d_1413.html Accessed May 30, 2021.

[33] Physical Properties Methanol, https://www.cetinerengineering.com/Properties.htm Accessed May 30, 2021.

[34] Physical Properties Maleic Anhydride, https://www.ashland.com/file_source/Ashland/Docu ments/Sustainability/rc%20maleic%20anhydride.pdf Accessed May 20, 2021.

[35] Thiourea Properties, https://en.wikipedia.org/wiki/Thiourea, Accessed May 20, 2021.

[36] Thiourea Properties, http://www.sciencemadness.org/smwiki/index.php/Thiourea *Accessed* May 20, 2021.

[37] Toluene Properties, http://www.homepages.ed.ac.uk/jwp/Chemeng/azeotrope/hetero.html Accessed June 2, 2021.

[38] Water/xylene azeotrope http://www.homepages.ed.ac.uk/jwp/Chemeng/azeotrope/hetero. html Accessed June 2, 2021.

[39] Xylene Properties, https://en.wikipedia.org/wiki/Xylene Accessed June 2, 2021.

[40] Dimethyl Amine, https://webwiser.nlm.nih.gov/substance?substanceId=204&identifier=N,N-Dimethylamine&identifierType=alias&menuItemId=82&catId=164 Accessed June 2, 2021.

[41] Dimethyl Amine, https://www.eastman.com/Pages/ProductHome.aspx?>product= 71104063&pn=Dimethylamine+Anhydrous+(DMA)+-+EU Accessed June 2, 2021.

[42] Ammonia Properties https://www.engineeringtoolbox.com/ammonia-d_1413.html Accessed June 12, 2021.

[43] Hydrochloric Acid Physical Properties https://www.chemicalbook.com/ChemicalProductProp erty_EN_CB7421538.htm Accessed June 15, 2021.

[44] Hydrochloric Acid Physical Properties https://www.oxy.com/ourbusinesses/chemicals/prod ucts/documents/hydrochloricacid/hydrochloric_acid_handbook.pdf Accessed June 15, 2021.

[45] https://www.pharmacompass.com Accessed June 12, 2020.

[46] Shell and Tube Heat Exchanger Data Sheet Template courtesy of CG Thermal LLC.

[47] Shell and Tube Heat Exchanger Data Sheet courtesy Power Plus International, https://www.pip.org.

[48] Data Sheets for Pumps, Courtesy https://www.pip.org June 25, 2021.

[49] API-610: Centrifugal Pump Data Sheet, American Petroleum Institute.

[50] Agitated Vessel Process Data Sheet https://www.chemengghelp.com/data-sheet-agitated-reactor/ Accessed June 21, 2021.

[51] Hussain, A. S. Office of Pharmaceutical Science CDER FDA, Pharmaceutical 6-Sigma Quality by Design, The 28th Annual Midwest Biopharmaceutical Statistical Workshop May 23-25, 2005, https://slideplayer.com/slide/6057519/ Accessed April 11, 2006.

[52] Tonne et al. USP 4464537, Preparation of Saccharin, BASF Aktiengesellschaft, August 7, 1984, Accessed September 30, 2020.

[53] Bhirud et al. US Patent 9,840,456, Process for Preparation of Dimethyl Fumarate, Glenmark Pharmaceuticals Limited, December 12, 2017 Accessed November 20, 2017.

[54] Dimethyl Sulphate Properties https://www.chemicalbook.com/ChemicalProductProperty_EN_CB9854316.htm, Accessed July 1, 2021.

[55] Caustic Soda Properties https://www.jsia.gr.jp/data/handling_01e.pdf Accessed July 6, 2021.

[56] Sulfuric Acid Properties https://www.chemicalbook.com/ChemicalProductProperty_EN_CB9675634.htm Accessed July 1, 2021.

[57] Downs, Pat Taking a right-first-time approach to equipment cleaning https://www.manufacturingchemist.com/news/article_page/Taking_a_right-first-time_approach_to_equipment_cleaning/161258 January 10, 2020 Accessed June 30, 2021.

[58] Benchmarking Shows Need to Improve Uptime, Capacity Utilization, https://www.pharmamanufacturing.com/articles/2007/144/ Sep 20, 2007, Accessed May 19, 2020.

[59] Validation of Cleaning Processes (7/93) https://www.fda.gov/validation-cleaning-processes-793Content current as of August 24, 2014 accessed July 5, 2021.

[60] Cleaning Pharma API Equipment, https://www.lechlerusa.com/en/resources/nozzle-insights/cleaning-pharma-api-equipment Accessed July 12, 2021.

[61] Toxicology, Wikipedia https://en.wikipedia.org/wiki/Toxicology Accessed July 5, 2021).

[62] Cleuvers, M. Aquatic Ecotoxicity of Pharmaceuticals Including the Assessment of Combination Effects, Toxicology Letters, 142, 2003, 185–194. Accessed July 5, 2021.

[63] Kienle, C. et Al. Effects of Treated Wastewater on the Ecotoxicity of Small Streams – Unravelling the Contribution of Chemicals Causing Effects, PLoS ONE, 14, 12, https://doi.org/10.1371/journal.pone.0226278, Accessed July 5, 2021.

[64] Larsson, D. G. J. et al Effluent from Drug Manufactures Contains Extremely High Levels of Pharmaceuticals, Journal of Hazardous Materials, 148, Issue 3, 30 September 2007, Pages, 751–755..

[65] Malhotra, G.: Pharmaceuticals, Their Manufacturing Methods, Ecotoxicology, and Human Life Relationship, Pharmaceutical Processing, November 2007, pgs. 24–26, Accessed August 10, 2009.

[66] Safety Data Sheet Omeprazole Sulfide https://www.caymanchem.com/msdss/13118m.pdf May 17, 2016 Accessed June 25, 2021.

[67] Omeprazole Sulfide Product Data sheet https://www.caymanchem.com/pdfs/18885.pdf Accessed June 25, 2021.

[68] Metformin Hydrochloride Safety Data Sheet https://www.caymanchem.com/msdss/13118m.pdf Accessed June 25, 2021.

[69] USP Metformin Hydrochloride and Levothyroxine Sodium http://www.pharmacopeia.cn/v29240/usp29nf24s0_m49795.html#:~:text=USP%20Monographs%3A%20Metformin%20Hydrochloride&text=Imidodicarbonimidic%20diamide%2C%20N%2CN%2D,calculated%20on%20the%20dried%20basis Accessed July 14, 2021.

Chapter 8
Road Map

All the earlier chapters dealt with the chemistry and chemical engineering considerations to create, scale-up, and commercialize processes for various active pharmaceutical ingredients. These chemistries are commercial and produce the needed API. It is valuable to review the history of the landscape again and create a roadmap for safe, economic, and optimal manufacturing processes. Manufacturing technology innovation will not only reduce the "E-factor" [1] of the API manufacturing, but also expand and protect public health of a significantly larger population [2]. Revenues and profits will increase through satisfying the needs of larger population as drugs will become more affordable. There are internal barriers [3] to lowering the "E-factors." By using the road map discussed in this chapter and throughout the book, these factors can be lowered.

8.1 Why manufacturing technology innovation has lacked in API manufacturing?

As indicated in the first chapter, most of the initial drugs were discovered at, what were then, the specialty/fine chemical manufacturing companies. The majority of small-molecule [4] active pharmaceutical ingredients (API) are fine/specialty chemicals that have a disease- or illness-curing value. From the onset, it was realized that they can be produced in existing processing equipment that was being used for the synthesis of more "traditional" industrial, specialty chemicals. Unit processes [5] and unit operations [6] that were practiced for chemical synthesis were applicable for API synthesis also.

Pharmaceutical companies in the last 100 years, besides creating new drugs, have not done much for manufacturing technology innovation. Their focus has been on "curiosity and speed" for the new drug or introducing generic drug treatments. Minimal or none of the "curiosity and speed" that has gone toward creating new treatments has percolated to manufacturing technology innovations.

No one has questioned the current API manufacturing strategies and practices, which are the same/similar from the beginning of the twentieth century. "Why innovation hasn't happened in pharma production?" is a very appropriate question. Pharma's opportunity cost alone makes an urgent case for innovation. Some estimates put the potential worldwide cost savings from efficiency improvement as high as $50 billion – equivalent to the cost of developing 80–90 new drugs every year [7]. These opportunity cost estimates do not include the savings realized from lowering pharma's "E-factor" [1].

https://doi.org/10.1515/9783110702842-008

With increasing global population, demand for drugs has increased. Most likely, the opportunity costs have increased. Doubling to $100 billion or more is not out of the realm of reality. In addition, the number of API manufacturing facilities (in-house and outsourced) has increased. This has led to higher "E-factor" [1] emissions from the API manufacturing facilities. Going forward, if manufacturing technologies are not improved, pharma's emissions will only increase. It is important to address this challenge. If any change is to take place, companies will have to be aggressive toward operational improvements and innovation. This is critical to control global climate change.

8.2 Why and how the current API manufacturing scenario developed?

There are reasons for the lack of manufacturing technology innovation in API manufacturing and their formulations. "It has not been necessary" is a misplaced answer. "Necessary" is a relevant term. Is it necessary for Pharma's profits or is it necessary for the benefit of all mankind?

A review and explanation of the current situation is easy and simple. It presents tremendous innovation opportunities. Some of these are examined. Chemists and chemical engineers have used their creativity and imagination to design and commercialize excellent processes for chemical manufacturing. They can do it for API manufacturing too.

It is necessary to review why the current scenario exists. Learnings from the current situation can be used to innovate API manufacturing practices. Figure 8.1 is a simple illustration of the progression of a product from the laboratory to commercial scale.

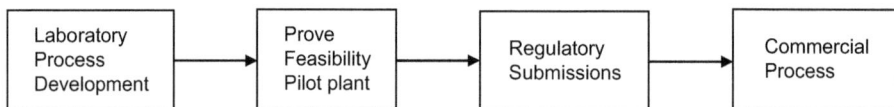

Fig. 8.1: Progression of API from laboratory to commercial product.

Laboratory equipment is used to demonstrate/prove the feasibility of the chemical synthesis route. Using the chemistry developed in the laboratory, chemists and chemical engineers develop, design, pilot, and commercialize a process that will produce the desired quality product from the onset. Yearly throughput at a site is the design basis for process equipment design. It dictates whether the commercial process will be a batch or a continuous process [8]. These definitions are succinct and well-defined. Altercations to established definitions for batch or continuous processes would be mocking the established science and engineering principles.

Calling a batch process a continuous process, as stated earlier, is an unacceptable practice. Too many entities (includes equipment vendors, pharma companies, consultants, regulators (FDA [Food and Drug Administration] and ASTM [9] [American Society for Testing and Materials], and ICH [10] [International Council for Harmonization]) have been lately doing that. Many claims of "continuous process" are cases of fake definitions [11], as "continuous process" is the latest buzz word.

Fine/specialty chemicals, due to the lack of production volume, are fitted in the existing processing equipment. They are generally batch processes. Due to similarities of synthesis chemistry and easy availability of processing equipment, API manufacturing adopted fine/specialty chemical manufacturing practices, since many of the APIs are produced at multiple sites, even when some have the production volume to have a single dedicated plant to meet the global need (Tab. 8.1).

Tab. 8.1: API dosage, global demand versus number of sites and API kg/year [12, 13].

Drug	Dosage (mg/day)	Population use estimate (million)	Total demand (kg/year)	Sites per API	API per plant (kg/year)
Omeprazole	40	1,092	2,184,000	87	25,100
Metoprolol	50	78	1,423,500	41	34,707
Hydrochlorothiazide	12.5	156	711,750	54	13,180
Metformin	1,000	175	63,875,000	77	829,545
Dimethyl fumarate	240	15.4	1,349,040	49	27,531
Levothyroxine	0.112	1,000	41,000	24	1,710

With multiple companies producing the same product, fitting processes in the existing equipment was a natural and an accepted practice [14, 15]. This practice of fitting processes using batch processes has continued.

Using batch processes for API manufacturing has served the global API needs very well. A fact remains that the equipment used has much lower productivity, compared to a process where the equipment is specifically designed for the product. The ability to fit the processes in existing equipment minimizes capital expenditures for new products.

The current commercial batch API manufacturing processes, in the simplest terms, are a larger version of the laboratory process. Laboratory syntheses are optimized and improved in the pilot plant and fitted in the existing equipment that was and is available. It can also be used for the manufacture of fine/specialty chemicals.

8.3 Cost of drugs

Drug prices were and are priced at the highest level that patients can afford to extend their life or stay well, rather than a commodity where prices are competitively priced. The current pricing works well in today's mutually subsidized healthcare systems and countries where distributors are guaranteed fixed margins. With every segment of the drug business (API manufacturing, formulations, and distribution) being profitable and comfortable with the current landscape, there is no pressure to innovate manufacturing practices, even if it improves affordability to a larger population [3].

With current practices, it is possible to have a good process, but it might not be the most efficient process, as the equipment is not designed for the process. Every process inefficiency cost is passed on to the patients, who do not know the drug cost component breakdown. Not many know the API cost contribution in the drug selling price. API cost contribution in the drug selling price of each dose is generally in single or fractional digit percentage [16]. Table 8.2 is an illustration for a few drugs. These are the best estimates. API manufacturers and others in the supply chain may not want to discuss these numbers.

Tab. 8.2: API cost percentage versus tablet sell price.

Drug	Metformin HCl	Ciprofloxacin	Levothyroxine	Atorvastatin
API sell price ($/kg)	4.00	25.00	4400.00	310.00
Inert excipients ($/kg) (@40%API cost)	1.60	10.00	1760.00	124.00
Conversion cost ($/kg) (@40%API cost)	1.60	10.00	1760.00	124.00
Profit (@ 40% above)	2.88	18.00	3168.00	223.20
Total ($/kg)	10.08	63.00	11,088.00	781.20
Dose (mg)	**500.00**	**500.00**	**0.112**	**20.00**
Selling price per tablet ex formulator ($)	0.005	0.032	0.001	0.016
Pharmacy				
Rite-Aid with insurance price/tablet	0.07	0.2	0.17	0.31
API cost % of sell price	2.86	6.25	0.29	2

Annual volume of most APIs (Tab. 8.1 and other chapters) is not large enough to have dedicated equipment and process designed for their production. Since new processing equipment cannot be justified for every API, fitting the chemistry, as stated earlier, in the existing equipment has been and is the chosen route. This is not ideal for many processes. Traditional batch production infrastructure is based on large reactors [17].

Every process does not fit well in every commercially available reactors tank, agitator, heat exchanger, and other processing equipment. Astute chemists and chemical engineers have very cleverly used the existing equipment and solvents to produce the needed quality products. Their best friends for this achievement are solvents. Processes are fitted in the existing equipment even when it is well-known that the batch processes are suboptimal [18] and are known to be the leading cause of high "E-factor" [1].

Solvents (water, single or multiple solvents or their combination) are the process facilitators. Even though they are recycled, they are still the leading cause of waste [19, 20]. If multiple, different solvents are used in a process and cannot be economically separated and/or reused; they are disposed while meeting the prevailing effluent regulations. In such cases, disposal is economic versus chemical or physical separation. As indicated in earlier chapters, besides water, every attempt has to be made to limit the solvent to one organic solvent.

Since the equipment used is not ideal for the processes, it necessitates extra cleaning and increased amount of solvent is needed for cleaning, which is a necessary part of the cGMP practice, and a must between product changes.

A combination of poorly-selected solvent and use of ill-designed equipment/process, at times, can be a hindrance to innovation. It is necessary to have a complete understanding of the mutual behavior of the physical and chemical properties of chemicals involved in every reaction. Their value has been discussed in earlier chapters and is reviewed later in this chapter.

API manufacturing facilities have low overall equipment utilization [21–25]. Excess batch manufacturing capacity [26] and too many plants [27] are producing the same API. Table 8.1 and similar tables in earlier chapters are an illustration of some APIs that are being produced using batch processes, even when some of them could be produced using continuous processes to meet the global demand. If done correctly, through properly designed plants, "E-factor" [1] for these and many other products can be significantly reduced.

The practice of fitting processes in existing equipment continues, as no one wants to invest in processing equipment that would not be used all the time. Actually, this practice is expanding in the recent years, as brand and generic API manufacturing is being outsourced to contract manufacturing organizations (CMO), where the processes are fitted in the available equipment. Instead of having the "best designed process," the intent is to have the "best fitted process" that will produce the desired quality product.

The art and science of creating and commercializing an optimum process for each chemistry is being lost. By outsourcing API manufacturing, the processes have become a commodity; one pot fits all syntheses, rather than specialty equipment, where perfection and quality are delivered from the onset. As discussed earlier, proper use of solvents can make this happen. There are ways to correct this and they are reviewed later in this chapter.

Unless there is long-term contract to manufacture an API, no contract manufacturing company will make an investment for dedicated equipment for a product, whose efficacy and need would become obsolete with the product. In addition, as has been discussed earlier, many of the APIs are produced at many factories; thus reducing the volume output at a single factory and this precludes any investment for low volume products. Any potential advantage that could be achieved through economies of scale is lost. Due to the unknown life of an API, companies have and will shrug from investing in dedicated equipment. Alternates that can bring the ownership to API process developers are possible and are reviewed later in this chapter. For this to happen, the current business model might have to be modified. Creativity and imagination would need to have a free reign.

Since new processing equipment cannot be justified for every new API product, fitting the chemistry, as stated earlier, in the existing equipment has been and will be the chosen route.

Quality by design (QbD) is a fundamental premise that each chemist and chemical engineer is taught to practice for any process they design and commercialize. If they do not practice and achieve this criterion, it suggests that their basic process design is flawed. They have to make it right from the onset.

Due to the mortar and pestle philosophy, that is, the batch process, even when the process is a QbD project, in-process testing is the established norm for API batch processes. This assures product quality through repeated analysis. This also complies with the FDA requirements [28]. API manufacturers want to make sure that each batch meets the established product specification. If a batch/es does/do not meet the specifications, it may be easier and cheaper to discard the batch/es. All of the additional costs were/are absorbed and passed on to the patients. These practices are still used. Since in-process intermediates are held for quality check, in-process storage tanks add to the idle equipment capacity [26], that is, lower asset utilization. For pharma, this is well documented [21–25].

The above discussion has been about the tradition of using existing batch processing philosophies. Current regulations are designed for batch processes. They will have to be modified for "continuous manufacturing". Since FDA staff does not have any/much development, design and commercialization experience in these processes, development of applicable regulations and promulgation will be a significant challenge.

8.4 Review of the laboratory process

It is necessary to understand and review some of the ways that can lead to better API manufacturing technologies and lower "E-factor" [1]. The process of use of excessive amount of solvent, "high E-factor," starts in the laboratory process development. Based on the different chemistries outlined in various patents (too many to cite), one can see that the solvent use, relative to solids, is high. Tables 8.3 [29] and 8.4 [30] are an illustration and analysis for the production of Paroxetine Hydrochloride and Paroxetine Mesylate, respectively.

If we review every patent filed for APIs and fine/specialty chemicals, similar stories of use of excessive solvent use appears over and over again. In fine/specialty chemical domain, product volumes are much larger than the typical volumes in the API domain. This is due to the milligrams dose needed to cure diseases (1 kg product can produce 1 million tablets of 1 mg each to serve many) versus fine/specialty use, as additives in kilograms. Since the API is produced at many (Tab. 8.1) sites, the value of economies of scale are lost and a larger version of the laboratory process gets commercialized in the existing processing equipment, resulting in high "E-factor" [1] emissions.

Once the chemistry has been optimized, after initial financial justification, it has to be tested and the processing conditions have to be defined. Learnings of the laboratory equipment teach us the unit processes [5] that are commercialized using established unit operations [6].

How processes get developed in the laboratory is well known. Unless all of the reactants are liquids, selected solvent/s are added to the reaction flask and the solid/liquid reactants are added until the mass can be agitated. Additional solvent/s might be necessary. Laboratory curiosities are translated to commercial realities. This continues until the process is defined. Since the laboratory chemistry development process, generally, cannot capitalize on the physical properties of the reactants and intermediates, the developed process in the laboratory is the best until scaled up.

If clinical trials prove the efficacy of a product, race and speed to commercialize the product matters more than having an economic process. Since the disease-curing value (selling price) of the drug is more critical than the cost of API, optimizing the process becomes irrelevant.

Based on the different chemistries outlined in various patents (too many to cite), one can see that the solvent use, relative to solids, is high.

Tab. 8.3: Paroxetine hydrochloride synthesis [29].

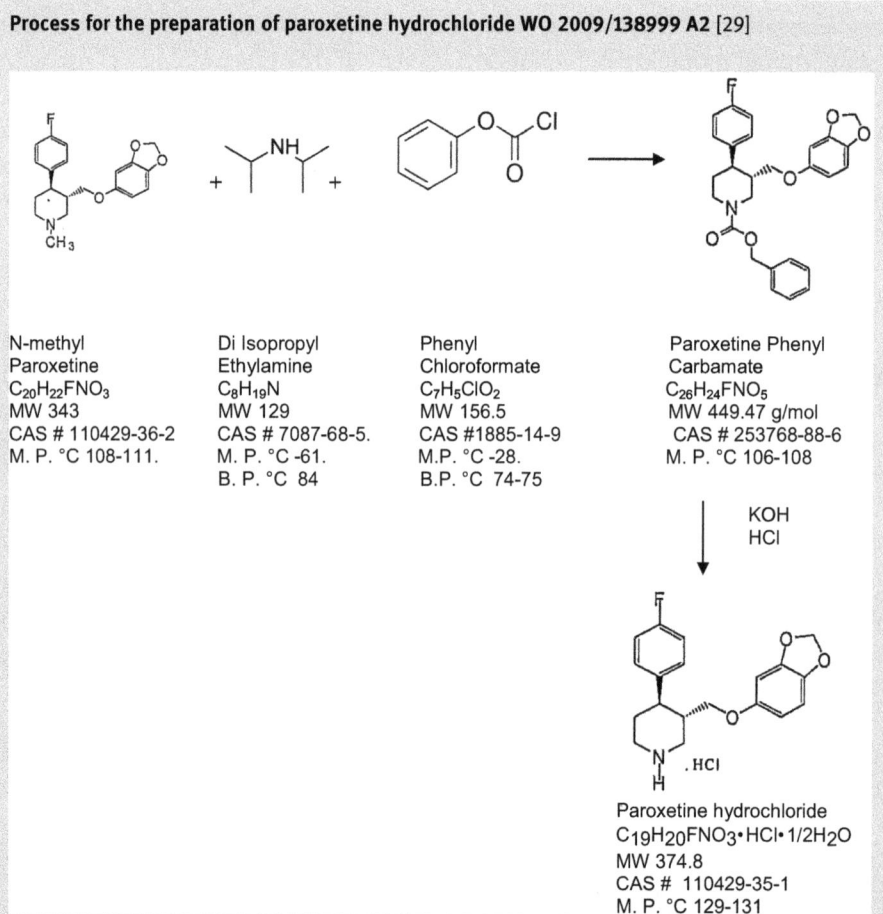

Process for the preparation of paroxetine hydrochloride WO 2009/138999 A2 [29]

N-methyl Paroxetine $C_{20}H_{22}FNO_3$ MW 343 CAS # 110429-36-2 M. P. °C 108-111.	Di Isopropyl Ethylamine $C_8H_{19}N$ MW 129 CAS # 7087-68-5. M. P. °C -61. B. P. °C 84	Phenyl Chloroformate $C_7H_5ClO_2$ MW 156.5 CAS #1885-14-9 M.P. °C -28. B.P. °C 74-75	Paroxetine Phenyl Carbamate $C_{26}H_{24}FNO_5$ MW 449.47 g/mol CAS # 253768-88-6 M. P. °C 106-108

KOH
HCl

Paroxetine hydrochloride
$C_{19}H_{20}FNO_3 \cdot HCl \cdot 1/2H_2O$
MW 374.8
CAS # 110429-35-1
M. P. °C 129-131

Example:
Process for paroxetine-N-phenyl carbamate

N-Methyl paroxetine (100 g) and toluene (900 mL) were added into the RB flask. Di-isopropyl ethyl amine (7.6 g) was added, under stirring, and the reaction mass was then heated to reflux temperature (110–112 °C). Phenyl chloroformate in toluene solution (57 g PCF in 100 mL toluene) was added and heated to reflux temperature. After the completion of the reaction, the reaction mass was washed with aqueous hydrochloric acid solution (100 mL) and extracted into toluene. The organic layer was washed with sodium bicarbonate solution (50 g in 1,000 mL) and then concentrated under reduced pressure to give a residue. The residue was treated with isopropyl alcohol (500 mL) to give paroxetine-N-phenyl carbamate.

Example:
Process for paroxetine hydrochloride:

Paroxetine-*N*-phenyl carbamate (125 g) and toluene (1,000 mL) were charged into the reaction mass. Potassium hydroxide flakes (81.6 g) were added and the reaction mass was heated to reflux temperature (110 °C). After the completion of the reaction, the mass was taken into aqueous sodium hydroxide solution (25 g of NaOH in 500 mL demineralized water) and extracted into toluene. The organic layer was separated and concentrated under reduced pressure to give a residue. The residue was dissolved in isopropyl alcohol (500 mL). IPA and HCl (53.2 g of HCl in 100 mL IPA) were charged to the above solution at 20–25 °C. The resulting solution was stirred, filtered, and washed with isopropyl alcohol to give paroxetine hydrochloride.

Paroxetine-*N*-phenyl carbamate (125 g) and toluene (1,000 mL) were charged into the reaction mass. Potassium hydroxide flakes (81.6 g) was added and the reaction mass was heated to the reflux temperature (110 °C). After the completion of the reaction, the mass was taken into 30 aqueous sodium hydroxide solution (25 g of NaOH in 500 mL DM water) and extracted into toluene. The organic layer was separated and concentrated under reduced pressure to give a residue. The residue was dissolved in isopropyl alcohol (500 mL). IPA. HCl (53.2 g of HCl in 100 mL IPA) were charged to the above solution at 20–25 °C. The resulting solution was stirred, filtered, and washed with isopropyl alcohol to give paroxetine hydrochloride.

Dry weight: 120 g

Analysis of the laboratory process:

In this example, an excessive amount of toluene is used. It is alright to demonstrate the reaction step feasibility in the lab. However, if this chemistry is to be scaled up, tested in the pilot plant, and the process commercialized, its stoichiometry and solvent use will have to be optimized. The theoretical mole ratio of phenyl chloroformate and *N*-methyl paroxetine is 1:1. In the suggested reaction, 25% excess of phenyl chloroformate is used.

Reactants (*N*-methyl paroxetine + phenyl chloroformate) make about 15% of the toluene/solid mix. This indicates that an excessive amount of toluene is being used. It will have to be significantly lowered for a commercial process.

Similarly, the conversion of paroxetine-*N*-phenyl carbamate to paroxetine also uses an excess of toluene. This is acceptable for laboratory experiments. However, for commercial operation, solvent use will have to be optimized and lowered. The equipment and process conditions will determine the optimum conditions. Isopropyl alcohol is the second solvent used.

Two separate solvent recovery systems would be needed if the process is to be commercialized. This adds complexity and cost.

Tab. 8.4: Paroxetine mesylate synthesis [30].

Process for the production of paroxetine USP **6686473** [30]

a) Paroxetine phenyl carbamate

The reaction vessel was charged with 17.4 kg of *N*-methyl paroxetine (containing approx. 5% of water) and 58 L of **toluene**. The reaction mixture was heated to reflux and the remaining water was removed by azeotropic distillation using a Dean-Stark receiver. Under reflux, 9.1 kg of phenyl chloroformate was added in 30 min and stirring was continued at reflux for 1.5 h. About 0.8 L of triethylamine was added and the mixture was heated under reflux for the next 30 min. The solution was concentrated to approx. 30 L volume by distillation at a diminished pressure (30 mbar) and cooled to ambient temperature.

b) Paroxetine-free base The reaction vessel was charged with 16 kg of potassium hydroxide and 45 L of **1-butanol,** and the mixture was heated under stirring to 80 °C. The solution from the preceding step was mixed with 33 L of 1-butanol and the combined solution was added to the above mixture under stirring. In approx. 20 min, the exothermic reaction brought the mixture spontaneously to reflux. Next, the supply vessels and pipes were rinsed with 5 L of butanol, which was combined with the reaction mixture. The reaction mixture was stirred under reflux for 2 h. Then, the mixture was cooled to approx. 95 °C and 45 L of water was added under stirring. The mixture was allowed to stand for phase separation and the lower phase was discarded. The upper phase was diluted with 13 L of toluene, and the combined solution was washed with 19 kg of 30% aqueous NaOH solution and 18 kg of 16% aqueous NaCl. The solvent was removed by distillation at reduced pressure, at 50 °C.

The crude product was dissolved in 70 L of toluene. 0.3 kg of Tonsil was added and the reaction mixture was filtered. Next, the reactor and the filter were rinsed with 5 L of toluene. The combined toluene solution was washed twice with 35 L of water and evaporated at reduced pressure, at 50–60 °C.

c) Paroxetine mesylate The product from step b) was mixed with 58 L of **ethanol** and heated to 60 °C. Methane sulfonic acid (4.9 kg) was added to the resulted solution under stirring. The solution was cooled to 25 °C and inoculated with a seed crystal. The mixture was cooled to 0 °C, stirred for 30 min, and the obtained solid was filtered. The filter cake was washed 2 times with 6 kg of cold ethanol. About 19.6 kg of wet paroxetine mesylate was obtained. After recrystallization from ethanol and drying, 14.5 kg of the dried product was obtained.

Analysis of the laboratory process:
Like paroxetine hydrochloride process, this process uses multiple solvents. Excess toluene is used as a solvent, thereby diluting the process. Butanol and ethanol are the two additional solvents used. Patent suggests removal of water from *N*-methyl paroxetine. This needs to be reviewed, as water has to be removed from the reaction mass before the phenyl chloroformate reaction.

Washing with 30% NaOH and 16% NaCl may be acceptable for a lab exercise, but their adoption in commercial process will suggest that the mutual physical properties have not been understood and exploited. For process simplification, it is necessary that the physical properties be exploited.

In both syntheses of paroxetine salts, excessive amounts of multiple solvents are used. That may be acceptable to prove the chemistry in the laboratory, but if they are to be commercialized, they will have excessive waste. They also add to the complexity of the whole process.

Table 8.5 is an estimate of global paroxetine hydrochloride needed. It is being produced at about 36 plants [31]. Since there are multiple plants, their "E-factor" [1] emissions will be high, and every potential benefit of economies of scale is completely lost. Like other APIs, it also presents an opportunity to lower its "E-factor" [1].

Tab. 8.5: Global paroxetine hydrochloride need.

Global population	Percent use	Use, global population	Dose mg @ 180 days/year	Total kg/year	Plants producing [29]
7,800,000,000	1	78,000,000	20	280,800	36

8.5 Why innovation is needed in pharma?

Discussion about innovation and simplification in API manufacturing has been going on for quite some time [32, 33] at FDA [34]. FDA's report in 2004 explained the current and the "desired state" of manufacturing very clearly.

> Excerpt of the Current State: Pharmaceutical manufacturing operations are inefficient and costly. Compared to other industrial sectors, the rate of introduction of modern engineering process design principles, new measurement and control technologies, and knowledge management systems is low. Opportunities for improving efficiency and quality assurance through an improved focus on design and control, from an engineering perspective, are not generally well recognized. For example, when discussions at the FDA Science Board and Advisory Committee for Pharmaceutical Science shed light on the current low efficiency and its cost implications (e.g., costs associated with manufacturing can far exceed those for research and development operations in innovator pharmaceutical firms) many at FDA had difficulty understanding this and common reactions were "how could this be possible?" or "this can't be true." Regulators and many in manufacturing operations express their frustration by suggesting that manufacturing is a "step-child" in this industry, and that there is no economic motivation (e.g., cost and price difference) for improvement. Other suggestions include a general lack of systems perspective, organizational barriers that inhibit exchange of knowledge, and the attitude that much of pharmaceutical formulation and process development is an "art." Some in pharmaceutical development suggest that there are very limited opportunities ("development time crunch") to realize and/or demonstrate the level of science underlying current formulation and process development efforts.

Excerpt of the Desired State:

> Improving the foundation of manufacturing science in our current manufacturing practices should be the primary basis for moving away from the corrective action "crisis" to continuous improvement. The "desired state" for pharmaceutical manufacturing in the 21st Century therefore emphasizes and aims to improve knowledge on design and understanding of product and processes. The "desired state" for pharmaceutical manufacturing in the 21st Century therefore emphasizes and aims to improve knowledge on design and understanding of product and processes.

Chemists and chemical engineers, who are development specialists for the API processes, need to put themselves in the operator's shoes and need to ask themselves: "If I had to operate the process in a manufacturing environment, how would I feel? Is this really the best that I can do?" [32]. Many of the answers become obvious. They would want each process to hum as a well-oiled machine and produce a quality product all the time. Unfortunately, batch processes, especially most API processes, fall short of that expectation.

It is interesting that not much progress in manufacturing innovation, to achieve what USFDA calls the "desired state", has taken place since 2004. All of the components (science and engineering) to significantly lower the "E-factor" [1] and manufacturing are taught and are available to all of the practitioners, but the industry, being tradition bound, has not exploited the methods to achieve what is possible.

Pharma manufacturing (API and their formulations) have not made much progress towards manufacturing technology innovation. It seems that the industry, in general, is going around in circles and will not have much of impact on process simplification. Regulators and equipment vendors, who might have relevant equipment for synthesis, have been pushing the API manufacturers and their formulators to adopt "continuous processing methods". However, neither, especially the regulators, have ever developed, designed, financially justified, or commercialized such processes for pharmaceuticals. With respect to equipment, API manufacturers have to justify the investment for any new equipment. With excess equipment capacity [21–26], it is going to be extremely difficult to justify any investment, if the equipment is not going to be producing a single product for about 7,000–8,000 h/year.

If pharma's most recent patents are any indication, they do not reflect much of "solvent conservation" as part of their API synthesis development process. It is very likely that the laboratory process development practices of excessive and multiple solvent use will continue, but a change in their scale-up and commercialization practices is definitely needed if pharma is to lower its "E-factor" [1] emissions.

Pharma has resisted any change in their manufacturing practices, once their product and process have been approved. Regulators have acknowledged that this resistance is due to regulatory uncertainty.

> Unfortunately, the pharmaceutical industry, generally, has been hesitant to introduce innovative systems into the manufacturing sector, for a number of reasons. One reason often cited is regulatory uncertainty, which may result from the perception that our existing regulatory system is rigid and unfavorable to the introduction of innovative systems. For example, many manufacturing procedures are treated as being frozen, and many process changes are managed through regulatory submissions. In addition, other scientific and technical issues have been raised as possible reasons for this hesitancy. Nonetheless, industry's hesitancy to broadly embrace innovation in pharmaceutical manufacturing is undesirable from a public health perspective. Efficient pharmaceutical manufacturing is a critical part of an effective U.S. health care system. The health of our citizens (and animals in their care) depends on the availability of safe, effective, and affordable medicine. [35]

Regulators claim that pharma's hesitancy toward "continuous improvement [36]" is unacceptable, but on the other hand, they have not done much to alleviate its own "red-tape" [37] practices, which could placate this hesitancy on the part of API manufacturers and their formulators. Their bureaucracy, which prolongs filings, approval, and making updates expensive and difficult, is well-known. Regulators need to practice what they preach and they have not done so. This is obvious from the prolonged time they take in approving new drug (NDA) and abbreviated new drug (ANDA) application filings. FDA does not publish the average time it takes to approve any "single" application, but publishes what they do in a fiscal year, which camouflages the real time per application. Regulators have to establish and improve their own current practices [38].

It would be beneficial for the pharmaceutical industry if the regulators practiced their current NDA (new drug application) and ANDA (abbreviated new drug application) filing requirements and processes in-house, to get a feel for what the industry has to follow for the respective approval. These would be reviewed by the fellow FDA employees as if they were filed by a company. These have to be blind tests. The time for each such process has to be published and used by FDA to simplify and reduce their current filing and approval times. They can use such work to set benchmark goals. Benefits of these will be faster introduction of brand and generic drugs to the market, which pharma companies would like to have, and it would be a global win.

When regulators make suggestions to the industry to improve manufacturing technologies, it seems they forget that for the adoption of any new manufacturing technology/methods, companies have to justify the investment. They have been pushing companies to innovate their manufacturing practices, but have not shown any leadership in innovating their own practices which will facilitate industry innovation [39, 40].

Every industry has innovated and many new ones have evolved through innovation. Cellular phones are an excellent example. However, pharma has been very passive with its manufacturing practices and needs a total "rethink" if it wants to lower its "E-factor" [1] emissions, make drugs affordable to a larger population, and, further, improve its profits. Pharma has an opportunity to innovate. It needs to shed its addiction to the old ways of manufacturing and implement improved methods [40, 41]. Pharma can choose "speed and curiosity" of the startup communities or Silicon Valley, or chose either of the following two ways to innovate manufacturing technologies. It has to make a selection. Continuous manufacturing of API and their formulations is very doable. All of the necessary fundamentals (science and engineering) and equipment are available and commercially used, except in the manufacture of API [9, 42–53].

In the continuous processes, none of the reaction process intermediates are isolated or purified for the subsequent reaction step. If they are, then the process is not a continuous process but a batch process. If the FDA and other regulators demand that the intermediates in a process be isolated and tested for defined characteristics and purity, then the process is no different from a batch process.

8.6 Transformational innovation/creative destruction

Transformational innovation [54] entails transition from the mode of operating that is known and secure, to one that is unknown and potentially chaotic. This requires doing something fundamentally different; a metamorphosis most organizations don't excel at. Pharma manufacturing has become extremely comfortable with its ways over the last one hundred years. For it to move away from "what it is" to "what it can be", it is going to be a challenge, as it has not faced any threat.

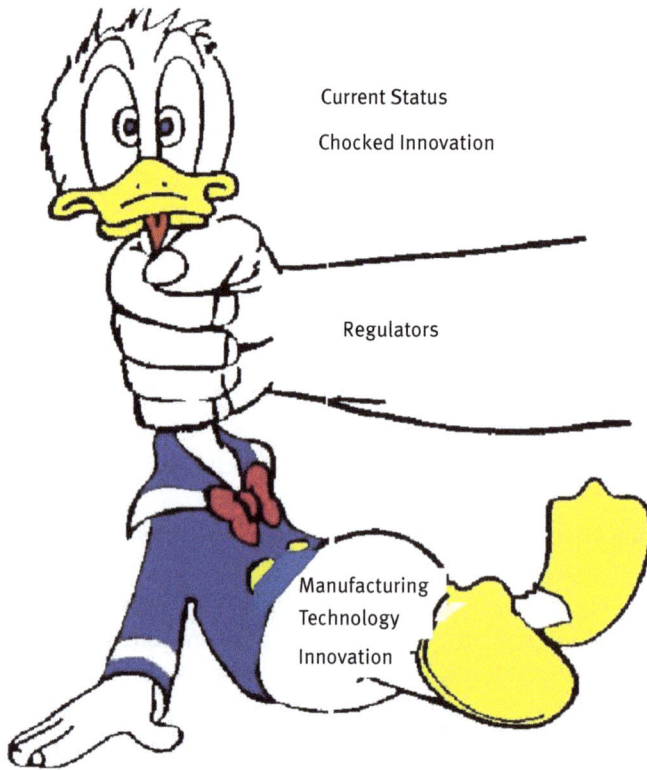

Fig. 8.2: Pharma's current state of manufacturing technology innovation.

Innovation via "creative destruction" [55], which is similar to transformational innovation, is disruptive, because it introduces products and services that change the business landscape by providing a dramatically different value proposition.

Championing innovation via "transformational innovation" or "creative destruction" involves going to war with all of the elements inside an organization that benefit from the status quo. Within the organization, it may involve persuading managers and staff to turn their working lives upside down and inside out, and not to do so slowly and grudgingly but quickly and enthusiastically. Such a shift will never be easy because it puts in question the existing strategies, jobs, careers, processes, brands, customers, and culture. For API manufacturing companies to consider such an innovation method will be difficult, as it has not faced any external threat.

8.7 Nondestructive creation

According to Edmund Phelps, a Nobel Laureate, "creative destruction" and "transformational innovation," in which free and uncoordinated contest of ideas generate growth, lead to chaos. Either of the above routes may not be suitable for pharma, as it would not be creating anything new, but causing more disruption. With either route, it is possible that the regulatory interference might also increase.

The concept of "nondestructive creation" [56, 57], introduced by Phelps, is another way for innovation. It takes the "whole village" for an improvement to be developed, adopted, and launched. It is "all inclusive" and would be the least disruptive way for any business to innovate. Team members, "the village" at a pharmaceutical manufacturing, would include chemists and chemical engineers, marketing, financial analysts, supply chain professionals, quality control, regulators, maintenance, and manufacturing, to think about and commercialize manufacturing processes [58].

The likelihood of success through "nondestructive creation" is the highest, and would be the least disruptive option, as all of the elements and components for its realization are part of every existing manufacturing and business process. Since it will be a collective effort and an extension of the current methods practiced differently, it will deliver results quickly and would be beneficial for the organization. With the success of this methodology, it is very likely that the current business model will improve significantly. It could influence and lessen the current migration of API manufacturing to, CMOs. It will bring the philosophies of "continuous improvement" [36] home to API manufacturing. A pride of creating excellent and innovative manufacturing processes, which is being lost as manufacturing is being outsourced, will come back. Like every initial effort for change, skeptics will be there and will have to be convinced of the values. Equipment vendors could resist and detest internalization of process design, as they could lose their current clout.

It is necessary to review the reasons for the current high solvent use landscape. Handling of solid raw materials, intermediates, and use of the available equipment size are the major causes. Solvents facilitate fitting of most unit processes and unit operations in the existing equipment, even if they are inefficient. In addition, since they can be recovered and recycled, methods to lower the solvent use are not a priority or explored. The current API costs are based on excessive solvent use. If the solvent changes from the current use levels, it does not impact much the drug selling price. The cost impact of the solvent on the selling price of the API or the selling price of the drug is in single-digit percentage (Tab. 8.2). Customers will not feel the impact of any conservation. The current API costs are based on excessive solvent use.

As reviewed earlier in this chapter and other chapters, solvents are process facilitators. Even though they are recycled, it does not mean that they can be and should be used in large quantities. In reality, the chemists and chemical engineers, due to the tradition of using the available processing equipment that is routinely used for chemical processing, are forced into using excessive amount of solvent. As a result, alternate processing options are not considered.

There are reasons and rationale for the high solvent use. A recap of the existing practices is necessary. It was, and still is, that with the help of solvent/s, organic synthesis reactions can be fitted in the existing equipment. This minimizes any investment for products. This practice continues:

1. Solvents are used as the medium to transfer the raw materials and intermediates, to and between the reactors. This allows fitting the process in the existing equipment.
2. Even if a solid raw material has a low melting point and is not a liquid at room temperature, it will be slurried to feed the reaction system. If the intermediates or a raw material to be added later in the reaction are solid, again they are slurried to make sure that the slurries can be pumped and easily agitated in the subsequent steps.
3. Excess solvent use also facilitates mixing and heat transfer in the current equipment.
4. Some raw materials could be fed as a melt. However, due to the low API annual production volume, a dedicated melt feed system would be expensive at any commercial plant. If it is used, it would have to comply with the cGMP cleaning requirements and may end up using additional solvent, thereby creating more solvent waste. Using a supplier provided molten liquid feed carrier would also be subject to FDA regulations and, in certain cases, that could be a challenge.
5. If any solid intermediate of the process is isolated, purified, and re-solubilized or re-slurried in a solvent for further processing, it will create additional waste. For a continuous process, such isolation would not be needed.

Chemists and chemical engineers by, their training, are seekers and experimenters who, by nature, create the simplest processes. They experiment and extrapolate the physical and chemical properties of the chemicals used and produced along with unit operations to create processes that are viable and safe.

Chapter 3 discusses value of the physical properties of raw materials, intermediates, understanding their solubilities, melting and boiling points, azeotropic behavior, mutual miscibility, and density differences. They are the crux of "nondestructive creation" [56, 57] and can be exploited to create excellent processes.

There are considerable ways and opportunities to improve batch and continuous API manufacturing. Companies will have to "think differently" [58–66] if they want to significantly lower their "E-factor" emissions. For API manufacturing simplification, it is time to exploit every possible opportunity.

The process of "nondestructive creation" [56, 57] in API process development has to start as soon as the efficacy of a molecule is acknowledged. The mission of all involved has to be to create an efficient process that will produce a quality product of the lowest cost, from the onset, with the lowest "E-factor" [1] emissions. The inclusion of lowest "E-factor" [1] emissions is a new or different value proposition for an organic synthesis process. This is an important consideration for each organic molecule synthesis in an effort to lower pharma's high organic emissions [67]. With the understanding of the root cause of high "E-factor" [1], it is distinctly possible to address the related issues and take the necessary remedial steps.

8.8 Why excess solvent is used?

Solvent use in large quantities, as has been discussed in earlier chapters, also begins at the first development step in the laboratory. A review of every patent indicates that if any of the reactants is a solid, a large excess of solvent will be used to produce a slurry or solution and, in general, the solid concentration would be about 15 to 50%. High use of solvent continues for most API for multiple reasons. The mortar and pestle philosophy persists in every step of API chemical synthesis.

If the chemists and the chemical engineers are constrained about the solvent use and given the freedom to exploit, they can create excellent processes with all the tools that are available. Their creativity and imagination have to be let loose. The methodology, technology, and knowledge to lower pharma's "E-factor" [1] have existed from the inception of chemical processing, but they have never been truly considered and applied to lower the solvent use. The requirement of using the existing equipment has stood in their way. API manufacturers will have to give away their deep-rooted and imbedded traditions [65]. Just the consideration or restriction of using smaller-sized equipment will set the path to reduce the amount of solvent use.

Another major benefit of such a restriction would be that many of the APIs that are produced using batch processes at multiple sites could be produced by continuous processes. In reality, the API manufacturing landscape could totally change. Many APIs that do not have the product demand for continuous production could be campaigned using modular plants for long production runs, emulating continuous processes. Chemists and chemical engineers have the necessary knowledge and background. These modular plants will be like the erectors sets that have been discussed earlier in Chapter 6. Pilot plant equipment can be used.

8.9 Continuous processing

The tradition in chemical process development has been that the development work starts in the laboratory. After the proof of safe and viable chemistry and all justifications, the process gets scaled-up and commercialized. In pharma manufacturing, due to the speed to commercialize the products, a totally different process development and commercialization tact has to be taken if continuous manufacturing has to be considered, even when the eventual volume of the product is not finalized. Learnings from the outlined process could also be applied to batch processes to reduce the product commercialization time. Process development for API manufacturing will have to be done differently.

Instead of the current practice of sequential process development and a need for the speed in product introduction, development has to be done in parallel, as described in Tab. 8.6. This can be modified to suit needs. This process will accelerate speed for product introduction. Compared to a batch process, the "E-factor" [1] for the same product produced by continuous process will be significantly lower.

Tab. 8.6: Current and proposed method of process development.

Current method:
A sequential process: Each process is followed after the first
Lab work →Pilot plant → Commercial

Proposed method:
Each of the steps to be done simultaneously.

Lab work	Pilot plant testing	Commercialization
Knowledge of chemistry Knowledge of the chemicals used and intermediates formed How the physical and chemical properties can be capitalized on to create a simple process in the low-volume pilot plant equipment?	Testing of the sequential addition/removal of different raw materials/intermediates using different small-volume processing vessels. This would simulate a continuous process. If and when the volume increases, the equipment used to test the process can become the dedicated process equipment for the product. This method can be ramped up and down to meet the global product is a brand product under patent or a generic when out of patent.	The smaller equipment could be considered as modular equipment that will not only be efficient, if properly designed, but also lower "E-factor" emissions and change the API manufacturing landscape.

The basic premise for every API development has to be that every laboratory process will be commercialized and each company has the necessary expertise for the required work. Parallel effort (proposed method, Tab. 8.6) could cut the commercialization time by as much as 50%. Such a development process will be a realignment of the current practices, and will be fast-paced. It is necessary to emphasize that the parallel process will significantly reduce the total process development and commercialization time. Such a pathway will also be a training ground for newbies. There will be naysayers within the companies.

The International Council for Harmonization of Technical Requirements for Pharmaceuticals, with its recent FDA Q7 [68], ICH Q11 [69] and ICH Q12 [70] guidances, make simplification of process development and commercialization, extremely difficult. These guidances can be considered as a hindrances. FDA will frown on such a connotation. Their focus needs to be on a quality product of a desired efficacy and characteristics, and not the details of who supplies the starting materials. For batch processes, there could be multiple suppliers for the same raw material. The FDA's connotation implies that the API manufacturers are ignorant of their raw material purchases and their impact on the product quality.

Process chemistries indicate the natural reaction mechanism for the highest yield. Developers have to define these conditions. In the laboratory synthesis, every reaction is conducted in sequential addition of various raw materials. If this is not followed and the reactants do not react properly, the end product will not have the desired yield and purity profile. Sequential addition of raw materials is the mainstay of every continuous process. Chemists and chemical engineers, by taking advantage of the mutual behavior of chemicals and the manipulation of operating conditions, can improve process yields. These, along with the manipulation of reaction kinetics and thermodynamics, result in creating an excellent manufacturing process.

Table 8.7 is the stoichiometric balance for Paroxetine hydrochloride [27]. Preparation of tables similar to Tab. 8.7 is necessary and a must if the process designers want to have control of the raw materials, solvent and raw material use, and the process and its operating parameters. If the yield is not as expected, effort needs to be made to improve the process. Such an analysis assists in developing and commercializing economic processes with low "E-factor" [1] emissions. Table 8.7 indicates that the yield is about 81.3% (=80/98.3) [29]. Compared to its theoretical yield, this presents opportunities.

Based on Tab. 8.7 one can conclude that the process uses an excess of toluene so that it can be processed in the laboratory flask. This will have to be evaluated and optimized for a commercial operation. The mole ratio of other chemicals (Phenyl chloroformate, KOH, and NaOH) indicates an opportunity to improve the stoichiometry of the process.

Figure 8.3 Is a schematic of the first step pf the paroxetine hydrochloride [29] process. Chemical engineers and chemists will quite understand the process. They would have to optimize it. Based on the total global demand, Tab. 8.5, a continuous process can be easily designed for a rate of about 30 kg/h.

Tab. 8.7: Stoichiometric balance for paroxetine hydrochloride [29].

	CAS #	Mol. wt	Gram	Moles	Mole ratio	Weight ratio (%)	Theoretical yield (g)
n-Methyl paroxetine	110429-36-2	343.39	100	0.29	1		
Di-isopropyl ethyl amine	7087-68-5	129	7.6	0.06	0.20	7.6	
Toluene	108-88-3	92	879	9.55	32.81	879	
Phenyl chloroformate	1885-14-9	156.5	57	0.36	1.25		45.58
Paroxetine phenyl carbamate	253768-88-6	449.47	125	0.28	1		
Toluene	108-88-3	92	879	9.55	34.36	703.2	
KOH	1310-58-3	56	81.6	1.46	5.24		
NaOH	7647-01-0	40	24	0.60	2.16		
IPA	67-63-0	60	393	6.55	23.55		
Paroxetine hydrochloride	78246-49-8	365.8	~75–85				98.30

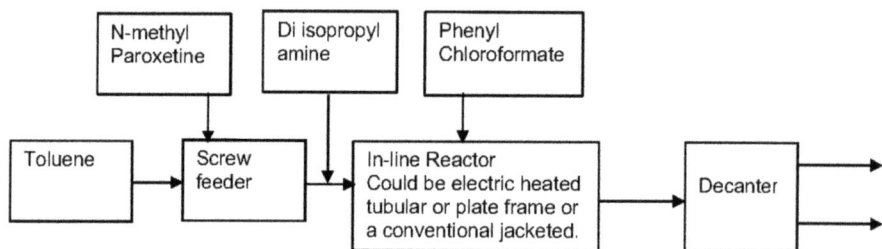

Fig. 8.3: Process flow diagram of a step for paroxetine hydrochloride.

A review of the laboratory process description suggests the use of excessive amount of toluene as the solvent. *N*-Methyl paroxetine concentration in toluene is ~11%. A catalytic quantity of an amine is added to the reaction mass, followed by a reaction with phenyl chloroformate. The reaction mass is converted to Paroxetine hydrochloride. The reaction mass is heated to about 110–112 °C. A catalytic quantity of di-isopropyl amine is added, with a subsequent addition of phenyl chloroformate. Di-isopropyl amine, being a liquid, can be metered in a defined stoichiometric ratio. The produced paroxetine phenyl carbamate is converted to paroxetine hydrochloride hemihydrate, with potassium hydroxide and hydrochloric acid. Paroxetine, the final product, is then isolated and purified.

Since, for every chemical reaction, there is a sequential addition of chemicals, it very clearly suggests that, if the feeds and stoichiometry are properly managed

and manipulated, paroxetine hydrochloride and many other APIs could be manufactured continuously using the equipment illustrated in Fig. 8. 3 or similar to it.
Some of the equipment is listed in Tab. 8.8. Most of the chemical engineers and
chemists are familiar with them and some of the equipment may be used in the
pilot plant scale-up.

Basically, the pilot plant equipment and ancillary equipment can be assembled
and very well used for test as commercial equipment. It can be a modular plant, designated and assembled by the maintenance and process engineers at a manufacturing company. It is doable and many of the fine/specialty producers have the
necessary equipment. Considerable thinking, design, and knowledge of the physical
properties, mutual behavior of chemicals, and reaction intermediates will have to be
incorporated. Process centricity [71] has to be the key. Chemists and chemical engineers are very well versed in this engineering and science.

8.10 Equipment

Two elements, high solvent use and larger-than-needed equipment, could be attrib-
uted to be the cause of high "E-factor" [1]. These could also be considered the cause
of API manufacturing not using continuous processes. For every API molecule process design, the use of currently available equipment in chemical reactions is the
first choice. As discussed earlier, excessive amounts of solvent [19, 20] has to be
used to fit the process in the existing equipment. It results in and is the cause of
high "E-factor" [1]. Since alternate equipment design and operating philosophies
are not considered for API manufacturing, batch processing has become the mainstay for most of the APIs.

To lower the solvent use, there are alternates. The equipment available in the
pilot plants is smaller than the conventionally used equipment. A properly designed
process that capitalizes on the physical and chemical properties and their mutual behavior can be used as a test case. Once such an equipment demonstrates the process
value, the use of similar other equipment can be extended to other products. The
equipment listed in Tab. 8.6 are suggestions. There can be different permutations
and combinations of the equipment listed to use and create an excellent processes
[52, 53]. Selection of the right equipment can simplify the chemistry also.

Tab. 8.8: Processing equipment for chemical processes.

Shell and tube, and plate and frame heat exchangers as reactors: These could be stainless steel,
Karbate, glass lined, PTFE, etc.
Pumps as mixer/agitator.
In-line dispersers as mixers/agitators.
Electrically heat-controlled tubular heaters.

Tab. 8.8 (continued)

Gravity decanters.
Screw conveyors as metering feeders for solid raw materials.
Venturi eductors for mixing and blending of solids.
Microreactors.

Except for microreactors, most of the equipment listed in Tab. 8.8 can be used in multiple applications. Electrical tubular heaters (discussed in Chapter 6) are moreversatile than the conventional heat exchangers, as the heat input can be modulated by plugging in electrical outlets, whereas, conventional heat exchangers would require an auxiliary heat source, for example, steam or hot oil system. Such heaters, by design, are modular. The use of each equipment has to be reviewed, and experienced chemical engineers can do that very well.

Microreactors have been in discussion for some years. Their applicability, due to fluid pathway, may be limited. Their use might be restricted to all-liquid or micronized solid systems only. Micronizing system might be needed to assure that the solids/liquid mix will not cause any plugging, and can be used for a variety of solids. Investment for such systems has to be justified. Considerable exploration of microreactors has been done in the laboratory, but not much is known about their commercial viability. Their economics could also interfere. Tubular electrical heaters and plate and frame heat exchangers can also be, in certain cases, used as microreactors. Since the heat input can be controlled, they can make an excellent case for a modular processing unit.

8.11 Regulations and technology

Elements of "nondestructive creation" [56, 57] for manufacturing technology and chemical synthesis innovation and simplification have been covered in different chapters. Most of the process developers and designers are not only familiar with the many elements but may have used them (Tab. 8.6) individually in the development and design of different processes. It is possible that they may have been used collectively also. Their ongoing adoption and application can lead to simplifying manufacturing and minimizing pharma's impact on "climate change" [72].

If the pharmaceutical industry wants to contribute to the sustainable development goals (SDG12) [73] and lower its "E-factor" [1] emissions that contribute to climate change, the current business model and their process inefficiencies need to be evaluated and remedied. The current fragmentation of API manufacturing and the thought of making appropriate improvements after the drugs have become generic will not have much of an impact on SDG12 [73] or "climate change" [72]. If any remedial steps to minimize the "E-factor" [1] impact are not taken from process inception, not much will change, and with increasing drug demand, the situation will worsen. Significant effort would be needed to remedy the current landscape.

Various regulatory guidances (old and new) [73] are used to assure drug safety and product quality repeatability. These, along with speed to market, have thwarted continuous improvement of the existing processes and any manufacturing technology innovation. The consequences of all of the above are manifested by pharma's highest "E-factor" [1] emissions in the chemical world. The economic impact of new drug regulations on "climate change" [71] is unknown and needs to be evaluated. It is equally critical that the toxicity of waste chemicals from API manufacturing and their formulation plants that end up in the waste streams be understood [75]. It may be necessary to have harmonization of the global environmental and safety [76] standards. All these, collectively, will hasten the progress toward the SDG12 goals and lower the "E-factor" [1] emissions. Climate change is not a cause of concern to Pharmaceutical industry.

Even with the pharmaceutical manufacturing being stuck in the stone age [77] and being impacted by different guidances and regulations, most of the API producers and their formulators are profitable, and pass every regulatory-based cost increase to patients. They have shrugged all of these situations and have ignored manufacturing technology innovation. Pharma's high "E-factor" [1] is generally considered a cost of doing business.

A complete understanding of the DNA (Fig. 8.4) of the process chemistries, the physical properties, and the product demand will lead to a selection of suitable equipment to create the right continuous process. In such processes, a proper process control technology assures that the produced product is within the established control limits. Quality is built/designed (QbD) [71, 78–80] in the manufacturing process and not analyzed in the process (QbA) [71, 78–80], a hallmark of batch process. Since the current regulatory focus is batch process-based, the industry will have to be doubly sure of the product quality. In continuous processes, in-process testing would not be necessary as the quality is built in the process design.

**Understanding the process DNA
we will result in a QbD process.**

QbA will be minimized/eliminated.

Fig. 8.4: Process DNA understanding.

Since companies will have a total command on the quality, this could be bewildering to the regulators as they have never seen such commercial processes.

Manufacturing technology improvements could alter the current model of many sites producing the same API molecules. Manufacturing of many APIs could be consolidated. Consolidation could lead to continuous manufacturing of many APIs. This would result in a significant reduction of "E-factor" [1] for pharmaceuticals. Large strides could be made toward meeting pharma's goals of SDG12 [73].

Conversation about conservation and ecology has been going on for a while.

> Invoking the sanctity of the free market to prove that present business practices are sound and constructive, and using it to rebut every charge of ecological malfeasance is, at its heart, dishonest. Historically, we have given industry great latitude for its miscalculations because there was no science sufficiently developed to inform society of industrialism's effects. One hundred years ago, industrial cities were coated with grime and cut off from the sun by permanent palls of smoke; the citizens were beset by disease; the very conditions under which workers toiled and died were inhumane and exploitative. These conditions had their analog in the industrial processes of waste and despoliation, and were the direct costs of the Industrial Revolution. It took many decades before an appreciation of the social and environmental damage spread beyond a small circle of Marxists and muckrakers to society as a whole. Today, businesspeople ready concede the abuses of the early days of this Revolution, but they do not wholly and genuinely acknowledge the more threatening abuses perpetuated by current practices. Troubling untruths lie uneasily with in a colossal economic system that denies what we all know while it continues to degrade our world, our society, and our bodies. Business economists can explain in detail the working of the modern corporation, its complex interrelation with financial markets, how its holdings might be valued on a discounted cash-flow basis, or the dynamics of global competitive advantages. These pronouncements and equations promise hope but they cannot explain – much less justify– the accelerating extinction of species, the deterioration of human health, the stress and anguish of the modern worker, the loss of our air, water, and forests. In short, they cannot explain the consequences of their actions. [81].

With the recent recognition by the US Department of Health and Human Service (HHS) that "climate change" is a public health issue [82], it should ask the USFDA and the US EPA to review the impact of their regulations on pharma manufacturing, as it has the highest "E-factor" [1].

Different words ("E-factor" emissions, "climate change," and net-zero) have been used to describe the same issue. To achieve the desired goal, each industry has to do its part. However, pharmaceuticals, being the largest emitter [1] in the chemical sector, will have to put in significant effort. As has been discussed in different chapters, pharma's current situation is due to multiple sites manufacturing many of the same APIs and their formulations [83]. Consolidation of many of the facilities will have to take place if the pharma industry is to lower its "E-factor" emissions.

Principles and methodologies of chemistry and chemical engineering discussed can also be applied to the manufacture of biotech drugs. With ever increasing development and commercialization of bio drugs, considerations and application of unit processes and unit operations is a must. They have been successfully applied in the production of industrial enzymes [84], foods and proteins [85]. Their application can result in process simplification, optimization and reducing equipment size and these can significantly reduce bio drugs environmental footprint and global warming.

As stated earlier in this and earlier chapters, it will take an effort and realignment of pharma's business model, and that will not be an easy task. The process can start with manufacturing technology innovation and it will facilitate the process. Companies and countries will resist the change from the current model, as the profitability, and consistent and reliable drug supply of the drugs, specially the generics, could be impacted. Such consolidation could raise "national security and supply," that is, "strategic vulnerability" issues. They will have to be sorted out.

Regulators will have to play their part, and that is going to more difficult than what the industry has to do (Fig. 8.2). While FDA and EPA review the impact of their regulations on pharma's manufacturing technology innovation, and this could take time, API manufactures and their formulators, independently and aggressively, should proceed to lower their "E-factor" emissions using "nondestructive creation" [56, 57] methods and strategies.

Peter Drucker said it right when he spoke, "Innovation can be systematically managed – if one knows where and how to look" [86]. Industry has all the knowledge and tools that it can use to API manufacturing and formulations, to the space age [87]. These tools and methods have been reviewed and discussed in different chapters throughout the book. Pilot plant equipment can be assembled and used as modular plants and tested. Done correctly, it will immediately lead to a reduction of solvent use. This will require minimal investment. Fear of failure would be a learning experience and result in better processes. Actually, some of these pilot plants could be used for full-scale continuous production and/or in batch campaigns to produce many different products. Another benefit of using pilot equipment, which is same or similar as modular plants, can be that they can be used in the production of drugs that are needed for national security.

Nondestructive creation [56, 57] is the **way** to improve pharma's manufacturing capabilities and technologies, which will result in improved "profitability through simplicity" and lower its environmental impact. Value of "nondestructive creation" is illustrated in Fig. 8.5.

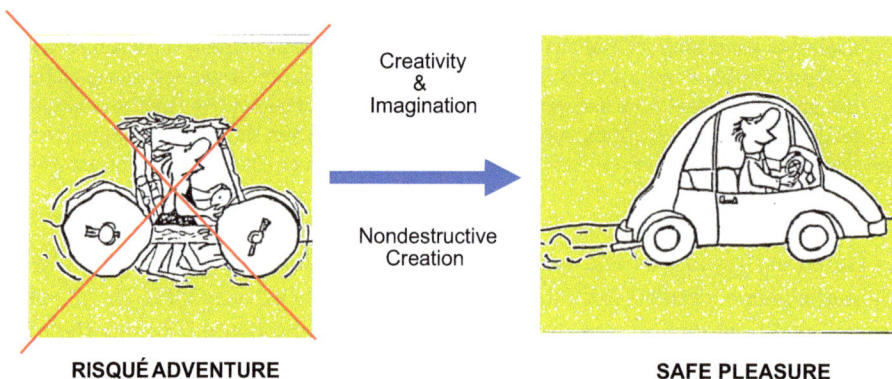

Creativity
&
Imagination

Nondestructive
Creation

RISQUÉ ADVENTURE   SAFE PLEASURE

Fig. 8.5: Value of nondestructive creation.

Abbreviations

API	Active pharmaceutical ingredient
FDA	Food and Drug Administration
ASTM	American Society for Testing and Materials
ICH	International Council for Harmonization of Technical Requirements for Pharmaceuticals for Human Use (Pun intended here: What a complex name for a simple organization)
E-factor	Emission factor
CMO	Contract manufacturing organization
QbD	Quality by design
QbA	Quality by analysis
IPA	Isopropyl alcohol
NDA	New drug application
ANDA	Abbreviated new drug application
SDG	Sustainable development goals
DNA	Deoxyribonucleic acid

References

[1]   Sheldon, R. A. The *E* factor 25 years on: the rise of green chemistry and sustainability, Green Chemistry https://pubs.rsc.org/en/content/articlelanding/2017/gc/c6gc02157c#!divAbstract 2017, 19, 18–43 Accessed February 17, 2021.

[2]   Roschangar, et al. Overcoming Barriers to Green Chemistry in the Pharmaceutical Industry – The Green Aspiration Level™, Green Chemistry 2014 pgs. 1–17 Accessed December 10, 2020.

[3]   Malhotra, G.: An Alternate Look at the Pharmaceutical World Revenues and Drug Affordability, Pages 2–5, CPhI 2017 Annual Report, Manufacturing Chemist https://www.man ufacturingchemist.com/news/article_page/An_alternate_look_at_pharmaceutical_world_rev

enues_and_drug_affordability/135165, Volume 88, Number 10, October 2017, Pg. 28-32 Accessed June 20, 2020.

[4] Small Molecule https://en.wikipedia.org/wiki/Small_molecule, Accessed July 29, 2021.

[5] Shreve, R. N. Unit Processes in Chemical Engineering, Industrial and Engineering Chemistry, 1954 Vol. 46, No. 4 pg. 672 Accessed June 22, 2020.

[6] Unit Operations, Wikipedia, https://en.wikipedia.org/wiki/Unit_operation Accessed March 11, 2021.

[7] Operations for the Executive Suite, McKinsey & Co. 2012 pg. 11 Accessed August 20, 2012.

[8] Spacy, J.: Batch vs. Mass Production https://simplicable.com/new/batch-production-vs-mass-production, February 3, 2017 Accessed January 3, 2021.

[9] Standard Guide for Application of Continuous Processing in the Pharmaceutical Industry https://www.astm.org/Standards/E2968.htm Accessed September 7, 2021.

[10] ICH guideline Q13 on continuous manufacturing of drug substances and drug products https://www.ema.europa.eu/en/documents/scientific-guideline/draft-ich-guideline-q13-continuous-manufacturing-drug-substances-drug-products-step-2b_en.pdf 29 July 2021Accessed September 7, 2021.

[11] Malhotra, G.: Batch, Continuous or "Fake/False" Continuous Processes in Pharmaceutical Manufacturing, Profitability through Simplicity https://pharmachemicalscoatings.blogspot.com/2017/07/batch-continuous-or-fakefalse.html July 20, 2017 Accessed July 10, 2021.

[12] Chapter 2, Table 2.8 Accessed July 4, 2021.

[13] Chapter 5, Accessed July 4, 2021.

[14] Malhotra, G.: Square Plug in A Round Hole: Does This Scenario Exist in Pharmaceuticals?, Profitability through Simplicity, https://pharmachemicalscoatings.blogspot.com/2010/08/square-peg-in-round-hole-does-this.html August 17, 2010.

[15] Malhotra, G.: Why Fitting a Square Plug in a Round hole is Profitable for Pharma and Most Likely Will Stay?, Profitability through Simplicity, https://pharmachemicalscoatings.blogspot.com/2014/08/why-fitting-square-plug-in-round-hole.html August 1, 2014.

[16] Malhotra, G.: Systematic Demystification of Drug Price Mystique and the Needed Creative Destruction, Profitability through Simplicity, https://pharmachemicalscoatings.blogspot.com/2019/10/systematic-demystification-of-drug.html, October 2, 2019 Accessed August 15, 2021.

[17] May, S. A. Journal of Flow Chemistry 7, pages 137–145 (2017) Accessed February 17, 2021.

[18] Doyle, et.al Mini-Monoplant Technology for Pharmaceutical Manufacturing, Organic Process Research and Development, 2020, 24, 10, 2169–2182, Accessed February 15, 2021.

[19] Li, et. al Green Chemistry for Chemical Synthesis, PNAS, September 9 2008, 105, 36, 13197–13202, Accessed September 7, 2020.

[20] Kansal, et. al. US 6,846,952 B2 Process for Manufacture of a 4-Bromo-2-Oxyimini Butyric Acid and its Derivatives, Lupin Limited January 25, 2005 Accessed January 29, 2008.

[21] Benchmarking Shows Need to Improve Uptime, Capacity Utilization, https://www.pharmamanufacturing.com/articles/2007/144/ Sep 20, 2007, Accessed May 19, 2020.

[22] Malhotra, G.: Continuous Pharmaceutical Processes and Their Demands, Contract Pharma, https://www.contractpharma.com/issues/2016-04-01/view_features/continuous-pharmaceutical-processes-and-their-demands/?userloggedin=true, April 5, 2016, Accessed May 15, 2020.

[23] Tyson, T.: Solving the Industry's Capacity Utilization Problem, https://www.pharmasalmanac.com/articles/solving-the-industrys-capacity-utilization-problem October 28, 2019 Accessed May 10, 2020.

[24] Malhotra, G.: Opportunities for Generic Pharma to Clear the Quality Stigma, https://pharmachemicalscoatings.blogspot.com/2019/05/opportunities-for-generic-pharma-to.html, May 23, 2019, Accessed April 25, 2020.

[25] Couper, J. R. et al. Chemical Process Equipment, Selection and Design, Third, Elsevier, ISBN: 9780123969590 2010.

[26] De. Souza, et.al Flow Processing as a Tool for API Production in Developing Countries, Journal of Flow Chemistry, 2017, 7, 3-4, pg., 146–150, Accessed January 23, 2020.

[27] Outpacing Change in Pharma, McKinsey & Co. 2009 pg. 9 Accessed August 20, 2012.

[28] Q7 Good Manufacturing Practice Guidance for Active Pharmaceutical Ingredients https://www.fda.gov/files/drugs/published/Q7-Good-Manufacturing-Practice-Guidance-for-Active-Pharmaceutical-Ingredients-Guidance-for-Industry.pdf September 2016 Accessed May 7, 2021.

[29] Konudula, et al. Matrix Laboratories Ltd. Process for the preparation of Paroxetine Hydrochloride WO 2009/138999 A2 Accessed August 16, 2012.

[30] Lemmens, et. al., Synthon BCT Technologies, LLC Process for the production of paroxetine USP 6,686,473 Accessed August 12, 2012.

[31] https://www.pharmacompass.com/active-pharmaceutical-ingredients/paroxetine-hydrochloride Accessed July 23, 2021.

[32] Malhotra, G., Less is More in API Process Development: Pharmaceutical Manufacturing, https://www.pharmamanufacturing.com/articles/2005/294/ July/August 2005, Pages 50–51.

[33] Malhotra, G.: API Manufacture-Simplification and PAT: Pharmaceutical Processing, https://www.researchgate.net/publication/220009740_API_Manufacture_-_Simplification_and_PAT November 1,2005, Pages 24–27.

[34] FDA The PAT Team and Manufacturing Science Working Group Report, http://docplayer.net/12621706-Innovation-and-continuous-improvement-in-pharmaceutical-manufacturing.html, Accessed September 1, 2005.

[35] PAT – A Framework for Innovative Pharmaceutical Development, Manufacturing, and Quality Assurance https://www.fda.gov/media/71012/download, September 2004 Accessed September 27, 2004.

[36] Continuous Improvement https://en.wikipedia.org/wiki/Continual_improvement_process Accessed August 28, 2021.

[37] https://en.wikipedia.org/wiki/Red_tape, Accessed August 23, 2021.

[38] Malhotra, G.: Impact of Regulations, Drug Manufacturing and Pharma Supply Chain (PBMs and allies) in Drug Shortages and Affordability Part 1 https://pharmachemicalscoatings.blogspot.com/2019/03/impact-of-regulations-drug.html March 8, 2018 Accessed August 23, 2021.

[39] Malhotra, G.: Simplified Roadmap for ANDA/NDA Submission and Approval will change Pharma Landscape, https://pharmachemicalscoatings.blogspot.com/2018/11/simplified-roadmap-for-andanda.html November 25, 2018 Accessed July 10, 2021).

[40] Malhotra, G. Pharmaceutical Manufacturing: Is It the Antithesis of Creative Destruction? Pharmaceutical Manufacturing https://www.pharmamanufacturing.com/assets/wp_downloads/pdf/Creative_Destruction_Malhotra.pdf, July 2008 Accessed August 19, 2021.

[41] Malhotra, G.: Pharma's future is putting innovations in the hands of innovators https://ubmemeaensoprod.s3.amazonaws.com/CPHI_EUROPE/cphi_annual_industry_report_2018.pdf August 28, 2018 Accessed September 21, 2018.

[42] Malhotra, G.: Batch or a Continuous Process: A Choice; Pharmaceutical Processing, March 2005, Pg. 16.

[43] Malhotra, G.: Alphabet Shuffle: Moving From QbA to QbD – An Example of Continuous Processing, Pharmaceutical Processing World, https://www.pharmaceuticalprocessingworld.com/alphabet-shuffle-moving-from-qba-to-qbd/, February 2009 pg. 12–13, Accessed February 19, 2019.

[44] Malhotra, G.: Hesitation In The Drive To A Continuous Pharmaceutical Manufacturing Process: Real or Imaginary? Pharmaceutical Processing, https://www.pharmaceuticalprocessingworld.

com/hesitation-in-the-drive-to-a-continuous-pharmaceutical-manufacturing-process-real-or-imaginary/ July 7, 2009 pg- 12–15,.

[45] Malhotra, G.: The Path Towards Continuous Processing, Pharmaceutical Processing World, https://www.pharmaceuticalprocessingworld.com/the-path-towards-continuous-processing/ August 2010, pgs. 16–20 Accessed February 19, 2019.

[46] Malhotra, G.: Is Continuous Processing in Pharma's Future? Profitability through Simplicity, https://pharmachemicalscoatings.blogspot.com/2012/07/is-continuous-processing-in-pharmas_3850.html July 24, 2012 Accessed February 19, 2012.

[47] Malhotra, G.: Continuous Process in Pharmaceutical Manufacturing: Considerations, Nuances and Challenges, Contract Pharma, https://www.contractpharma.com/issues/2015-06-01/view_features/continuous-process-in-pharmaceutical-manufacturing-considerations-nuances-and-challenges June 2, 2015, Accessed February 19, 2012.

[48] Malhotra, G.: My perspective for Pharmaceutical Manufacturing Technologies/Processes and Continuous Improvements, CPhI 2015, http://epcotint.com/Pharmaevolution/GirishMalhotra September2015CPhIreport.pdf.

[49] Malhotra, G.: Continuous Pharmaceutical Processes and Their Demands, Contract Pharma, https://www.contractpharma.com/issues/2016-04-01/view_features/continuous-pharmaceutical-processes-and-their-demands/ pgs. 37–40, April 5, 2016.

[50] Malhotra, G.: Strategies for Improving Batch or Creating Continuous Active Pharmaceutical Ingredient (API) Manufacturing Processes, Profitability through Simplicity, https://pharmachemicalscoatings.blogspot.com/2017/03/strategies-for-enhancing-active.html March 20, 2017.

[51] Malhotra, G.: Reality and Un-Reality: Continuous Processing in Pharmaceutical Manufacturing, Contract Pharma, https://www.contractpharma.com/issues/2017-04-01/view_features/reality-and-un-reality-continuous-processing-in-pharmaceutical-manufacturing/ April 3, 2017.

[52] Malhotra, G.: Chapter 4 "Simplified Process Development and Commercialization" in "Quality by Design-Putting Theory into Practice" co-published by Parenteral Drug Association and DHI Publishing© February 2011.

[53] Malhotra, G.: Chemical Process Simplification: Improving Productivity and Sustainability John Wiley & Sons, February 2011.

[54] TRANSFORMATIONAL INNOVATION: A JOURNEY BY NARRATIVE http://www.stevedenning.com/slides/Innovation-pt2-Final-Feb2-05.pdf June 1, 2005 Accessed January 29, 2008.

[55] Schumpeter, J. A. Creative Destruction" from Capitalism, Socialism and Democracy, New York, Harper, 1975, [orig. pub. 1942], pp. 82–85, accessed January 31, 2008.

[56] Hubbard, G. Nondestructive Creation https://www.strategy-business.com/article/07203?pg=0, May 29, 2007 (originally published by Booz & Company) Accessed December July 14, 2008.

[57] Hubbard, G: Nondestructive creation: Entrepreneurship and management research in the study of growth, Volume 30, Issue 4, July–August 2008, pp. 595–602, July 14, 2008.

[58] Malhotra, G.: The Good, The Bad, The Ugly (1) Complexities of Pharmaceutical Manufacturing, Profitability through Simplicity, https://pharmachemicalscoatings.blogspot.com/2018/04/the-good-bad-ugly-1-complexities-of.html April 9, 2018, Accessed January 12, 2020.

[59] Malhotra, G.: Process Simplification and The Art of Exploiting Physical Properties, Profitability through Simplicity, https://pharmachemicalscoatings.blogspot.com/2017/03/process-simplification-and-art-of.html, March 10, 2017.

[60] Malhotra, G.: Chemical Process Simplification: Improving Productivity and Sustainability John Wiley & Sons, February 2011.

[61] Malhotra, G.: Could Software Technologies be the "Creative Destructionist" for Pharmaceuticals and Chemicals? Profitability through Simplicity, https://pharmachemicalscoat ings.blogspot.com/2015/08/could-software-technologies-be-creative.html, August 31, 2015.

[62] Malhotra, G.: Considerations to Simplify Organic Molecule (API) Manufacturing Processes: My perspective, Profitability through Simplicity, https://pharmachemicalscoatings.blogspot. com/2019/04/considerations-to-simplify-organic_77.html April 20, 2019.

[63] Malhotra, G.: Quick Review of Chemicals Related Process Development, Design and Scale up Considerations, Profitability through Simplicity, https://pharmachemicalscoatings.blogspot. com/2018/11/quick-review-of-chemicals-related.html, November 10, 2018.

[64] Malhotra, G.: Innovation In Pharmaceuticals: What Would It Take & Who is Responsible? https://pharmachemicalscoatings.blogspot.com/2017/11/innovation-in-pharmaceuticals-what.html, November 28, 2017.

[65] Malhotra, G.: Impact of 21CFR 314.70 on The Life of a Pharmaceutical Manufacturing Process, Profitability through Simplicity, https://pharmachemicalscoatings.blogspot.com/2015/08/im pact-of-21cfr-314670-on-life-of.html, August 7, 2015.

[66] Malhotra, G.: My perspective for Pharmaceutical Manufacturing Technologies/Processes and Continuous Improvements, CPhI Report 2015 http://epcotint.com/Pharmaevolution/Girish MalhotraSeptember2015CPhIreport.pdf.

[67] Malhotra, G.: Active Pharmaceutical Ingredient Manufacturing (API) and Formulation Drive to NET ZERO (Carbon Neutral)? https://pharmachemicalscoatings.blogspot.com/2021/04/ac tive-pharmaceutical-ingredient.html April 29, 2021 Accessed May 1, 2021.

[68] Q7 Good Manufacturing Practice Guidance for Active Pharmaceutical Ingredients, https:// www.fda.gov/files/drugs/published/Q7-Good-Manufacturing-Practice-Guidance-for-Active-Pharmaceutical-Ingredients-Guidance-for-Industry.pdf September 2016 Accessed August 28, 2021.

[69] Q11 Development and Manufacture of Drug Substances (Chemical Entities and Biotechnological/Biological Entities) Questions and Answers Guidance for Industry, February 2018 https://www.fda.gov/media/103162/download Accessed August 28, 2021.

[70] ICH guideline Q12 on technical and regulatory considerations for pharmaceutical product lifecycle management, 4 March 2020 https://www.ema.europa.eu/en/documents/scientific-guideline/ich-guideline-q12-technical-regulatory-considerations-pharmaceutical-product-lifecycle-management_en.pdf Accessed May 27, 2021.

[71] Malhotra, G.: Process Centricity is the Key to Quality by Design https://pharmachemicalscoat ings.blogspot.com/2010/04/process-centricity-is-key-to-quality-by.html, April 6, 2010 Accessed January 3, 2011.

[72] Climate Change https://en.wikipedia.org/wiki/Climate_change Accessed September 8, 2021.

[73] Sustainable Development Goals, Chemicals and waste: https://sdgs.un.org/topics/chemi cals-and-waste Accessed September 5, 2021.

[74] US FDA Guidances (Drugs) https://www.fda.gov/drugs/guidance-compliance-regulatory-information/guidances-drugs Accessed August 31, 2021.

[75] Malhotra, G.: Pharmaceuticals, Their Manufacturing Methods, Ecotoxicology, and Human Life Relationship, Pharmaceutical Processing, https://www.researchgate.net/publication/ 220009804_Pharmaceuticals_Their_Manufacturing_Methods_Ecotoxicology_and_Human_ Life_Relationship November 2007 Pgs. 18–23 Accessed January 12, 2008.

[76] Malhotra, G.: Can Uniform Safety, Health and Effluent and Manufacturing Standards Create Process Technology Innovation and Competition in Pharmaceuticals? Profitability through Simplicity, https://pharmachemicalscoatings.blogspot.com/2017/01/can-uniform-safety-health-and-effluent.html January 10, 2017 Accessed September 5, 2021.

[77] Stone Age https://en.wikipedia.org/wiki/Stone_Age Accessed August 31, 2021.

[78] Malhotra, G.: Pharmaceuticals: What is Holding Back Quality By Design, Profitability through Simplicity, https://pharmachemicalscoatings.blogspot.com/2009/06/in-pharmaceuticals-what-is-holding-qbd.html June 12, 2009 Accessed August 12, 2021.

[79] Malhotra, G.: Alternate Interpretation of Pharmaceutical TLAs: Three Letter Acronyms, https://pharmachemicalscoatings.blogspot.com/2010/03/alternate-interpretation-of.html March 16, 2010 Accessed August 12, 2021.

[80] Malhotra, G.: Who Is/Or Should Be Driving Pharma's Manufacturing Car: Regulators or the Regulated? Profitability through Simplicity, https://pharmachemicalscoatings.blogspot.com/2015/12/who-isor-should-be-driving-pharmas.html December 7, 2015 Accessed August 21, 2021.

[81] Paul Hawken -The Ecology of Commerce, https://p2infohouse.org/ref/31/30233.pdf HarperBusiness a Division of HarperCollins Publishers, 1993 Accessed 1996 Accessed September 15, 2021.

[82] Climate Change to Be Treated as Public-Health Issue https://www.wsj.com/articles/climate-change-to-be-treated-as-public-health-issue-11630315800 Accessed August 31, 2021.

[83] Malhotra, G.: Impact of Regulations, Manufacturing and Pharmaceutical Supply Chain (PBMs) on Drug Shortages and Affordability Part 2: https://pharmachemicalscoatings.blogspot.com/2019/04/impact-of-regulations-manufacturing-and.html April 3, 2019 Accessed August 31, 2021.

[84] Industrial Enzymes, https://en.wikipedia.org/wiki/Industrial_enzymes Accessed July 12, 2022.

[85] Tripathi, N. K., Shrivastava A. Recent Developments in Bioprocessing of Recombinant Proteins: Expression Hosts and Process Development, Frontiers of Bioengineering and Biotechnology, https://doi.org/10.3389/fbioe.2019.00420 20 December 2019, Accessed January 12, 2022.

[86] Drucker, P. F. The Discipline of innovation, Harvard Business Review May-June 1985, pg. 67–72 Accessed June 10, 1990.

[87] Space Age https://en.wikipedia.org/wiki/Space_Age Accessed July 10, 2021.

Index

https://doi.org/10.1515/9783110702842-009

Lightning Source UK Ltd.
Milton Keynes UK
UKHW050934120822
407214UK00003B/13

9 783110 702828